Biology of Extracellular Matrix

Volume 15

Extracellular matrix (ECM) biology, which includes the functional complexities of ECM molecules, is an important area of cell biology. Individual ECM protein components are unique in terms of their structure, composition and function, and each class of ECM macromolecule is designed to interact with other macromolecules to produce the unique physical and signaling properties that support tissue structure and function. ECM ties everything together into a dynamic biomaterial that provides strength and elasticity, interacts with cell-surface receptors, and controls the availability of growth factors. Topics in this series include cellular differentiation, tissue development and tissue remodeling. Each volume provides an in-depth overview of a particular topic, and offers a reliable source of information for post-graduates and researchers alike.

All chapters are systematically reviewed by the series editor and respective volume editor(s).

"Biology of Extracellular Matrix" is published in collaboration with the American Society for Matrix Biology and the International Society for Matrix Biology.

Kirsi Rilla
Editor

Extracellular Vesicles as Matrix Messengers

Springer

Editor
Kirsi Rilla
Institute of Biomedicine
University of Eastern Finland
Kuopio, Finland

ISSN 0887-3224 ISSN 2191-1959 (electronic)
Biology of Extracellular Matrix
ISBN 978-3-031-68810-2 ISBN 978-3-031-68811-9 (eBook)
https://doi.org/10.1007/978-3-031-68811-9

This Springer imprint is published by the registered company Springer Nature Switzerland AG
The registered company address is: Gewerbestrasse 11, 6330 Cham, Switzerland

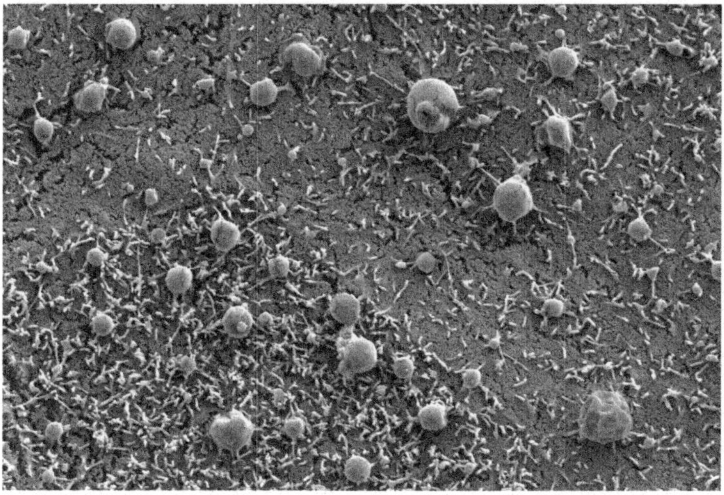

Scanning electron microscopic image of EVs attached to the surface of tumor cells

Preface

The cells of our body release a wide variety of factors to the extracellular space to signal, regulate, and build the extracellular matrix (ECM). One of the most fascinating of those factors are extracellular vesicles (EVs). Those lipid membrane-covered particles have now emerged as a new powerful means of near and distant intercellular communication between cells and their extracellular environment of solid tissues and body fluids.

This volume in the Biology of Extracellular Matrix series summarizes the current knowledge on EVs as integral and active components of the extracellular matrix. It covers topics such as EV interplay between ECM molecules and how cancer cells modify their microenvironment by releasing EVs. Furthermore, it will be discussed how EVs can mediate tissue repair and regeneration by modulating matrix degradation, cross-linking of matrix proteins, and matrix calcification. Furthermore, examples of the role of EVs in disease states related to matrix biology such as cancer and arthritis, as well as their potential as therapeutic tools and biomarkers in those disease states, will be covered. Finally, some examples of specific methodology for EV analysis will be introduced.

The book is organized into eleven chapters. The first two chapters introduce the role of EVs as matrix components, messengers, and modulators, Chaps. 3–8 address the role of ECM and EVs in synovial fluid, adipose tissue, vitreous body, and retina of the eye and their role in tumor progression, inflammation, and other disease states related to these tissues with specific emphasis on the role of the extracellular matrix. Chapters 9–11 give some examples of specific methodology to study EVs in the context of the ECM.

Chapter 1 by Vincent Anto and his colleagues acts as an introduction to the exciting world of EVs and their role as components and modulators of the extracellular matrix. Furthermore, the chapter describes matrix-bound nanovesicles, a novel type of EVs that resides within the ECM. This chapter also discusses the therapeutic potential of EVs and their clinical implications.

Chapter 2 by Giuseppina Poppa and others addresses the role of EVs in matrix degradation and remodeling, giving examples of their role in the extracellular environment in health and disease. Several aspects of EV–ECM interaction under healthy and pathological conditions are covered in this chapter.

Chapter 3 focuses on the role of EVs as modulators of the tumor microenvironment. The authors, Nao Nishida-Aoki and Takahiro Ochiya, summarize the roles of cancer EVs in remodeling ECM in tumors, highlighting how EVs influence ECM secretion by stromal cells, directly modify ECM structures, and impact tumor progression. The chapter emphasizes that the interaction between EVs and the ECM in the tumor microenvironment is crucial for developing novel diagnostic and therapeutic strategies to improve cancer treatment efficacy.

In Chapter 4 by Rafał Szatanek and others, evidence is provided on the role of the tumor tissue HA and HA–TEV complexes in modulation of monocytes, macrophages, and myeloid-derived suppressor cell activities during cancer development and progression. The chapter also presents the possible forms of HA-targeted therapies as potentially promising approach in the future treatment of cancer.

Chapter 5 concentrates on the role of EVs in inflammation. The chapter by Saara Hämälistö and Lujain AlGhadir summarizes the current view of EVs in promoting basic processes of innate and adaptive immunity. The authors also discuss the inflammatory conditions such as autoimmune diseases and antitumor immune responses where EVs are reported to take part.

Chapter 6 by Enrico Ragni and Laura de Girolamo presents the role of EVs in the synovial fluid, a specific lubricating ECM component of the joints. The chapter describes the current understanding of the role of EVs in joints, the current knowledge of their role in disease states of the joints, and their possible use as biomarkers in these diseases.

Chapter 7 presents a broad discussion of the role of EVs in the adipose tissue microenvironment. The chapter provides a comprehensive overview of current research on adipose tissue EVs, including their composition and molecular content. The authors Tamara Camino, Nerea Lago-Baameiro, and Maria Pardo discuss the potential impact of EVs on metabolic regulation, obesity-associated inflammatory diseases, and the extracellular matrix (ECM), which is a key component in the development of obesity and its comorbidities.

In Chap. 8 Stef Bosch and others review and discuss the current knowledge of EVs in the posterior eye segment, particularly their existence and role in the vitreous humor, a specific ECM material in the eye. The rather unknown area of ocular kinetics of EVs and potential factors regulating it will be also discussed in detail. The authors also discuss the potential use of EVs as retinal therapeutics.

Chapter 9 by Paul Debashish and others introduces a methodology to investigate mechanical properties of single EVs using atomic force microscopy (AFM). The authors show that AFM is not only an imaging tool of EVs that delivers high-resolution images, but the label-free AFM force spectroscopy could also be valuable to study the mechano-responses of single EVs. They propose that the multi-parametric AFM structure-mechanical property analysis has important implications for EV biology research and clinical applications.

Chapter 10 by Khirul Islam and Janne Leivo addresses the aspect of glycosylation landscape of EVs and its impact on biological functions, along with the analytical tools used to study them. In addition, the most widely used carbohydrate-specific natural and synthetic affinity scaffolds currently used to study the glycosylations on EVs are described.

Chapter 11 by Heikki Kyykallio and Kirsi Rilla is focused on the challenges and advances of imaging of EVs especially in the 3D environment. The chapter discusses the basics and the current advances of imaging techniques and discusses the challenges of EV imaging in the 3D environment. The authors also describe examples for most commonly used labeling methods, imaging techniques, and in vitro and in vivo models for EV imaging as components of the ECM.

I would like to thank all contributors for their valuable contribution to this book. The authors to this book are both pioneers in the field, offering their unique insights and expertise, and junior researchers with new inspiration for future discoveries and innovations in the field. Their collective efforts provide a rich, multidimensional perspective on the role of EVs as matrix messengers. I also express my gratitude to Prof. Nikos Karamanos for providing me the opportunity to act as a guest editor for this Series, as well as for valuable support and comments. I also thank Series Editor, Dr. Alina Cepraga, Associate Editor, and the Editorial Board Members for their revision and valuable comments in the whole process of preparing the book.

In conclusion, EVs serve as essential mediators in ECM dynamics, influencing tissue structure, repair, and disease progression, and thus play a fascinating role in the context of the extracellular matrix. This book is a great opening to introduce the field for EV biology especially for extracellular matrix research community. The topics covered in this volume provide an important context for understanding the role of extracellular vesicles in matrix biology. Welcome to the world of EVs, where tiny messengers make enormous impacts!

Kuopio, Finland Kirsi Rilla

Contents

Chapter 1
Manipulating the Matrix: Role of Extracellular Vesicles

Vincent P. Anto, Kaylee M. Montanari, George S. Hussey, and Stephen F. Badylak

Abstract Extracellular vesicles (EVs) are membrane-bound vesicles ubiquitously secreted by nearly all cell types, containing bioactive cargo that profoundly influences cellular behavior. One mechanism through which EVs exert their effects is by modulating the extracellular matrix (ECM). Originally viewed as a mere physical scaffold, the ECM plays a pivotal role in orchestrating cellular interactions. This chapter explores the ability of EVs to regulate the ECM in both physiological conditions and various disease states. Additionally, this chapter will delve into the discovery and functional significance of a novel type of EV, the matrix-bound nanovesicle, which resides within the ECM itself. Lastly, the clinical implications and therapeutic potential of EVs will be explored given their profound influence on cellular function and ECM modulation.

Vincent P. Anto and Kaylee M. Montanari contributed equally to this work and are co-first authors.

V. P. Anto
McGowan Institute for Regenerative Medicine, University of Pittsburgh, Pittsburgh, PA, USA

Department of Surgery, School of Medicine, University of Pittsburgh, University of Pittsburgh Medical Center Presbyterian Hospital, Pittsburgh, PA, USA

K. M. Montanari
McGowan Institute for Regenerative Medicine, University of Pittsburgh, Pittsburgh, PA, USA

G. S. Hussey
McGowan Institute for Regenerative Medicine, University of Pittsburgh, Pittsburgh, PA, USA

Department of Pathology, School of Medicine, University of Pittsburgh, University of Pittsburgh Medical Center Presbyterian Hospital, Pittsburgh, PA, USA

S. F. Badylak (✉)
McGowan Institute for Regenerative Medicine, University of Pittsburgh, Pittsburgh, PA, USA

Department of Surgery, School of Medicine, University of Pittsburgh, University of Pittsburgh Medical Center Presbyterian Hospital, Pittsburgh, PA, USA

Department of Bioengineering, University of Pittsburgh, Pittsburgh, PA, USA
e-mail: badysx@upmc.edu

K. Rilla (ed.), *Extracellular Vesicles as Matrix Messengers*, Biology of Extracellular Matrix 15, https://doi.org/10.1007/978-3-031-68811-9_1

1.1 Introduction

The initial description of extracellular vesicles (EVs) dates back to 1967, when Peter Wolf observed the presence of small particulate matter in freshly ultracentrifuged plasma, coining it "platelet dust" (Wolf 1967). Unbeknownst at the time, this "dust" was later identified as platelet-derived EVs. Dr. Wolf's discovery confirmed the work performed by Drs. Chargaff and West had previously noted a particulate fraction in ultracentrifuged blood with high clotting potential, suggesting "minute breakdown products of blood corpuscles" (Chargaff and West 1946). However, it was not until 1971 that the term "extracellular vesicle" made its first appearance in the scientific literature, when Aaronson et al. demonstrated that eukaryotic algae could produce extracellular membrane-bound vesicles (Aaronson et al. 1971). Over the subsequent decade, research on EVs gained traction within the scientific community, revealing their pervasive impact on both normal homeostatic functions and pathological conditions. In 2011, the International Society of Extracellular Vesicles (ISEV) was formed and officially adopted the term "extracellular vesicle" to encompass any naturally released or secreted particle from cells, delimited by a lipid bilayer (Lötvall et al. 2014).

1.2 EV Biology, Classification, and Biogenesis

EVs represent a broad range of secreted membrane vesicles that are produced by all cell types. These vesicles, upon biogenesis, are secreted into the extracellular space and transmit important cargo to other cells. Producing lipid-bound vesicles is a highly evolutionarily conserved process (Van Niel et al. 2018) that is employed by most organisms including bacteria and plants as well as in vitro cell and organoid cultures (Abdollahi 2021; Toyofuku et al. 2019; Baldrich et al. 2019). EVs are typically classified based on their biogenesis, cellular origin, size, cargo, and function (Fig. 1.1).

Standard EV subdivisions are microvesicles (MVs), exosomes, and apoptotic bodies. These subtypes have different protein, lipid, and nucleic acid contents based on their cellular origin and relative biogenesis. Elucidating the characterization of a specific EV is a challenging task in the EV field, as cellular and membrane-bound markers often overlap between groups; thus, lipidomics and proteomics, EV size, membranous and intracellular protein expression, cargo, and cellular origin all must be taken into consideration to classify an EV (Doyle and Wang 2019). There is a lack of a standardized EV isolation technique in the field, and different isolation methods on the same population of EVs result in variable properties. Nevertheless, EVs can still be categorized, and work in the field continues to expose distinctions between traditional, novel, and disease-associated classes of EVs. Upon secretion from cells, EVs circulate extracellularly and are either endocytosed by cells through fusion with the plasma membrane and subsequently delivery of cargo or interact

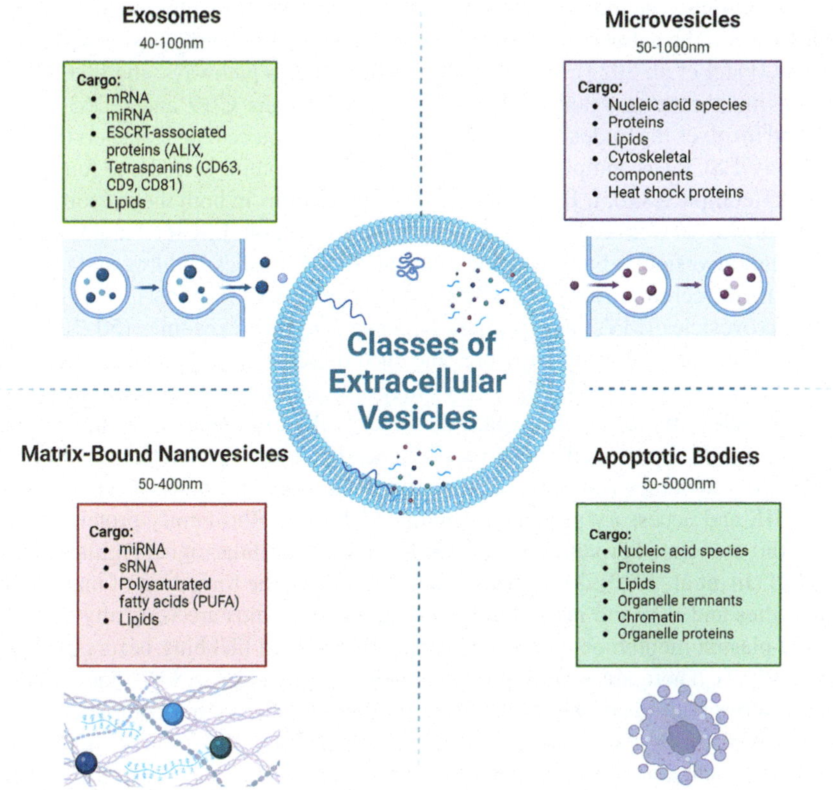

Fig. 1.1 Classes of Extracellular Vesicles and Common Cargo: Exosomes, microvesicles (MV), matrix-bound nanovesicles (MBV), and apoptotic bodies all carry similar cargo including a lipid, protein, and nucleic acid. Evidence suggests each type of vesicle goes through cargo sorting, selectively including and excluding proteins, nucleic acids, lipids, and other molecules

directly with cells via ligand binding and intracellular cascade activation. EVs deliver protein, lipid, and nucleic acids to the receiving cell, directing downstream signaling cascades inducing cellular responses.

Exosomes have been a focus of investigation for several decades and have implications in tissue homeostasis, disease pathogenesis, drug delivery, and prognosis. Exosomes are typically 40–100 nm in diameter and can be derived from virtually all biological fluids. They have a unique biogenesis that imparts an impressive array of membrane-bound cellular markers that enable their classification (Simons and Raposo 2009). Exosomes are produced and secreted by cells through an endosomal pathway, giving exosomes a protein cargo that is unique to the protein interactions along the route to secretion. The biogenesis of exosomes begins with the plasma membrane inwardly budding to form an early endosome, which transitions into a late endosome. The late endosome matures into a multivesicular body which itself invaginates to produce intraluminal vesicles (ILV) inside the multivesicular body.

Once the multivesicular body fuses with the cell plasma membrane, the ILVs are released into the extracellular environment and are subsequently designated as exosomes (Patel et al. 2023). ILV formation relies on two pathways: the first is endosomal membrane reorganization involving tetraspanins CD9 and CD63 and the second involves the Endosomal Sorting Complexes Required for Transport (ESCRT) pathway. ESCRT is composed of approximately 20 proteins sorted into four protein ESCRT complexes (0, I, II, III) that have vital functions in both the exosome secretion pathway as well as the cargo sorting before secretion. ESCRT I and II are critical for multivesicular body budding and ESCRT III finishes budding and cleavage, resulting in secretion-ready ILV (Burtenshaw et al. 2022) (Fig. 1.2).

Microvesicles (MVs) are typically larger in size than exosomes (50–1000 nm), are more heterogeneous in structure, and shed directly from the plasma membrane through fission (Patel et al. 2023). MVs may also be referred to as ectosomes, shedding vesicles, microparticles, and oncosomes. Like exosomes, MVs traffic membrane proteins, lipids, and nucleic acids to nearby cells to generate a rapid cellular response. Exosomes and MVs share similar biogenesis machinery such as ESCRT I, II, III, and accessory proteins including ARRDC1, Rho family proteins, VPS4, and Rho-associated protein kinase (ROCK) and extracellular signal regulator kinase (ERK) (Jin et al. 2022). MV genesis does not involve the formation of multivesicular bodies and ILV, and instead, proteins and nucleic acids are vertically trafficked to the plasma membrane at the site where MV outward blebbing occurs (Tricarico et al. 2017). There are several proteins involved with roles in MV biogenesis and cargo sorting, although the mechanisms of MV budding are not completely understood. ADP ribosylation factor 6 (ARF6) is a GTP-binding protein that recycles cellular proteins to selectively incorporate and exclude cargo. Imperative for biogenesis, ARF6 is a key regulator of MV shedding through activation of ERK and localization of myosin light chain kinase at the head of budding vesicles, resulting in contractile machinery and ROCK activation (Muralidharan-Chari et al. 2009).

Exosomes and MVs are constantly being produced by normal, viable cells while apoptotic bodies are a distinct class of EVs released by dying cells during the last stage of apoptosis. Apoptotic bodies were once thought of as simply debris of dying cells, but it has since been well established they have important roles in cell–cell communication. During apoptotic dismantling, remnants of the apoptotic cell such as condensed chromatin, cytosol, DNA, and degraded proteins, are packed into large (50–5000 nm) membranous vesicles (Yu et al. 2023). The formation of apoptotic bodies is a highly regulated process and important for recruiting phagocytes for clearance. Apoptotic bodies form by the initialization of "membrane blebbing" caused by the rupturing of the plasma membrane regulated by several kinases (i.e., ROCK) and actomyosin-mediated contracting. Once the nucleus invaginates and blebbing occurs, membrane protrusions appear which can either be microtubule spikes, apoptopodia, or beaded apoptopodia (Caruso et al. 2019). Each of these protrusions has distinct molecular and cytosolic mechanistic pathways. Caspases are activated either through the intrinsic pathway, controlled by the BCL-2 protein which results in the permeabilization of the mitochondrial outer membrane, or the extrinsic pathway, involving transmembrane receptor interactions transmitting

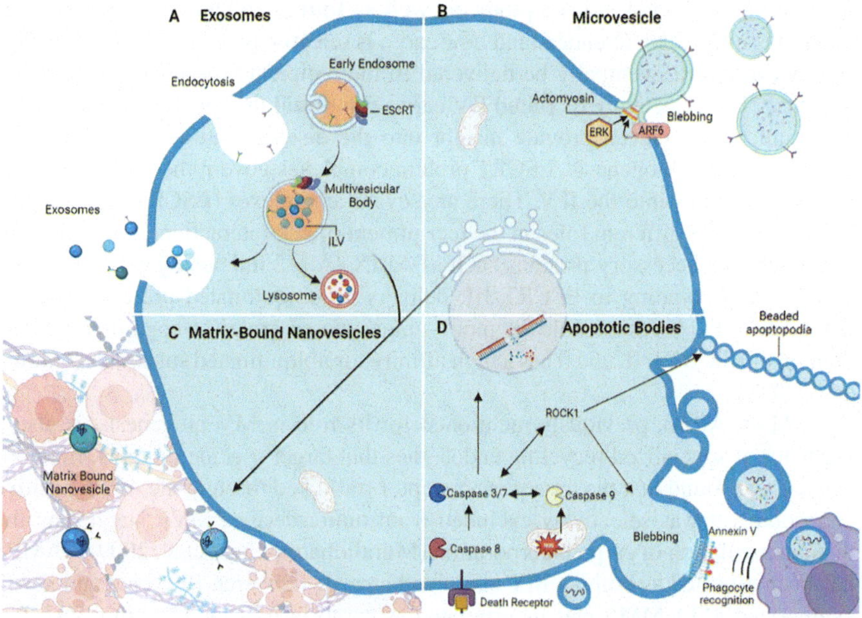

Fig. 1.2 Biogenesis of the major classes of extracellular vesicles. (**a**) Exosome biogenesis occurs through the endosomal pathway. Early endosomes form and go through stepwise sorting and processing mechanisms using ESCRT machinery 0, I, II, and III (Patel et al. 2023). Intraluminal vesicles (ILV) form within the multivesicular body (also called early endosomes) upon membrane invagination before further cargo sorting into lysosomes or late endosomes. Late endosomes fuse with the plasma membrane to release the ILV into the extracellular fluid, now denoted as exosomes (Simons and Raposo 2009). (**b**) Microvesicle (MV) formation involves similar machinery used in exosome biogenesis, involving ESCRT and accessory proteins, but lacks ILV formation. Instead, cargo is vertically trafficked to plasma membrane where membrane blebbing occurs. ARF6 has been found to be a regular MV formation for its role in the assembly of actomyosin localization and ERK activation. ARF6 also has roles in cargo sorting (Tricarico et al. 2017). (**c**) Matrix-Bound Nanovesicle (MBV) biogenesis remains elusive, however, emerging evidence suggests that MBV are secreted in concert with fibrillar matrix components (Frantz et al. 2010). (**d**) Apoptotic bodies are formed either through intrinsic or extrinsic stimuli. Extrinsic apoptosis starts with the ligand binding of a "death receptor" which begins a caspase cascade, leading to nuclear condensation and DNA condensation. Intrinsic stimuli (cell stress, DNA damage, hypoxia, etc.) can lead to mitochondrial outer membrane permeabilization (Nössing and Ryan 2023) and activation of caspase 9, which in turn activates effector caspase 3/7. Once caspase 3/7 is activated, ROCK1 is cleaved and increases actomyosin contraction, inducing membrane blebbing and formation of apoptopodia (Dixson et al. 2023). Phosphatidylserine translocates to the outer leaflet of the vesicle lipid bilayer, binding to Annexin V, a marker for phagocytic recognition (Gurung et al. 2021)

external death signals inside the cell. After signaling cascades and the involvement of several procaspase and caspase enzymes, effector caspase 3/7 is activated in both pathways to form apoptotic bodies (Yu et al. 2023).

Exosomes, MVs, and apoptotic bodies are all enclosed by a lipid membrane derived from the parent cell membrane and have various nucleic acid and protein cargos that are preferentially sorted during biogenesis. EV protein content depends on both the origin cell transcriptome as well as sorting complex selective recruitment. It is imperative to understand how cargo is selectively organized into vesicles, as this cargo will eventually be delivered to the recipient cell inducing a specific response. The selectively recruited EV cargo can reveal the role of EVs in cellular communication or even provide insight into the disease state of the origin cell. During exosome biogenesis, ESCRT protein complexes govern the sorting of ubiquitinated proteins into the ILV. There are several complexes (ESCRT I, II, III, and IV) that all have different roles in the recruitment of and interaction with each other and additional accessory proteins such as ALIX (Alg-2 interacting protein X) and VPS4. ALIX, binding to ESCRT III, delivers un-ubiquitinated protein cargos to ILVs. ESCRT 0 is responsible for mono-ubiquitinated protein recognition, while a domain of ESCRT I, II, and 0 has a high affinity for ubiquitinated substrates (Gurung et al. 2021).

In MVs, ARF6, previously mentioned for its role in MV biogenesis, is also a regulator of specialized recycling endosomes that target specific proteins. In tumor cells, it was found that the transferrin receptor trafficked through the ARF6 positive endosome yet was selectively excluded from tumor-derived MVs, suggesting the distinct regulation of cargo incorporation (Muralidharan-Chari et al. 2009). VAMP3 has been reported as both an MV cargo and a cargo deliverer. Recycled and newly synthesized MT1-MMP was incorporated as cargo in nascent MVs in a VAMP3-dependent mechanism. The authors suggested that MT1-MMP is stored in reserves for quick release. It is likely different cells and EVs derived from those cells use both distinct and redundant protein sorting mechanisms (Muralidharan-Chari et al. 2009). Another study explored how cellular stress, specifically hypoxia, increases MV shedding. RAB22 colocalizes with reservoirs of transglutaminase-2 at locations of MV bud formation; it was found that knockdown of RAB22A, typically upregulated in hypoxic conditions, cancels out the hypoxia-induced increase in MV shedding, indicating RAB22A recruits protein cargo to sites of MV release (Wang et al. 2014).

Apoptotic bodies, much larger than exosomes and MVs, contain protein and RNA, but also may have lipid, DNA, and organelle remnant cargo. Importantly, apoptotic bodies contain information derived from dying cells. Most apoptotic bodies are phagocytosed by macrophages through the recognition of the distinct translocated phosphatidylserine to the outer leaflet of the vesicle lipid bilayer. The phosphatidylserine then binds to Annexin V, a recognition marker for phagocytic cells (Akers et al. 2013). Apoptotic bodies carry the remains of dead cells including proteins from the nucleus, mitochondria, and plasma membrane, as well as lipids and a wide range of whole or fragmented nucleic acids (mRNA, miRNA, rRNA, DNA, etc.). Apoptotic bodies can horizontally transfer these leftovers to neighboring cells. Although it was previously believed that apoptotic bodies acquired cargo in a stochastic manner, it has since been proven that there is regulation of apoptotic body cargo sorting. A large percentage of apoptotic bodies were found to contain RNA and not DNA, or DNA and not RNA, suggesting the selective packaging of biomolecules into apoptotic vesicles (Halicka et al. 2000). Apoptotic bodies also contain autoantigens, suggesting their role in antigen presentation. A newly discovered mechanism of apoptotic body formation, termed "beads-on-a-string," found membrane protrusions extend from the apoptotic cell suggesting the selective incorporation or exclusion of nuclear material (Dixson et al. 2023; Nössing and Ryan 2023). Nuclear contents were depleted in apoptotic body-enriched fractions while proteins with roles with signal transduction, cell maintenance, and transportation, were augmented. Nuclear DNA and components were excluded from beaded apoptopodia protrusions but present in apoptotic cell bodies. Acidic organelles and mitochondria were highly enriched in the apoptotic cell body and occasionally tracked in beaded apoptopodia. These data collectively suggest there are selective organization mechanisms at play during apoptosis and apoptotic vesicle biogenesis (Atkin-Smith et al. 2015).

Aside from utilizing nucleic acid species to distinguish distinct EV subsets or to act as biomarkers for various disease states which will later be discussed in detail, an EV's nucleic acid profile reflects the cell type and disease state of the source cell as well as the EV isolation method used (Crescitelli et al. 2013; Van Deun et al. 2014; Ciferri et al. 2021). EVs generally have been shown to contain a wide array of full-length and fragmented RNA species including mRNA, non-coding RNA, small nuclear RNA, long non-coding RNA, and transfer RNA. The wide range and variability of EV RNA content make it difficult to establish functionality, thus many are trying to understand the biological impact of RNA species contained in vesicles. One vesicle may contain hundreds of RNA species, adding to the complexity of the EV RNA characterization. Small RNA molecules are extremely susceptible to degradation, making EV-mediated small RNA analysis advantageous, as an EV confers protection against enzymatic destruction (Schulz-Siegmund and Aigner 2021).

A comprehensive study delved into the repertoire of RNA species across five distinct cell lines, revealing that parental cells predominantly harbored small RNAs, with a notable emphasis on microRNAs (miRNAs) (Sork et al. 2018). In contrast, EVs exhibited minimal and highly variable levels of small RNAs across cell lines. The investigation unveiled a moderately high correlation between the abundance of miRNAs in vesicles and their parent cells. However, notable differences were observed among cell lines, as exemplified by the substantial enrichment of miRNAs in C2C12 cells compared to HEK EVs, despite the absence of a detectable difference in their miRNA-related proteomes. This study serves as a valuable illustration of the intricate nature of understanding the mechanisms and purpose of EV nucleic acids (Sork et al. 2018). The observed variability and inconsistencies in EV nucleic acid content, influenced by factors such as cell type, disease state, timing, EV subset, and other confounding variables, highlight the complexity of unraveling the biology of EVs. It is acknowledged that numerous transcriptional and translational components contribute to the resultant EV proteome and transcriptomic. The scientific community remains actively engaged in efforts to uncover the biological mechanisms governing EV RNA incorporation and to further explore the theranostic potential of EVs.

1.3 Matrix-Bound Nanovesicles, the Messenger in the Matrix

Once considered merely a static structural support, extensive research has unveiled the crucial role of the extracellular matrix (ECM) in preserving tissue homeostasis. The ECM is now acknowledged as a dynamic network of both functional and structural macromolecules arranged in tissue-specific conformations (Frantz et al. 2010). In addition to foundational proteins such as collagen, elastin, fibronectin, and proteoglycan, the ECM boasts a myriad of growth factors, cytokines, and bioactive molecules that intricately govern cellular behavior (Brown and Badylak 2014). A recent revelation in ECM composition is the matrix-bound nanovesicle (MBV) (Huleihel et al. 2016) (Fig. 1.2).

The assessment of double-stranded DNA commonly serves as a metric for evaluating the decellularization of biologic scaffolds. Huleihel et al. conducted studies to explore nucleic acid content following the decellularization of various commercially available extracellular matrix (ECM) scaffolds (Huleihel et al. 2016). An intriguing revelation from these investigations was the identification of a small population of nuclease-protected nucleic acids encapsulated within vesicular bodies. These minute vesicular bodies were characterized as matrix-bound nanovesicles (MBVs), measuring 50–400 nm, and were observed tightly associated within the collagen matrix under electron microscopy. Complete ECM digestion allowed for the separation of MBVs from the collagen network. Notably, these newly discovered vesicles lacked common exosome surface markers (CD63, CD81, CD9, and HSP70), suggesting that they represented a distinct entity and were not merely exosomes entrapped in the extracellular matrix (Huleihel et al. 2016).

In addition to their physical location within the extracellular matrix, MBVs exhibit distinct disparities in lipid content and cargo when compared to exosomes. Recent studies have unveiled a unique lipid profile for MBVs in comparison to other EVs (Hussey et al. 2020). Notably, MBVs harbor a significant proportion of polyunsaturated fatty acids (PUFAs), with enrichments in phosphatidylinositol, phosphatidylglycerol, phosphatidylserine, and bis-monoacylglycerophosphate. These PUFAs likely serve as precursors to signaling molecules crucial for the immunomodulatory effects of MBVs. Hydrolysis of these phospholipids by phospholipases can yield lysophospholipids, bioactive molecules intricately involved in inflammatory and repair processes. The lipoprotein lipase (LPL) profiles of MBVs differ from both EVs and parent cells, suggesting that these LPL molecules may play a role in influencing the transfer of vesicular cargo to cellular targets. Furthermore, the free fatty acids resulting from PUFA metabolism, such as arachidonic acid, docosahexaenoic acid, and docosapentaenoic acid, are notably enriched in MBVs. This lipid profile indicates that MBVs may function as a reservoir for these lipid mediators, potentially activated by phospholipases in response to the inflammatory context of the extracellular environment.

MiRNA contained within EVs functions as posttranslational gene regulators via binding to target mRNA and altering translation (O'Brien et al. 2018). Hussey et al. demonstrated distinct miRNA signatures in MBVs compared to exosomes in a 3T3 fibroblast model (Hussey et al. 2020). In their study, liquid-phase exosomes and MBVs integrated into the matrix, both derived from the same parent cells, were harvested and subjected to next-generation miRNA sequencing. Half of miRNAs were found to be differentially expressed in the MBVs compared to exosomes. Overall, miRNA profiles of the exosomes and MBVs compared to the parental cell miRNA were clearly distinct.

In concordance with analogous phenomena observed in other extracellular vesicles, MBVs exhibit distinctive cargo profiles contingent upon their tissue of origin. Investigative efforts by Turner et al. encompassed a comparative analysis of MBVs derived from diverse biological sources, including porcine urinary bladder, dermis, small intestinal submucosa, liver, and heart ECM (Turner et al. 2022). This study evaluated cytokine arrays and miRNA content within MBVs originating from these varied tissues. Although certain cargo constituents were conserved across different sources, discernible distinctions were identified that were specific to the tissue of origin. Particularly noteworthy were the outcomes of small RNA sequencing, which revealed cargo elements intricately linked to the functional attributes of the source tissues. For instance, MBVs derived from liver tissue manifested cargo characterized by its association with synthetic and metabolic functionalities. This cargo repertoire encompassed the expression of miRNAs pertinent to endoplasmic reticulum processing, growth factor production, and glucose metabolism. MBVs sourced from cardiac tissue exhibited a distinct miRNA expression profile indicative of oxidative stress, aligning with the heightened metabolic activity inherent in this particular source tissue.

MBVs play a pivotal role in influencing the biological activities of the extracellular matrix (ECM). Extracellular matrix scaffolds actively facilitate the structural remodeling of both healthy and injured tissues through the orchestrated release of bioactive molecules, recruitment of progenitor cells, and modulation of the immune response toward an anti-inflammatory, pro-reparative phenotype (Swinehart and Badylak 2016). A prevailing hypothesis posits that MBVs are released concomitantly with the degradation of ECM scaffolds, thereby contributing substantively to these crucial biological activities (Huleihel et al. 2017). Recent investigations have methodically examined the biological effects of isolated MBVs. Notably, decellularized tissue scaffolds consistently exhibit a propensity to transition inflammatory M1 macrophages toward an anti-inflammatory M2 phenotype both in vitro and in vivo. MBVs have similarly been observed to induce M2 macrophage phenotypes and harbor interleukin-33 (IL-33), a recognized mediator of M2 macrophage polarization (Huleihel et al. 2017; Hussey et al. 2019). The potential translation of the anti-inflammatory effects of MBVs into clinical applications has been explored using a rodent model of pristane-induced rheumatoid arthritis (Crum et al. 2022). Systemic administration of MBVs demonstrated comparable improvements in inflammatory arthritis when juxtaposed with the standard methotrexate treatment, albeit without the broad immunosuppressive effects associated with the latter.

Furthermore, MBVs have exhibited neuroprotective effects in various experimental models. Faust et al. demonstrated that MBVs derived from urinary bladder matrix (UBM) are internalized by hippocampal neurons, positively influencing neurite growth to a greater extent than UBM hydrogel alone (Faust et al. 2017). In a corresponding study, MBVs were found to regulate retinal ganglion neurite growth in a model of ischemic axon degeneration (van der Merwe et al. 2019). These findings collectively underscore the multifaceted therapeutic potential of MBVs in diverse biological contexts.

1.4 EVs Effects on the ECM

1.4.1 Calcification

One of the initial instances identifying the modulation of the extracellular matrix (ECM) by EVs is observed in the process of calcification (Fig. 1.3). Matrix vesicles (MtVs) represent a specific subclassification of EVs originating from osteoblasts that play a pivotal role in matrix mineralization processes (Liu et al. 2018). These vesicles are characterized by the presence of particular osteogenic components such as phosphatases, calcium, and inorganic phosphate, in addition to the standard cargo found in EVs. Notably, the inorganic constituents within MtVs have the capability to undergo precipitation, forming hydroxyapatite crystals. Subsequently, these crystals can grow and disrupt the vesicle membrane, leading to the formation of mineralized nodules (Hasegawa et al. 2017). The mineralized nodules established through this process come into contact with the collagen present in the matrix, potentially initiating collagen mineralization. Alternative theories regarding mineralization mechanisms propose that MtVs may bind to the collagen matrix through calcium channel-forming annexins and alkaline phosphatase, thereby serving as a source of amorphous calcium phosphate (Wu et al. 1991; Wuthier and Lipscomb 2011; Kirsch et al. 1997).

Apart from direct roles in mineralization, specialized EVs function to promote osteogenic differentiation. Bone homeostasis is regulated by the precise balance of bone formation and resorption by osteoblasts and osteoclasts respectively. Historically the functions of these two critical cell types were thought to be controlled by direct contact, cytokines, and growth factors. EVs present an alternative method of communication via the exchange of EVs with surface ligands or EV cargo (Ansari et al. 2021). Various studies have provided evidence of EVs modulating RANK-RANKL interactions critical to bone remodeling. After periods of bone resorption, osteoclasts undergo apoptosis releasing large amounts of apoptotic bodies that promote osteogenesis via reverse RANKL signaling (Morrell et al. 2018; Huynh et al. 2016; Yuan et al. 2018).

EVs have been investigated in pathologic extraosseous calcification, in addition to their role in physiologic skeletal processes (Bakhshian Nik et al. 2017; Qin et al. 2021). A prime example of this process is vascular calcification, a component of advanced atherosclerosis with profound clinical consequences. The accumulation of calcium compounds in vascular tissue leads to arterial stiffening, a major

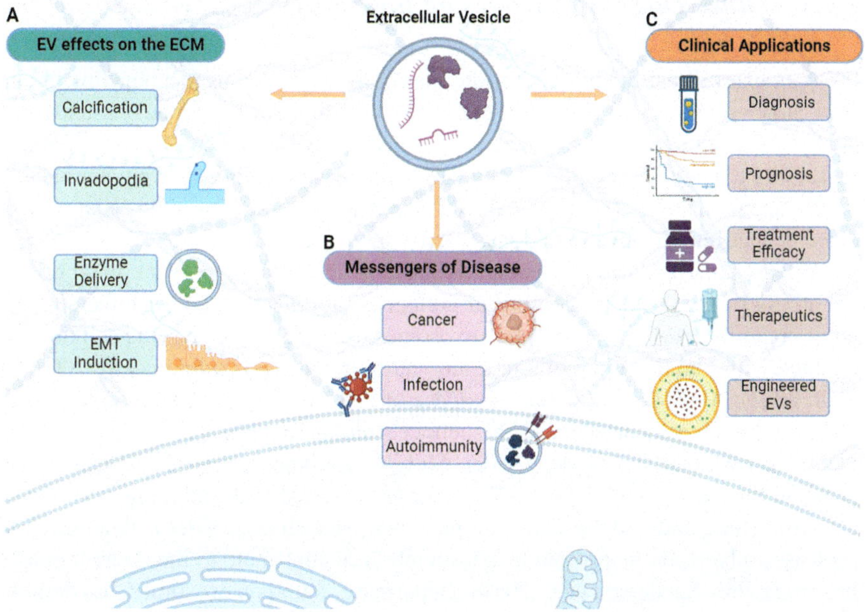

Fig. 1.3 Extracellular vesicles as messengers of the matrix: EVs hold great potential in various applications in the medical and research field. (**a**) EVs can affect the organization and behavior of the ECM by harboring osteogenic molecules and inducing calcification, stimulating invadopodia in cancer cells, delivering enzymes to remodel the matrix, and inducing the epithelial to mesenchymal transition (EMT), especially in terms of cancer cell migration. (**b**) EVs can act as messengers of disease through their alteration of cell phenotype. Pathogen-derived EVs serve as warnings to immune cells, displaying pathogen-associated molecular patterns (PAMPs) that can be recognized by pattern recognition receptors (PRR) on surveilling cells. Evidence suggest EVs are associated with autoimmune disease, carrying peptides and antigens that may contribute to the initiation/progression of diseases such as rheumatoid arthritis and inflammatory bowel disease. (**c**) EVs are under further investigation for their potential use for diagnosis, prognosis, treatment efficacy, therapeutics, and targeted, engineered EVs

component of cardiovascular morbidity. Under normal physiologic conditions, vascular smooth muscle cells (VSMC) EVs contain inhibitors of vascular calcification, fetuin A, and osteoprotegerin as major examples (Touyz and Montezano 2009). However, in settings of injury or stress VSMC shift toward a synthetic osteoblastic phenotype with the release of calcified exosomes depleted in fetuin A and membranes enriched in annexin and phosphatidyl serine (Kapustin et al. 2015). These osteogenic exosomes serve as a nidus for calcification within the extracellular matrix (Zazzeroni et al. 2018). Similarly, vascular endothelial cells release matrix vesicles containing bone morphogenic protein 2 (BMP2), stimulating calcification (Buendía et al. 2015). Both osteogenic MVs and exosomes have been found in precalcified vessels and it is theorized that their release during vascular injury may catalyze vessel calcification (Kapustin et al. 2015).

1.4.2 Other Direct Influences on the Matrix

EVs can exert direct modulation of the matrix through various mechanisms. Some EVs carry proteases and glycosidases capable of remodeling the matrix (Sanderson et al. 2019). In the context of cancer, many EVs derived from cancer cells leverage this ability to enhance cell migration, thereby increasing their invasive potential and facilitating metastasis (Franzen et al. 2015). Subsequent sections in this chapter will delve deeper into this phenomenon.

Proteases, particularly matrix metalloproteinases (MMPs), play a crucial role in the ability of EVs to degrade fibrillar collagens, fibronectins, and members of the laminin family. MMP-14, present on exosomes from many tumorigenic cell lines, is credited with playing a key role in tumor cell migration (Hakulinen et al. 2008; Li et al. 2022). ADAM-17, a potent sheddase of cell surface molecules implicated in conditions such as heart disease, Alzheimer's, inflammation, and cancer, has been identified on exosomes (Gooz 2010). Mobile exosomes carrying ADAM-17 can promote the shedding of ADAM-17 substrates at distal sites, including cytokines, growth factors, and cellular adhesion molecules (Groth et al. 2016). In addition to proteases, glycosidases present in EV membranes contribute to the degradation of proteoglycans (Sanderson et al. 2019). Heparinase, found on exosomes, has demonstrated biological activity. Heparin sulfate proteoglycans sequester biosignaling molecules by anchoring them to the ECM. EVs modulate ECM signaling by degrading proteoglycans, releasing these molecules, which likely play a significant role in processes such as angiogenesis, inflammation, and tumor migration (Bandari et al. 2018).

Other membrane-bound enzymes identified in EVs include collagenases, elastases, and lysyl oxidase-like-2 (LOXL2) (Genschmer et al. 2019; de Jong et al. 2016). EVs derived from activated neutrophils exhibit notable increases in collagen and elastin degradation compared to EVs from inactivated neutrophils via increased surface neutrophil elastase (Genschmer et al. 2019). Activated neutrophil-derived exosomes from lung disease donors were able to induce emphysema in previously healthy murine lungs via ECM degradation (Genschmer et al. 2019). This effect was not seen in quiescent neutrophil exosomes. LOXL2, responsible for collagen crosslinking, is present when endothelial cells are grown in hypoxic conditions. When exosomes from endothelial cells cultured under hypoxic conditions were incubated with collagen gel, crosslinking was increased resulting in twofold higher stiffness compared to exosomes from cells cultured under normoxic conditions (de Jong et al. 2016). Mesenchymal-derived stem cell exosomes have been shown to downregulate LOXL2 and subsequent liver fibrosis via enrichment in a master microRNA (miR-27b-3p) (Cheng et al. 2023). This presents a possible mechanism of disease and therapeutic target for fibrogenic conditions such as idiopathic pulmonary fibrosis and liver fibrosis (Barry-Hamilton et al. 2010).

1.4.3 Indirect Matrix Effects

EVs show the capacity to indirectly impact the extracellular matrix (ECM) by influencing cellular behavior. A noteworthy illustration of this phenomenon is evident in cancer cell invadopodia, specialized membrane protrusions housing enzymes responsible for ECM degradation, thereby facilitating cancer cell invasion (Hoshino et al. 2013). Studies by Hoshino et al. used a carcinoma cell line with a knockdown of RAB27a, a crucial factor for exosome formation, and showed a substantial reduction in invadopodia production and lifespan. Conversely, the introduction of exogenous exosomes to these cells resulted in an increased formation of invadopodia (Hoshino et al. 2013). It is plausible that exosomes harbor essential growth factors and signaling molecules essential for invadopodia formation and subsequent modification of the matrix.

EVs contribute to alterations in cell phenotype, ultimately influencing the matrix. Epithelial–mesenchymal transition (EMT) is one such process by which exosomes can alter cells, which can lead to matrix modification. EMT entails a transformation from an epithelial architecture with tight cell connections to mesenchymal properties characterized by weakened intercellular adhesion, loss of apical-basal polarity, and secretion of matrix metalloproteinases (MMPs) (Lamouille et al. 2014; Gonzalez and Medici 2014). This phenomenon is observed in processes such as wound healing, organ fibrosis, and cancer (Haensel and Dai 2018; López-Novoa and Nieto 2009; Huang et al. 2022). EVs contain inducers of EMT, such as TGF-β, TNF-α, β-catenin, annexin A2, caveolin-1, and MMPs (Xu et al. 2009; Wang et al. 2013; Huang and Feng 2017; Lin et al. 2019; Zhu et al. 2022). Phenotypic shifts induced by EVs have been identified as promoters of tumor growth. An illustrative example involves the utilization of EVs derived from breast cancer, wherein Song et al. demonstrated the stimulation of desmoplastic reprogramming in adipose-derived stem cells. This process, in turn, facilitated the creation of a proangiogenic tumorigenic environment (Song et al. 2017). The findings underscore the role of EVs in orchestrating phenotypic alterations that alter the extracellular matrix and tumor microenvironment, ultimately fostering conditions conducive to enhanced tumor growth.

1.5 Messengers of Disease

1.5.1 Infection

The ubiquitous presence and diverse functionalities of EVs contribute significantly to the fundamental pathogenesis of various disease states. EVs play a crucial role in modulating immune function through the transfer of cargo to cells, a phenomenon evident in both bacterial pathogenesis and autoimmune inflammatory conditions. Gram-negative bacteria, for instance, release exosomes, also known as outer membrane vesicles (OMVs). These vesicles contain pathogen-associated molecular pattern molecules (PAMPs), serving as alerts to immune cells upon recognition by pattern recognition receptors (PRRs) such as Toll-like receptors (TLRs) (Tan et al. 2018). An example of a PRR is Nucleotide Oligomerization Domain 1 (NOD1), which recognizes bacterial peptidoglycan present in OMVs released by Gram-negative bacteria like *Pseudomonas aeruginosa* (Tan et al. 2018; Kaparakis et al. 2010). NOD1 activation triggers an innate immune response cascade, including the upregulation of nuclear factor-κB (Kaparakis et al. 2010). Even in the absence of an outer membrane, Gram-positive bacteria also produce EVs. Protease and toxins contained within *Staphylococcus aureus* EVs have the ability to potentiate the immune response and further the pathogenicity of *S. aureus* (Kim et al. 2012; Wang et al. 2021).

1.5.2 Autoimmunity

Given the compelling body of evidence implicating EVs in the modulation of the immune response, it is predictable that they are implicated in autoimmune diseases. EVs possess the capability to carry major histocompatibility complexes I and II, endowing them with the capacity to participate in antigen presentation and activation of immune cells (Barros et al. 2018). One autoimmune disease of particular relevance is rheumatoid arthritis (RA), characterized by synovial inflammation and concurrent collagen destruction. Notably, synovial fluid obtained from individuals with RA shows a heightened concentration of EVs that contain content such as citrullinated peptides that play a role in the inflammatory manifestations of the condition (Cloutier et al. 2013; Withrow et al. 2016; Foers et al. 2020).

Exosomes have been identified as a potential source of autoantigenic proteins. Exosomes derived from salivary glands, for instance, harbor ribonucleoproteins Ro (SS-A), Lupus La protein (SS-B), and the Smith antigen (Sm) (Kapsogeorgou et al. 2005). These proteins are intrinsically associated with autoimmune conditions such as systemic lupus erythematosus (SLE) and Sjögren's syndrome. The presence of such autoantigenic cargo in exosomes suggests their potential role in the pathogenesis and immune dysregulation observed in these autoimmune disorders. Exosomes play a pivotal role in shaping the immune cascade associated with inflammatory bowel disease (IBD). Central to the pathogenesis of IBD is the maintenance of the integrity of the intestinal epithelial barrier, which governs the intricate interactions between luminal contents and the immune system (Antoni et al. 2014). The dysregulation of this delicate equilibrium, responsible for clearing pathogenic bacteria and fostering immune tolerance toward commensal bacteria, is implicated in the etiology of IBD. Exosomes, through their influence on wound repair pathways, can significantly impact the permeability of the intestinal epithelial barrier.

EVs released by epithelial cells, particularly those harboring Annexin A1, play a significant role in the reparative processes of the epithelium during inflammation. This influence is mediated through a mechanism that engages formyl peptide receptors and NOX1 (Leoni et al. 2013). Notably, within the context of IBD, there is evidence of an increased release of Annexin A1-containing EVs during inflammation (Leoni et al. 2015). This augmentation in EV release may potentially contribute to the observed dysregulation in the intestinal epithelial barrier. In addition to the role of Annexin A1, neutrophil-mediated inflammatory membrane damage is facilitated by the production of oxidative radicals through the action of myeloperoxidase (MPO). EVs serve as carriers for MPO, delivering it to epithelial cells in the intestine, thereby perpetuating the inflammatory response (Slater et al. 2017; Zhang et al. 2018).

Moreover, intestinal EVs, including exosomes, actively contribute to immune tolerance by engaging with dendritic cells. EVs featuring αvβ6 are internalized by dendritic cells, thereby initiating processes associated with transforming growth factor-beta (TGF-β) and regulatory T cells, as elucidated by Chen et al. (Chen et al. 2011). Furthermore, exosomal miRNA has been identified as a key modulator capable of promoting inflammatory reactions through the NF-κB signaling pathway, subsequently leading to the downstream production of proinflammatory cytokines. For instance, miR-16 downregulates adenosine A2A receptor inhibition of NF-κB, resulting in an increased expression of inflammatory cytokines (Tian et al. 2016). These multifaceted interactions underscore the diverse roles of exosomes in the pathogenesis of IBD.

1.5.3 Cancer

Although EV mechanisms implicated in cancer were discussed in preceding sections their relevance to cancer biology cannot be overstated. EVs released by cancer cells, integral to their malignant potential, play a significant role in matrix degradation through the enzymatic activities of matrix metalloproteinases (MMPs) and glycosidases. Notably, nasopharyngeal cancer cells exhibit heightened MMP-13 expression, leading to increased matrix degradation and enhanced infiltration of cancer cells (Shan et al. 2018). Enzymes situated on the surface vesicles target proteoglycans and glycoproteins, contributing to degradation processes that augment cell migration. The consequential release of active signaling molecules from these degradatory processes influences local cellular behavior and spurs tumor cell migration. The tumor microenvironment undergoes significant alterations due to the impact of EVs on matrix components and cell interactions (Tao and Guo 2020). Under hypoxic conditions, EVs can elevate cellular MMP expression, thereby increasing cell migration and invasion. Additionally, EVs can reduce E-cadherin expression, consequently promoting tumor growth and migration (Shan et al. 2018; Aga et al. 2014).

The pre-metastatic niche refers to the site conducive to potential cancer cell metastasis, aligning with the “seed and soil” hypothesis. This hypothesis posits that circulating tumor cells (the “seed”) necessitates a conducive environment (the “soil”) for colonization. Tumors release EVs that prime the microenvironment for tumor growth (Peinado et al. 2017). Melanoma-derived EVs influence bone marrow progenitor cells toward a pro-tumor growth phenotype (Peinado et al. 2017). These exosomes carry activated MET-related signaling proteins that induce tumor cell proliferation and migration (Peinado et al. 2012). Furthermore, EVs derived from melanoma orchestrate extracellular matrix (ECM) modulation, thereby promoting tumor invasion, angiogenesis, and priming sentinel lymph nodes for tumor migration (Peinado et al. 2017). EVs released from breast cancer, leukemia, and pancreatic cancer cell lines contain extracellular matrix metalloproteinase inducer (EMMPRIN) (Redzic et al. 2013; Hood et al. 2011). Treatment with these respective EVs upregulates MMP-9 and interleukin-6 (IL-6) across all cell lines (Redzic et al. 2013). Moreover, a positive feedback cycle is observed, wherein cells exposed to EMMPRIN-containing EVs produce a greater number of EMMPRIN EVs (Redzic et al. 2013). The CD44-hyaluronic axis is an important adhesion interaction for tumor-derived EV and ECM crosstalk. CD44-positive tumor exosomes have been detected in several cancer types as key modulators of the extracellular matrix by supplying proteases, enabling cell motility and migration, promoting proliferation, and inducing apoptosis resistance (Mu et al. 2021).

1.6 Clinical Applications

EVs inherently alter the composition of the ECM by regulating the synthesis or degradation of matrix molecules through the transfer of proteins, nucleic acids, and cargo. The fundamental role of EVs in cellular regulation and disease is closely tied to their composition, prompting the exploration of differences between EVs derived from healthy and diseased cells. Thanks to the high biological stability conferred by the lipid bilayer of EVs, numerous studies have delved into the potential use of EVs as diagnostic markers in various cancers such as breast, liver, colon, and others (Sandfeld-Paulsen et al. 2016; Su et al. 2021). The ease of collecting EVs from body fluids like urine, saliva, or serum makes them particularly suitable for diagnostic and prognostic purposes. Consequently, most studies focus on circulating fluid-residing EVs, with a majority being exosomes. The application of EVs in cancer diagnostics is multifaceted, encompassing tumor diagnosis, prognosis prediction, and assessing response to treatment. Similarly, the diagnostic potential of EVs has been explored in various other diseases beyond cancer.

1.6.1 Diagnosis

Many studies seek to find the upregulation or downregulation of distinct EV species, distinguished by surface markers, nucleic acids, or protein composition in cancer patients or cell lines compared to healthy controls. Sandfeld-Paulsen et al. extracted exosomes from plasma isolated from over 500 patients and found three distinct markers (CD151, CD171, tetraspanin 8) were enriched in cancer patient exosomes independent of disease stage (Sandfeld-Paulsen et al. 2016). Another found 144 phosphorylated proteins significantly elevated in breast cancer patients (Chen et al. 2017) while in a different highly impactful study, proteoglycan GPC1-positive circulating cancer exosomes had a 100% correlation with KRAS mutations in pancreatic tumors, suggesting the use of GPC1 exosomes as a biomarker for early pancreatic cancer. Circulating EVs are promising diagnostic markers as they are often detectable before expensive and inaccurate MRI-based screenings (Melo et al. 2015). EV-derived miRNA is one of the most promising biomarkers, as several studies have shown the unique EV-derived miRNA content of benign versus malignant tumors. Apart from the use of EVs as diagnostic tools in several types of cancer, there are completed and ongoing clinical trials in other pathologic states including but not limited to myocardial infarction, chronic inflammation, and Parkinson's Disease (Ciferri et al. 2021). The identification of circulating extracellular vesicle microRNA 21-5p (miR-21-5p), derived from beta cells (insulin-producing cells), has been proposed as a potential marker for the development of type 1 diabetes. Elevated levels of this EV-specific miR-21-5p have been observed in hyperglycemic mice and children experiencing the onset of type 1 diabetes. Ongoing studies are increasingly exploring the role of these specific EVs in the context of type 1 diabetes (Lakhter et al. 2018).

1.6.2 Prognosis

The potential to acquire EVs through non-invasive methods, along with their compositions reflecting those of parental cell lines, positions them as promising prognostic markers. EVs, being able to permeate the blood-brain barrier, may have important implications in brain cancers with miRNA-21 as one of the most investigated small RNA species for its involvement in glioma. miRNA-21 expression level was found to be inversely proportional to the survival of glioblastoma patients and was delivered to microglia by glioma cell-derived EVs, increasing microglia proliferation (Yang et al. 2014; Abels et al. 2019). The relative levels of nucleic acids can reveal cancer severity. For example, the levels of exosome miR-373 were significantly higher in triple-negative breast cancer and estrogen and progesterone-negative tumors compared to luminal carcinomas and hormone-positive tumors respectively (Eichelser et al. 2014). One study analyzed several tumor models that have a high rate of metastasis to specific sites and found that exosome-derived integrins direct an organ-specific exosome biodistribution in mouse models. Breast and lung cancer cell line-derived exosomes were preferentially uptaken by cells in the lungs, suggesting exosomes have a role in organ-specific metastasis. This study linked exosome integrin expression with tissue organotropism and metastasis, revealing correlations between distinct exosome-derived integrin levels before metastasis in patients with breast cancer and liver cancer, suggesting a possible method to predict likely sites of metastasis (Hoshino et al. 2015).

1.6.3 Response to Treatment

Like the diagnostic and prognostic applications of EVs, correlations can also be drawn between EV protein and nucleic acid composition and treatment efficacy. In patients with metastatic melanoma, high expression of PD-1 and CD28 on dendritic cell-derived exosomes correlated with improved progression-free survival and a prolonged survival rate after IPI treatment (Tucci et al. 2018). A study using EV long RNA sequencing determined a diagnostic signature able to detect stage I/II breast cancer with high accuracy and also found EV methylsterol monooxygenase 1 (MSMO1) mRNA could distinguish pathological complete response in 58 patients receiving neoadjuvant treatment. MSMO1 was highly enriched in the non-pathological complete response group and was found to reduce the phosphorylation of AKT and mTOR of breast cancer cells. Silencing of MSMO1 sensitized highly metastatic tumor cells to chemotherapy drugs and increased apoptosis rates, demonstrating exosome MSMO1 can act as a non-invasive prediction tool for neoadjuvant treatment success (Su et al. 2021).

1.6.4 Therapeutics

EVs have emerged as a promising avenue for therapeutic intervention, given their diverse functional effects that impact both body homeostasis and disease states. Mesenchymal stem cells (MSCs), widely used in regenerative medicine due to their differentiation capacity and local paracrine effects, face limitations such as biodistribution challenges, poor tissue engraftment, transient life cycle, tumorigenic effects, instability during extended in vitro expansion, and difficulty in long-term storage for off-the-shelf therapies. In contrast, EVs present a compelling alternative to MSCs (Kou et al. 2022).

EVs show inherent stability, a lower likelihood of immune rejection in allogenic administration, and an inability to directly form tumors (Phinney and Pittenger 2017). Immortalized cells can secrete EVs in quantities sufficient for production needs (Phinney and Pittenger 2017; Xunian and Kalluri 2020). Notably, EVs preserve their bioactivity due to their lipid membrane structure, which allows them to remain functional even after prolonged storage at −80 °C. This remarkable stability extends to multiple freeze-thaw cycles, ensuring their activity remains intact over time (Xunian and Kalluri 2020). EVs also possess an encapsulation capability, enabling the loading of specific drugs for transport to target cells (Piffoux et al. 2017; Cheng et al. 2019; Ge et al. 2014). Additionally, EVs possess an encapsulation capability, enabling the loading of specific drugs for transport to target cells (Piffoux et al. 2017). Their ability to traverse the blood–brain barrier further enhances their therapeutic potential (Ramos-Zaldívar et al. 2022). Recent studies propose that many therapeutic effects traditionally attributed to MSC administration may be primarily attributable to EVs alone (Ramos-Zaldívar et al. 2022; Li et al.

2020; Keshtkar et al. 2018). This realization underscores the potential of EVs as a promising therapeutic modality, offering advantages in terms of safety and applicability (Li et al. 2020; Keshtkar et al. 2018).

Not surprisingly, over the past decade, EVs have been extensively explored for therapeutic applications across a spectrum of conditions. EVs have been evaluated as a method for immunomodulation, cancer treatment, drug delivery vehicles, and tissue repair (Dai et al. 2020; Herrmann et al. 2021). The therapeutic potential of EVs as messengers is expansive, and we are only beginning to uncover the breadth of their capabilities. A significant aspect of EVs' remarkable therapeutic potential lies in their capacity to promote regeneration and restoration of the extracellular matrix. As detailed in earlier sections, EVs directly modulate the extracellular matrix (ECM) by influencing structural molecules (employing MMPs, glycosidases, and collagenases) and indirectly by impacting cells involved in matrix modulation. Exosomal therapies can harness this ability to influence the matrix. For instance, exosomes derived from mesenchymal stem cells have demonstrated the ability to reduce scar formation in murine models by increasing TGF-B and altering matrix metalloproteinase expressions, thus modulating collagen type III to type I ratios (Wang et al. 2017). Similar efficacy in mitigating disease through ECM alterations was observed by Woo et al. in an osteoarthritis model, where MSC EVs facilitated the synthesis of type 2 collagen while inhibiting the expression of matrix metalloproteinases (MMPs), including MMP-1, MMP-3, MMP-13, and ADAMTS-5, thereby slowing the progression of articular cartilage degeneration inherent to the disease (Woo et al. 2020).

The most recent addition to the domain of EVs, matrix-bound nanovesicles (MBVs), has demonstrated clinical efficacy in modulating inflammatory conditions and fostering tissue regeneration. MBVs have been identified within commercially available biological scaffolds employed in wound repair (Huleihel et al. 2016). These bioactive entities can be liberated from ECM products through enzymatic digestion. Labeled MBVs derived from a urinary bladder matrix scaffold exhibited uptake in a myoblast cell line (Huleihel et al. 2016). Notably, MBVs originating from the scaffold exerted immunomodulatory effects by promoting an M2 macrophage phenotype and stimulating neurite growth (van der Merwe et al. 2019). It is postulated that a substantial portion of the observed biologic activity associated with ECM scaffolds may be attributed to the actions of MBVs. It is not only plausible, but likely, that MBVs play a significant role in the augmented remodeling observed in the context of wound repair utilizing biologic scaffolds. Subsequent investigations are warranted to assess the in vivo release of MBVs from scaffolds and to design scaffolds with optimized MBV cargo.

1.6.5 Engineering EV Therapies

Due to their virtually boundless therapeutic potential, the processes involved in isolating, characterizing, and potentially engineering EVs have become subjects of significant research interest. A crucial focus lies in elucidating targeting mechanisms that can facilitate the precise delivery of exosomes to specific organs, tissues, or even individual cells, thereby enhancing their efficacy as a therapeutic avenue. Furthermore, the creation of exosomes with tailored cargo holds promise for augmenting their biologic potency and directing mechanistic effects while minimizing nonspecific impacts.

Various methods are employed in the bioengineering of exosomes, including the modification of target genes into cells through plasmid-based approaches with the coexpression of EV membrane proteins. In a more passive approach, cargo modification can be accomplished with the coincubation of small molecules with exosomes and methods such as electroporation to drive these molecules into the vesicles (Liang et al. 2021). Small therapeutic drugs and miRNAs have been incorporated into exosomes using these passive packaging techniques. EVs can be modified at the cell level or post-EV genesis through biochemical or physical modifications. Scientists can strategically choose the parent cell line, manipulate cell culture, or genetically engineer founder cells to overexpress miRNA, growth factors, and molecules depending on the role of the EV messenger in the context of the therapeutic goal.

Manipulation of cell culture conditions is one of the most straightforward and accessible methods of engineering EVs. MSC EVs are known to mediate MSC and innate immune cell crosstalk, triggering macrophage polarization from an M1 to M2-like phenotype making them a desirable source for EV engineering (Lo Sicco et al. 2017). Hypoxic conditioning of cultured MSCs increased the production of EVs enriched in miRNAs associated with the stages of the healing process and there was an increase in M2 markers CD206, Arg1, and Ym1, and a decrease in M2 markers Nos2 upon cardiotoxin-induced skeletal muscle injury in both normoxic and hypoxic cell culture conditions (Lo Sicco et al. 2017).

Researchers have demonstrated creativity in engineering specific peptide-expressing EVs, hybrid EVs, drug-loaded EVs, and more, employing unique manufacturing methods. Manipulating reversible protein–protein interactions using optogenetics, or the use of light to control cell activity, Choi et al. engineered NF-κB inhibitor-loaded exosomes and ameliorated sepsis-induced organ inflammation and injury accompanied by decreased inflammatory cytokine secretion and improved mouse survival (Choi et al. 2020). To enhance membrane targeting, strategies involve the fusion of known membrane proteins with proteins designed to target specific tissues or cells. Exosomes expressing cardiac-targeting peptide (CTP)-Lamp2b on the surface of the lipid bilayer membrane were increasingly delivered to cardiac but not kidney cells in vitro and in vivo there was a 15% increase of the engineered EV delivered to the hearts of mice (Kim et al. 2018). Another study explored EV-mediated drug delivery using a hybrid exosome composed of macrophage exosome and synthetic liposome, showing greater cancer cell toxicity when loaded with cancer drug doxorubicin (Rayamajhi et al. 2019).

One limitation of using engineered EVs is their rapid clearance after administration. To tackle this challenge, one study developed collagen-binding EVs, by chemically incorporating the collagen-binding peptide SILY which would in turn create EV-ECM complexes to enhance homing duration. In mouse hind limb ischemia, SILY-populated EVs significantly suppressed inflammatory conditions and improved muscle regeneration and vascularization compared to non-engineered EVs (Hao et al. 2022). Interestingly, MBV, which are naturally adhered to ECM constituents like collagen, have yet to be explored for their long in situ retention. It is likely that MBV could naturally achieve the desired outcome of extended residence time, exemplifying the need for future studies to commence the investigation of engineering solid-phased MBV-like vesicles. As advancements continue in the development of methods to create targeted exosomes with precisely defined cargo, the therapeutic utility of these engineered EVs is poised to expand significantly. This ongoing progress holds promise for unlocking new dimensions in therapeutic interventions.

1.7 Conclusions and Future Perspectives

EVs are ubiquitous entities intricately involved in all cellular processes, exerting pivotal control over the extracellular matrix in both physiological states and pathological conditions. Among the diverse classes of EVs, MBVs emerge as a distinct subclass, residing within the extracellular matrix.

Despite significant advancements in understanding how EVs modulate the ECM and govern cell behavior, there remains a substantial gap in knowledge regarding their regulatory mechanisms, targeting, and cellular uptake. The nascent field of MBVs is particularly challenging, with their biological characteristics still largely unexplored. Extensive research is required to characterize and comprehend the biogenesis, markers, cargo, and biological effects of MBVs. Despite these challenges, the unique attributes of EVs offer an unparalleled avenue for both therapeutic interventions and diagnostic breakthroughs.FundingThe University of Pittsburgh holds a Physician-Scientist Institutional Award from the Burroughs Wellcome Fund. This award supports VPA.

References

Aaronson S, Behrens U, Orner R, Haines TH (1971) Ultrastructure of intracellular and extracellular vesicles, membranes, and myelin figures produced by Ochromonas danica. J Ultrastruct Res 35(5):418–430

Abdollahi S (2021) Extracellular vesicles from organoids and 3D culture systems. In: Biotechnology and bioengineering. Wiley, pp 1029–1049

Abels ER et al (2019) Glioblastoma-associated microglia reprogramming is mediated by functional transfer of extracellular miR-21. Cell Rep 28(12):3105–3119.e7

Aga M et al (2014) Exosomal HIF1α supports invasive potential of nasopharyngeal carcinoma-associated LMP1-positive exosomes. Oncogene 33(37):4613–4622

Akers JC et al (2013) Biogenesis of extracellular vesicles (EV): exosomes, microvesicles, retrovirus-like vesicles, and apoptotic bodies. J Neurooncol 113(1):1–11

Ansari S et al (2021) Matrix vesicles: role in bone mineralization and potential use as therapeutics. Pharmaceuticals (Basel) 14(4)

Antoni L, Nuding S, Wehkamp J, Stange EF (2014) Intestinal barrier in inflammatory bowel disease. World J Gastroenterol 20(5):1165–1179

Atkin-Smith GK et al (2015) A novel mechanism of generating extracellular vesicles during apoptosis via a beads-on-a-string membrane structure. Nat Commun 6:7439

Bakhshian Nik A, Hutcheson JD, Aikawa E (2017) Extracellular vesicles as mediators of cardiovascular calcification. Front Cardiovasc Med 4:78

Baldrich P et al (2019) Plant extracellular vesicles contain diverse small RNA species and are enriched in 10- to 17-nucleotide "tiny" RNAs. Plant Cell 31(2):315–324

Bandari SK et al (2018) Chemotherapy induces secretion of exosomes loaded with heparanase that degrades extracellular matrix and impacts tumor and host cell behavior. Matrix Biol 65:104–118

Barros FM, Carneiro F, Machado JC, Melo SA (2018) Exosomes and immune response in cancer: friends or foes? Front Immunol 9

Barry-Hamilton V et al (2010) Allosteric inhibition of lysyl oxidase-like-2 impedes the development of a pathologic microenvironment. Nat Med 16(9):1009–1017

Brown BN, Badylak SF (2014) Extracellular matrix as an inductive scaffold for functional tissue reconstruction. Transl Res 163(4):268–285

Buendía P et al (2015) Endothelial microparticles mediate inflammation-induced vascular calcification. FASEB J 29(1):173–181

Burtenshaw D et al (2022) Exosomal composition, biogenesis and profiling using point-of-care diagnostics-implications for cardiovascular disease. Front Cell Dev Biol 10:853451

Caruso S et al (2019) Defining the role of cytoskeletal components in the formation of apoptopodia and apoptotic bodies during apoptosis. Apoptosis 24(11–12):862–877

Chargaff E, West R (1946) The biological significance of the thromboplastic protein of blood. J Biol Chem 166(1):189–197

Chen X et al (2011) Intestinal epithelial cell-derived integrin αβ6 plays an important role in the induction of regulatory T cells and inhibits an antigen-specific Th2 response. J Leukoc Biol 90(4):751–759

Chen IH et al (2017) Phosphoproteins in extracellular vesicles as candidate markers for breast cancer. Proc Natl Acad Sci USA 114(12):3175–3180

Cheng Y, Zeng Q, Han Q, Xia W (2019) Effect of pH, temperature and freezing-thawing on quantity changes and cellular uptake of exosomes. Protein Cell 10(4):295–299

Cheng F et al (2023) Mesenchymal stem cell-derived exosomal miR-27b-3p alleviates liver fibrosis via downregulating YAP/LOXL2 pathway. J Nanobiotechnol 21(1):195

Choi H et al (2020) Exosome-based delivery of super-repressor IκBα relieves sepsis-associated organ damage and mortality. Sci Adv 6(15):eaaz6980

Ciferri MC, Quarto R, Tasso R (2021) Extracellular vesicles as biomarkers and therapeutic tools: from pre-clinical to clinical applications. Biology 10(5)

Cloutier N et al (2013) The exposure of autoantigens by microparticles underlies the formation of potent inflammatory components: the microparticle-associated immune complexes. EMBO Mol Med 5(2):235–249

Crescitelli R et al (2013) Distinct RNA profiles in subpopulations of extracellular vesicles: apoptotic bodies, microvesicles and exosomes. J Extracell Vesicles 12:2

Crum RJ et al (2022) Immunomodulatory matrix-bound nanovesicles mitigate acute and chronic pristane-induced rheumatoid arthritis. NPJ Regen Med 7(1):13
Dai J et al (2020) Exosomes: key players in cancer and potential therapeutic strategy. Signal Transduct Target Ther 5(1):145
de Jong OG, van Balkom BW, Gremmels H, Verhaar MC (2016) Exosomes from hypoxic endothelial cells have increased collagen crosslinking activity through up-regulation of lysyl oxidase-like 2. J Cell Mol Med 20(2):342–350
Dixson AC, Dawson TR, Di Vizio D, Weaver AM (2023) Context-specific regulation of extracellular vesicle biogenesis and cargo selection. Nat Rev Mol Cell Biol 24(7):454–476
Doyle LM, Wang MZ (2019) Overview of extracellular vesicles, their origin, composition, purpose, and methods for exosome isolation and analysis. Cells 8(7):727
Eichelser C et al (2014) Increased serum levels of circulating exosomal microRNA -373 in receptor-negative breast cancer patients. Oncotarget 5(20):9650–9663
Faust A et al (2017) Urinary bladder extracellular matrix hydrogels and matrix-bound vesicles differentially regulate central nervous system neuron viability and axon growth and branching. J Biomater Appl 31(9):1277–1295
Foers AD et al (2020) Proteomic analysis of extracellular vesicles reveals an immunogenic cargo in rheumatoid arthritis synovial fluid. Clin Transl Immunol 9(11):e1185
Frantz C, Stewart KM, Weaver VM (2010) The extracellular matrix at a glance. J Cell Sci 123(Pt 24):4195–4200
Franzen CA et al (2015) Urothelial cells undergo epithelial-to-mesenchymal transition after exposure to muscle invasive bladder cancer exosomes. Oncogenesis 4(8):e163
Ge Q et al (2014) miRNA in plasma exosome is stable under different storage conditions. Molecules 19(2):1568–1575
Genschmer KR et al (2019) Activated PMN exosomes: pathogenic entities causing matrix destruction and disease in the lung. Cell 176(1–2):113–126.e15
Gonzalez DM, Medici D (2014) Signaling mechanisms of the epithelial-mesenchymal transition. Sci Signal 7(344):re8
Gooz M (2010) ADAM-17: the enzyme that does it all. Crit Rev Biochem Mol Biol 45(2):146–169
Groth E et al (2016) Stimulated release and functional activity of surface expressed metalloproteinase ADAM17 in exosomes. Biochim Biophys Acta 1863(11):2795–2808
Gurung S, Perocheau D, Touramanidou L, Baruteau J (2021) The exosome journey: from biogenesis to uptake and intracellular signalling. Cell Commun Signal 19(1):47
Haensel D, Dai X (2018) Epithelial-to-mesenchymal transition in cutaneous wound healing: where we are and where we are heading. Dev Dyn 247(3):473–480
Hakulinen J et al (2008) Secretion of active membrane type 1 matrix metalloproteinase (MMP-14) into extracellular space in microvesicular exosomes. J Cell Biochem 105(5):1211–1218
Halicka HD, Bedner E, Darzynkiewicz Z (2000) Segregation of RNA and separate packaging of DNA and RNA in apoptotic bodies during apoptosis. Exp Cell Res 260(2):248–256
Hao D et al (2022) Engineered extracellular vesicles with high collagen-binding affinity present superior in situ retention and therapeutic efficacy in tissue repair. Theranostics 12(13):6021–6037
Hasegawa T et al (2017) Ultrastructural and biochemical aspects of matrix vesicle-mediated mineralization. Jpn Dent Sci Rev 53(2):34–45
Herrmann IK, Wood MJA, Fuhrmann G (2021) Extracellular vesicles as a next-generation drug delivery platform. Nat Nanotechnol 16(7):748–759
Hood JL, San RS, Wickline SA (2011) Exosomes released by melanoma cells prepare sentinel lymph nodes for tumor metastasis. Cancer Res 71(11):3792–3801
Hoshino D et al (2013) Exosome secretion is enhanced by invadopodia and drives invasive behavior. Cell Rep 5(5):1159–1168
Hoshino A et al (2015) Tumour exosome integrins determine organotropic metastasis. Nature 527(7578):329–335

Huang Z, Feng Y (2017) Exosomes derived from hypoxic colorectal cancer cells promote angiogenesis through Wnt4-induced β-catenin signaling in endothelial cells. Oncol Res 25(5):651–661
Huang Y, Hong W, Wei X (2022) The molecular mechanisms and therapeutic strategies of EMT in tumor progression and metastasis. J Hematol Oncol 15(1):129
Huleihel L et al (2016) Matrix-bound nanovesicles within ECM bioscaffolds. Sci Adv 2(6):e1600502
Huleihel L et al (2017) Matrix-bound nanovesicles recapitulate extracellular matrix effects on macrophage phenotype. Tissue Eng Part A 23(21–22):1283–1294
Hussey GS et al (2019) Matrix bound nanovesicle-associated IL-33 activates a pro-remodeling macrophage phenotype via a non-canonical, ST2-independent pathway. J Immunol Regen Med 3:26–35
Hussey GS et al (2020) Lipidomics and RNA sequencing reveal a novel subpopulation of nanovesicle within extracellular matrix biomaterials. Sci Adv 6(12):eaay4361
Huynh N et al (2016) Characterization of regulatory extracellular vesicles from osteoclasts. J Dent Res 95(6):673–679
Jin Y et al (2022) Extracellular signals regulate the biogenesis of extracellular vesicles. Biol Res 55(1):35
Kaparakis M et al (2010) Bacterial membrane vesicles deliver peptidoglycan to NOD1 in epithelial cells. Cell Microbiol 12(3):372–385
Kapsogeorgou EK, Abu-Helu RF, Moutsopoulos HM, Manoussakis MN (2005) Salivary gland epithelial cell exosomes: a source of autoantigenic ribonucleoproteins. Arthritis Rheum 52(5):1517–1521
Kapustin AN et al (2015) Vascular smooth muscle cell calcification is mediated by regulated exosome secretion. Circ Res 116(8):1312–1323
Keshtkar S, Azarpira N, Ghahremani MH (2018) Mesenchymal stem cell-derived extracellular vesicles: novel frontiers in regenerative medicine. Stem Cell Res Ther 9(1):63
Kim MR et al (2012) Staphylococcus aureus-derived extracellular vesicles induce neutrophilic pulmonary inflammation via both Th1 and Th17 cell responses. Allergy 67(10):1271–1281
Kim H et al (2018) Cardiac-specific delivery by cardiac tissue-targeting peptide-expressing exosomes. Biochem Biophys Res Commun 499(4):803–808
Kirsch T et al (1997) Annexin V-mediated calcium flux across membranes is dependent on the lipid composition: implications for cartilage mineralization. Biochemistry 36(11):3359–3367
Kou M et al (2022) Mesenchymal stem cell-derived extracellular vesicles for immunomodulation and regeneration: a next generation therapeutic tool? Cell Death Dis 13(7):580
Lakhter AJ et al (2018) Beta cell extracellular vesicle miR-21-5p cargo is increased in response to inflammatory cytokines and serves as a biomarker of type 1 diabetes. Diabetologia 61(5):1124–1134
Lamouille S, Xu J, Derynck R (2014) Molecular mechanisms of epithelial-mesenchymal transition. Nat Rev Mol Cell Biol 15(3):178–196
Leoni G et al (2013) Annexin A1, formyl peptide receptor, and NOX1 orchestrate epithelial repair. J Clin Invest 123(1):443–454
Leoni G et al (2015) Annexin A1-containing extracellular vesicles and polymeric nanoparticles promote epithelial wound repair. J Clin Invest 125(3):1215–1227
Li Y, Altemus J, Lightner AL (2020) Mesenchymal stem cells and acellular products attenuate murine induced colitis. Stem Cell Res Ther 11(1):515
Li X et al (2022) Chemoresistance transmission via exosome-transferred MMP14 in pancreatic cancer. Front Oncol 12:844648
Liang Y, Duan L, Lu J, Xia J (2021) Engineering exosomes for targeted drug delivery. Theranostics 11(7):3183–3195
Lin CJ et al (2019) The paracrine induction of prostate cancer progression by caveolin-1. Cell Death Dis 10(11):834
Liu M, Sun Y, Zhang Q (2018) Emerging role of extracellular vesicles in bone remodeling. J Dent Res 97(8):859–868

Lo Sicco C et al (2017) Mesenchymal stem cell-derived extracellular vesicles as mediators of anti-inflammatory effects: endorsement of macrophage polarization. Stem Cells Transl Med 6(3):1018–1028

López-Novoa JM, Nieto MA (2009) Inflammation and EMT: an alliance towards organ fibrosis and cancer progression. EMBO Mol Med 1(6-7):303–314

Lötvall J et al (2014) Minimal experimental requirements for definition of extracellular vesicles and their functions: a position statement from the International Society for Extracellular Vesicles. J Extracell Vesicles 3:26913

Melo SA et al (2015) Glypican-1 identifies cancer exosomes and detects early pancreatic cancer. Nature 523(7559):177–182

Morrell AE et al (2018) Mechanically induced Ca(2+) oscillations in osteocytes release extracellular vesicles and enhance bone formation. Bone Res 6:6

Mu W et al (2021) Exosomal CD44 cooperates with integrin α6β4 to support organotropic metastasis via regulating tumor cell motility and target host cell activation. Engineering 7(10):1413–1423

Muralidharan-Chari V et al (2009) ARF6-regulated shedding of tumor cell-derived plasma membrane microvesicles. Curr Biol 19(22):1875–1885

Nössing C, Ryan KM (2023) 50 years on and still very much alive: 'apoptosis: a basic biological phenomenon with wide-ranging implications in tissue kinetics'. Br J Cancer 128(3):426–431

O'Brien J, Hayder H, Zayed Y, Peng C (2018) Overview of microRNA biogenesis, mechanisms of actions, and circulation. Front Endocrinol 9

Patel NJ, Ashraf A, Chung EJ (2023) Extracellular vesicles as regulators of the extracellular matrix. Bioengineering 10(2):136

Peinado H et al (2012) Melanoma exosomes educate bone marrow progenitor cells toward a pro-metastatic phenotype through MET. Nat Med 18(6):883–891

Peinado H et al (2017) Pre-metastatic niches: organ-specific homes for metastases. Nat Rev Cancer 17(5):302–317

Phinney DG, Pittenger MF (2017) Concise review: MSC-derived exosomes for cell-free therapy. Stem Cells 35(4):851–858

Piffoux M et al (2017) Extracellular vesicle production loaded with nanoparticles and drugs in a trade-off between loading, yield and purity: towards a personalized drug delivery system. Adv Biosyst 1(5):1700044

Qin Z et al (2021) A narrative review of exosomes in vascular calcification. Ann Transl Med 9(7):579

Ramos-Zaldívar HM et al (2022) Extracellular vesicles through the blood–brain barrier: a review. Fluids Barriers CNS 19(1):60

Rayamajhi S, Nguyen TDT, Marasini R, Aryal S (2019) Macrophage-derived exosome-mimetic hybrid vesicles for tumor targeted drug delivery. Acta Biomater 94:482–494

Redzic JS et al (2013) Extracellular vesicles secreted from cancer cell lines stimulate secretion of MMP-9, IL-6, TGF-β1 and EMMPRIN. PLoS One 8(8):e71225

Sanderson RD, Bandari SK, Vlodavsky I (2019) Proteases and glycosidases on the surface of exosomes: newly discovered mechanisms for extracellular remodeling. Matrix Biol 75–76:160–169

Sandfeld-Paulsen B et al (2016) Exosomal proteins as diagnostic biomarkers in lung cancer. J Thorac Oncol 11(10):1701–1710

Schulz-Siegmund M, Aigner A (2021) Nucleic acid delivery with extracellular vesicles. Adv Drug Deliv Rev 173:89–111

Shan Y et al (2018) Hypoxia-induced matrix Metalloproteinase-13 expression in exosomes from nasopharyngeal carcinoma enhances metastases. Cell Death Dis 9(3):382

Simons M, Raposo G (2009) Exosomes - vesicular carriers for intercellular communication. Curr Opin Cell Biol 21(4):575–581

Slater TW et al (2017) Neutrophil microparticles deliver active myeloperoxidase to injured mucosa to inhibit epithelial wound healing. J Immunol 198(7):2886–2897

Song YH et al (2017) Breast cancer-derived extracellular vesicles stimulate myofibroblast differentiation and pro-angiogenic behavior of adipose stem cells. Matrix Biol 60-61:190–205
Sork H et al (2018) Heterogeneity and interplay of the extracellular vesicle small RNA transcriptome and proteome. Sci Rep 8(1):10813
Su Y et al (2021) Plasma extracellular vesicle long RNA profiles in the diagnosis and prediction of treatment response for breast cancer. NPJ Breast Cancer 7(1):154
Swinehart IT, Badylak SF (2016) Extracellular matrix bioscaffolds in tissue remodeling and morphogenesis. Dev Dyn 245(3):351–360
Tan K, Li R, Huang X, Liu Q (2018) Outer membrane vesicles: current status and future direction of these novel vaccine adjuvants. Front Microbiol 9:783
Tao S-C, Guo S-C (2020) Role of extracellular vesicles in tumour microenvironment. Cell Commun Signal 18(1):163
Tian T et al (2016) MicroRNA-16 is putatively involved in the NF-κB pathway regulation in ulcerative colitis through adenosine A2a receptor (A2aAR) mRNA targeting. Sci Rep 6(1):30824
Touyz RM, Montezano AC (2009) Vascular smooth muscle cells sense calcium: a new paradigm in vascular calcification. Cardiovasc Res 81(2):237–239
Toyofuku M, Nomura N, Eberl L (2019) Types and origins of bacterial membrane vesicles. Nat Rev Microbiol 17(1):13–24
Tricarico C, Clancy J, D'Souza-Schorey C (2017) Biology and biogenesis of shed microvesicles. In: Small GTPases. Taylor & Francis, pp 220–232
Tucci M et al (2018) Serum exosomes as predictors of clinical response to ipilimumab in metastatic melanoma. Onco Targets Ther 7(2):e1387706
Turner NJ et al (2022) Matrix bound nanovesicles have tissue-specific characteristics that suggest a regulatory role. Tissue Eng Part A 28(21–22):879–892
van der Merwe Y et al (2019) Matrix-bound nanovesicles prevent ischemia-induced retinal ganglion cell axon degeneration and death and preserve visual function. Sci Rep 9(1):3482
Van Deun J et al (2014) The impact of disparate isolation methods for extracellular vesicles on downstream RNA profiling. J Extracell Vesicles 18:3
Van Niel G, D'Angelo G, Raposo G (2018) Shedding light on the cell biology of extracellular vesicles. Nature Publishing Group, pp 213–228
Wang H et al (2013) Epithelial-mesenchymal transition (EMT) induced by TNF-α requires AKT/GSK-3β-mediated stabilization of snail in colorectal cancer. PLoS One 8(2):e56664
Wang T et al (2014) Hypoxia-inducible factors and RAB22A mediate formation of microvesicles that stimulate breast cancer invasion and metastasis. Proc Natl Acad Sci USA 111(31):E3234–E3242
Wang L et al (2017) Exosomes secreted by human adipose mesenchymal stem cells promote scarless cutaneous repair by regulating extracellular matrix remodelling. Sci Rep 7(1):13321
Wang X, Koffi PF, English OF, Lee JC (2021) Staphylococcus aureus extracellular vesicles: a story of toxicity and the stress of 2020. Toxins (Basel) 13(2)
Withrow J et al (2016) Extracellular vesicles in the pathogenesis of rheumatoid arthritis and osteoarthritis. Arthritis Res Ther 18(1):286
Wolf P (1967) The nature and significance of platelet products in human plasma. Br J Haematol 13(3):269–288
Woo CH et al (2020) Small extracellular vesicles from human adipose-derived stem cells attenuate cartilage degeneration. J Extracell Vesicles 9(1):1735249
Wu LN, Genge BR, Lloyd GC, Wuthier RE (1991) Collagen-binding proteins in collagenase-released matrix vesicles from cartilage. Interaction between matrix vesicle proteins and different types of collagen. J Biol Chem 266(2):1195–1203
Wuthier RE, Lipscomb GF (2011) Matrix vesicles: structure, composition, formation and function in calcification. Front Biosci (Landmark Ed) 16(8):2812–2902
Xu J, Lamouille S, Derynck R (2009) TGF-beta-induced epithelial to mesenchymal transition. Cell Res 19(2):156–172

Xunian Z, Kalluri R (2020) Biology and therapeutic potential of mesenchymal stem cell-derived exosomes. Cancer Sci 111(9):3100–3110
Yang CH et al (2014) MicroRNA-21 promotes glioblastoma tumorigenesis by down-regulating insulin-like growth factor-binding protein-3 (IGFBP3). J Biol Chem 289(36):25079–25087
Yu L et al (2023) Apoptotic bodies: bioactive treasure left behind by the dying cells with robust diagnostic and therapeutic application potentials. J Nanobiotechnol 21(1):218
Yuan FL et al (2018) Osteoclast-derived extracellular vesicles: novel regulators of osteoclastogenesis and osteoclast-osteoblasts communication in bone remodeling. Front Physiol 9:628
Zazzeroni L, Faggioli G, Pasquinelli G (2018) Mechanisms of arterial calcification: the role of matrix vesicles. Eur J Vasc Endovasc Surg 55(3):425–432
Zhang X et al (2018) Metaproteomics reveals associations between microbiome and intestinal extracellular vesicle proteins in pediatric inflammatory bowel disease. Nat Commun 9(1):2873
Zhu Y et al (2022) Exosomal MMP-1 transfers metastasis potential in triple-negative breast cancer through PAR1-mediated EMT. Breast Cancer Res Treat 193(1):65–81

Chapter 2
Extracellular Matrix Remodeling in Physiological and Pathological Conditions: Insight into Extracellular Vesicles Contribution

Giuseppina Poppa, Ilaria Giusti, and Vincenza Dolo

Abstract The extracellular matrix (ECM) is a dynamic, acellular framework that maintains homeostasis and supports the preservation of tissues' structural integrity. From early embryonic stages forward, the ECM organizes itself through a variety of changes, including degradation, synthesis, and rearrangement of matrix elements. Although the mechanisms that underlie these changes are strictly controlled, the ECM can undergo structural and functional alterations that result in diseases or even promote the growth of tumors. Several elements mediate communication between the cells that make up the tissue and the matrix, with extracellular vesicles (EVs) playing a significant role. EVs are spherical particles surrounded by a phospholipid bilayer, secreted by all living organisms and present in all organs, tissues, and biological fluids, both in physiological and in pathological conditions. In this review, we will present a comprehensive understanding of the matrix's structure and the roles in which it is engaged, as well as the implications of EVs. Several aspects of EVs–ECM interaction under healthy and pathological settings will be covered, with an emphasis on EVs' function in matrix degradation.

Giuseppina Poppa and Ilaria Giusti contributed equally to the chapter writing.

G. Poppa · I. Giusti · V. Dolo (✉)
Department of Life, Health and Environmental Sciences, University of L'Aquila, L'Aquila, Italy
e-mail: giuseppina.poppa@graduate.univaq.it; ilaria.giusti@univaq.it; vincenza.dolo@univaq.it

K. Rilla (ed.), *Extracellular Vesicles as Matrix Messengers*, Biology of Extracellular Matrix 15, https://doi.org/10.1007/978-3-031-68811-9_2

2.1 Introduction

Cellular communication is an important process for cellular homeostasis maintenance and tissue organization, as well as for the progression of diseases, including cancer. This communication is often mediated by soluble factors, or by extracellular vesicles (EVs) (van Niel et al. 2022).

EVs are a group of spherical particles surrounded by a phospholipid bilayer, heterogeneous in size and biogenesis, secreted by all cells, and ubiquitary in the organism. They have lipids, proteins, DNA, and different types of RNA (miRNA, lncRNA, etc.) in their cargo, which can be released from the donor cell into the recipient cell or extracellular space. Given their key role in cell-to-cell communication, current studies identify their involvement not only in physiological processes but also in various diseases (Théry et al. 2018). However, to move from one cell to another, EVs must cross an acellular barrier: the extracellular matrix (ECM).

The first studies on this highly dynamic structure date back to the 1980s on cartilage and bone matrices, leading to the discovery of the different groups of proteins that make up the matrix. The advancement of molecular biology techniques, genetics, -omics studies and bioinformatics subsequently allowed a broader characterization of the ECM's components. The results of these studies have led to the clustering of more than a thousand genes coding for ECM proteins, collectively referred to as "the matrisome"; the latter can be divided into three categories: (i) the central matrisome, characterized by genes coding for around 300 proteins including glycoproteins, proteoglycans, and collagen; (ii) the matrix-associated proteins; and (iii) the secreted factors and ECM regulatory factors (Hynes and Naba 2012; Naba et al. 2016).

The ECM constitutes a support structure for tissues and likewise contributes to all normal processes of cellular homeostasis through its remodeling, consisting of its deposition, chemical-structural modification, proteolytic degradation, and physical remodeling of ECM components (Winkler et al. 2020). All these steps are tightly regulated; alterations that occur generate excessive degradation or overproduction of the matrix, resulting in disease outbreaks or cancer promotion and progression (Theocharis et al. 2019). Metalloproteinases (MMPs) and cathepsins are the main proteolytic enzymes able to remodel the ECM by mediating the production or degradation of matrix components. It can happen that the degradation of ECM also leads to the release of bioactive matrix fragments, known as matrikines, which can also drive the reorganization of the ECM (Karamanos et al. 2019).

In this scenario, EVs can play a dual role: passive, as they can merely diffuse through the ECM from the donor cell to the recipient cell; active in relation to the ECM, interacting with its various structural components, and modulating its deposition and degradation (Nawaz et al. 2018; Rilla et al. 2019; Sariano et al. 2023). The aim of this review is to provide an overview of the EVs-ECM relationship and its impact on physiological and pathological conditions. Before this, an overview will be given of both the composition of the matrix and its role in normal homeostatic processes and pathological conditions, as well as the general characteristics and role of EVs. The role of EVs in the modulation of ECM will then be addressed, with a special focus on matrix degradation and the molecules involved in this process. Finally, potential therapeutic applications and the use of EVs as biomarkers for ECM will be discussed.

2.2 Composition and Function of ECM

Tissues and organs are composed of cells and an interposed component between them, known as the extracellular matrix (ECM).

ECM is commonly defined as a dynamic noncellular structure, consisting of molecules that build up to form a 3D scaffold which in turn undergoes constant structural changes, thus regulating cellular functions and homeostasis. Even if ECM can be made up of specific molecules—which provide to the matrix different mechanical, viscoelastic, structural, and functional properties—depending on the type of tissue considered, it shares a basic structure and components with all of them (Karamanos et al. 2021).

The main components of the ECM are summarized in Table 2.1.

2.2.1 ECM in Health

Despite the initially assumed role of ECM was merely scaffolding, its involvement in development and homeostasis regulated by the remodeling processes it constantly undergoes is now widely recognized (Theocharis et al. 2012; Soans and Norden 2021).

Early in embryogenesis, several studies in chicken embryos have shown how the ECM undergoes changes that lead to the epithelial–mesenchymal transition (EMT). In this process, molecules including fibronectin and laminin are involved, which support cell migration during gastrulation and neural crest formation; during cardiac formation, versican and periostin both contribute to cell migration through binding to other ECM proteins (collagen, fibronectin, integrins) and are involved, in deposition of new matrix and promotion of EMT, respectively (Zito 2012). During tendon development in the limbs of murine embryos, matrix proteoglycans act as connection points between the epidermis and cartilage (Acuna et al. 2018). The ECM also acquires a key role in the function and development of the musculoskeletal system: it regulates both the processes of proliferation and the differentiation of myoblasts during myogenesis and the migration, adhesion, proliferation, and differentiation of satellite cells during skeletal muscle development (Zhang et al. 2021).

The interaction between cells and ECM is also critical for normal skin development and homeostasis, as well as wound healing processes (Tracy et al. 2016; Kleiser and Nyström 2020). In the case of a skin wound, the body carries out a highly regulated process in several events in which soluble factors, blood elements,

Table 2.1 Main ECM components

ECM component	Subcomponents	Functions	References
Collagens	Collagen type I–XXVIII	Structural support and integrity, tissue rigidity, cell signaling, adhesion	Ricard-Blum and Ruggiero (2005), Kadler et al. (2007), Zhang et al. (2021)
Proteoglycans	Heparan Sulfate Proteoglycans (HSPGs), syndecans, betaglycans, glypican, perlecan, hyalectans, aggrecan, versican, brevican, Small Leucine-rich Proteoglycans (SLRPs)	Viscoelastic properties, tissue hydration and maintenance of osmotic pressure, trapping of growth factors, contributors to the correct organization of collagen	Couchman and Pataki (2012), Iozzo and Schaefer (2015)
Glycosaminoglycans	Hyaluronic acid (HA), Heparin (Hep), Heparan Sulfate (HS), Chondroitin Sulfate (CS), Dermatan Sulfate (DS), Keratan Sulfate (KS)	Cell adhesion and signaling through interaction with growth factors, chemokines, cytokines, and adhesion molecules; main participants in tissue repair and regeneration, embryogenesis, and maintenance of homeostasis	Shi et al. (2021a), Ricard-Blum and Perez (2022), Zappe et al. (2022)
Laminin	LAMA 1-5; LAMB 1-3; LAMC 1-3	Structural integrity of the basement membrane through interaction with other matrix proteins; signaling, cellular adhesion, differentiation, and gene expression by binding to cell surface receptors.	Aumailley (2013), Guldager Kring Rasmussen and Karsdal (2016), Hohenester (2019)
Fibronectin		Mechanical tension, cell adhesion through integrin binding, tissue repair by binding growth factors, chemokines, and other matrix components	To and Midwood (2011), Dalton and Lemmon (2021)
Elastic fibers	Elastin	Tissue elasticity	Baumann et al. (2021)

components of the ECM, and cells are co-involved. Fibroblasts produce a temporary ECM, predominantly consisting of fibrin, plasma fibronectin, vitronectin, and platelets, during the first phase of re-epithelialization. Keratinocytes, which are tightly bonded to the basal membrane the area adjacent to the wound site, move to the temporary scaffold site to deposit new ECM proteins such as tenascin C, laminin, and cellular fibronectin. Their creation commences the matrix maturation and remodeling phase, with the scaffold being converted from fibrin–fibronectin to a reticulum mostly composed of proteoglycans and collagen (Rousselle et al. 2019; Moretti et al. 2022; Huang et al. 2022).

All the processes described above involve morphological changes in the ECM known as "matrix remodeling." Indeed, from the earliest stages of development, the ECM is constantly subjected to modifications that include reorganization, renewal, and degradation of the molecules of which it is composed; these processes ensure the structural and functional integrity of tissues, maintain cellular homeostasis, and regulate responses to damage. The underlying mechanisms are highly regulated and involve several factors, including matrix metalloproteinases (MMPs).

MMPs are a class of proteolytic enzymes that are generally classified depending on their activity, substrate specificity, and structural organization in collagenases, gelatinases, stromelysin, matrilysin, ADAM (A Disintegrin and Metalloproteinase), and ADAMTS (A Disintegrin and Metalloproteinase with Thrombospondin Motif). The MMPs family is essential for physiological activities, such as development and tissue repair (Cui et al. 2017). For example, one of the first steps in the wound-healing process is the establishment of an inflammatory state, which draws several inflammatory cells to the wound site. MMPs are known to promote leukocyte and chemokine activity at the wound site: in wounded mucous membranes, for example, the epithelial cells at the wound edge release MMP-7 which, acting on the matrix, establishes a local chemokine gradient that controls neutrophil influx and activation; then, MMP-7 acts by degrading adherent junctions, facilitating cell migration to promote the re-epithelialization phase. In skin wounds, on the other hand, MMP1-mediated alteration of collagen promotes keratinocyte migration to the wound site (Parks et al. 2004; Caley et al. 2015). MMP-2 and MMP-9 also act on growth factors in the ECM and their receptors: in particular, they allow the release of TGF-β and VEGF and modulate their activity in the inhibition or promotion of angiogenesis during wound healing processes (de Almeida et al. 2022).

An active role of MMPs during tissue development and morphogenesis has also been suggested, although many of the mechanisms underlying these processes are not fully understood. In mouse models, for example, it has been seen how MMPs, in conjunction with hormonal activity, are involved in the branching changes of the mammary gland from puberty to pregnancy; the underlying mechanism involves the degradation of the ECM, mainly mediated by MMP-2 and MMP-3, to allow invasion and branching of the mammary ducts (Wiseman et al. 2003). Similarly, in tadpole models of gut formation, during metamorphosis, MMP-11, following stimulation by thyroid hormones, is secreted by fibroblasts and acts by influencing epithelial cell apoptosis and connective tissue growth in favor of this tissue formation (de Almeida et al. 2022).

In addition to MMPs, an important role in matrix remodeling is played by cathepsins, which represent a family of lysosomal proteolytic enzymes that exert their activity at a slightly acidic pH. Although, to date, their activity is mostly associated with a broad spectrum of pathological conditions, cathepsins are also important for various physiological functions in the body. Just as an example, cathepsins, especially the cysteine class, play a key role in maintaining bone tissue homeostasis;

indeed, during bone resorption, osteoclasts release cathepsin K, which is responsible for the degradation of type I collagen and elastin (Fonović and Turk 2014; Vizovišek et al. 2019).

Hence, MMPs and cathepsins can be considered among the main modulators of ECM remodeling. A dysregulation in their activity is often associated with the development of pathological conditions such as fibrosis, inflammatory diseases, and cancer progression.

2.2.2 *ECM in Disease*

The breakdown of the ECM can result either in an overproduction of its component molecules or in its abnormal degradation. In the first case, we can consider the wounds: very often their healing results in the formation of hypertrophic or keloid scars. While hypertrophic scars result from increased production of collagen (I and III), fibronectin, laminin, periostin, and tenascin, keloids are the result of an increased expression of collagen (I, III, and VI), fibronectin, periostin, tenascin, some proteoglycans, and inhibition of hyaluronic acid, dermatopontine, and decorin (Ikeda et al. 2009; Sidgwick and Bayat 2012; Hohenester 2019). Fibrosis is a further condition in which there is an excess of ECM production, which influences tissue properties, especially in terms of stiffness, and alters function, leading to disease and often death. In cardiac fibrosis there is an increase in the deposition of collagen fibers and incorporation of matricellular proteins; these simultaneously act on the TGF-β signaling pathway for the differentiation of cardiac fibroblasts into myofibroblasts (Li et al. 2018; Liu et al. 2021). Similarly, in pulmonary fibrosis, excessive matrix deposition, resulting mainly from an overproduction of collagen, contributes to the differentiation and activation of fibroblasts, impacting the mechanical properties of the tissue (Deng et al. 2020). Fibrotic-related matrix stiffness can also occur during aging processes, as a result of an increase in intramolecular bonds between fibrillar collagens and elastin (Jones et al. 2018).

On the other hand, dysregulated matrix degradation has been found in degenerative diseases such as osteoarthritis, in which the accumulation of reactive oxygen species (ROS) contributes to chondrocyte apoptosis and exacerbated matrix degradation (Loeser 2009; Lotz and Caramés 2011), and in cancer, that will be discussed in more detail in the following section.

In the same way as physiological processes, MMPs and cathepsins do influence ECM activity, promoting the onset of disease.

ADAMs have been identified as key players in several diseases, including cardiomyopathies. In a mouse model study, the downregulation of ADAM15 leads to an interruption in the remodeling of the ECM of the left ventricle, due to a loss of fibrillar collagen and reduced expression of proteins necessary for the meshwork of collagen, thus contributing to the development of myocardial infarction (Chute et al. 2022). In murine wound model on skin, the decrease in MMP-1 leads to an accumulation of collagen, promoting the formation of skin scars: the increase in

TGF-β1 activated the Smad signaling pathway and consequently caused an increased level of connexin 43, resulting in the downregulation of the Erk1/2 phosphorylation, thus impairing the expression of MMP-1, which in turn is decreased; the reduction of MMP-1 levels matches an increased in collagen III expression, suggesting a relationship with excessive ECM deposition and fibrosis development (Li et al. 2020).

In scar tissues where keloid formation occurs, however, several MMPs (1, 3, and 9) were elevated in fibroblasts adjacent to wound sites. It should be remembered that fibroblasts are responsible for the deposition of collagen and ECM both under physiological conditions and in the presence of wounds; consequently, high expressions of MMPs generate hyperstimulation of fibroblastic activity, which in this context causes keloid formation (Andrews et al. 2016). In vitro studies on tissues of patients with keloids have shown a possible correlation between fibroblasts and activated immune cells in lesions: specifically, high expressions of ADAM17 and ADAM10 in fibroblasts were found, which stimulate the infiltration of inflammatory cells, which, once activated, seem to increase the functionality and proliferation of fibroblasts through a mechanism that is not yet fully understood (Rath et al. 2021).

In osteoarthritis, fibroblast growth factor (FGF) is involved in several pathways leading to matrix degradation: FGF2, for example, by binding to the perlecan, upregulates the transcription of MMP-1 and -13 in articular chondrocytes, promoting cartilage degradation; also, by upregulating the transcription of the factors AP1 and RUNX2 it stimulates the expression of ADAMTS5, which is also fundamental in the pathogenesis of osteoarthritis as it is involved in the degradation of the articular cartilage matrix aggrecan (Xie et al. 2020).

In intervertebral disc degeneration, high miR-21 expression levels promote, via the PTEN/Akt/mTOR signaling pathway, the inhibition of autophagy and the overregulation of MMP-3 and -9, which act by degrading type II collagen and aggrecan in nucleus pulposus (Wang et al. 2018).

Dysregulation of cathepsins, on the other hand, has been found in several diseases, including cardiovascular diseases, and more recently in renal fibrosis (Wu et al. 2018). While in the former they are mainly related to ECM proteolysis, in the fibrosis increased cathepsin levels are related to an overproduction of ECM. In a study in murine and in vitro models, it was shown how in hydrophrenosis-related renal fibrosis, high cathepsin S levels cause an increase in ECM production through increased expression of TGF-β1, α-SMA, increased phosphorylation of SMAD2 and SMAD3 and decreased expression of E-Cadherin (Yao et al. 2019).

ECM in Cancer Alterations in the normal ECM remodeling processes can result in the development of tumors and regulate various hallmarks of cancer (Pickup et al. 2014; Bertero et al. 2019; Najafi et al. 2019; Raskov et al. 2023). Broadly speaking, a few key features and stages in which ECM is involved and for which it plays a key role in tumor promotion and progression can be identified.

First, the different composition of the tumor matrix gives the tumor greater stiffness than normal tissue. This is due to an intensified deposition and cross-linking of certain molecules, mainly collagen, followed by fibronectin, hyaluronic acid, and tenascin C, and to increased interactions between matrix molecules acting as ligands and membrane surface receptors, including integrins (Najafi et al. 2019). The increased deposition of ECM is mainly due to stromal cells—mainly fibroblasts—which, recruited by tumor cells, also assume a tumor-associated and activated phenotype called CAF (Cancer-Associated Fibroblasts).

The ECM activity mediated by tumor and stromal cells also promotes the establishment of an inflammatory state, e.g., by upregulating TGF-β, which draws cytokines and chemokines to the site of the tumor lesion, increasing the inflammatory state (Acerbi et al. 2015; Marozzi et al. 2021). As the tumor spreads to nearby sites, the collagen fibers of the matrix reorganize to stretch and spread with the tumor cells.

In the case of tumors developing metastases, changes in the ECM remodeling are necessary in order to allow tumor cells to "break" the basement membrane, invade and intravasate into the circulation, extravasate, and establish themselves at sites distant from the primary tumor (Kai et al. 2019; Paolillo and Schinelli 2019; Winkler et al. 2020).

The deposition of new ECM components, mainly represented by fibronectin, periostin, and tenascin-C, contributes to the formation of the pre-metastatic niche, which precedes the establishment of tumor cells at the new site (Liu and Cao 2016). However, it should be considered that depending on the site of the establishment of the pre-metastatic niche, the composition of the ECM will differ, supporting the concept of trophism in metastases (Hebert et al. 2020). On the other hand, the ECM -to allow tumor cells to invade and intravasate—rearranges itself into rigid collagen bundles. The interaction of this new ECM with the circulating fibrin and platelets forms a kind of protective barrier for the circulating tumor cells against shear stress and thus ensures their survival. Once extravasated, the tumor cells encounter a previously established pro-tumorigenic environment that will allow the tumor to proliferate at the new site; alternatively, however, the tumor cells may enter a transient dormant state and develop metastasis later (Kai et al. 2019). Even in the latter case, the role of the ECM was found to be crucial: for example, once the tumor cells reach the new site, they can remodel the matrix by forming a niche rich in type III collagen which has been seen to support the dormant state (Di Martino et al. 2021); fibronectin production may also be involved in the transition from the dormant to the proliferative state of metastatic cancer cells in a different manner (Barkan et al. 2008). In addition, ECM can also reactivate dormant tumor cells; for example, in metastatic breast cancer, the proliferative growth of dormant cells is induced at least in part by fibronectin and mediated by integrin β1 and the reorganization of the cytoskeleton (Barkan et al. 2008). Related to the awakening of dormant tumor cells and tumor progression are also matrikines: these represent peptide cytokines produced as a result of the fragmentation of ECM proteins and are also related to ECM remodeling during aging and in cancer (Jariwala et al. 2022). Laminin, for example, is a glycoprotein whose fragmentation has been associated with lung inflammation and the awakening of dormant metastatic cells in mice, leading to the exposure of an α3β1 integrin-binding epitope that causes the activation of the FAK/ERK/MLCK/YAP proliferative signaling pathway in tumor dormant cells (Albrengues et al. 2018).

Among all the processes potentially leading to the development and progression of cancer, it is now widely recognized that the enzymatic activity performed by MMPs also plays a role in this scenario. While this activity is carried out through the regulation of the activation of various growth factors, or through the regulation of immune responses, MMPs are also involved in the degradation of matrix components, facilitating tumor cell metastasis and angiogenesis (Jabłońska-Trypuć et al. 2016; Mittal et al. 2016; Roy et al. 2020). In vitro studies on chondrosarcoma have shown how connective tissue growth factor (CTGF) by binding to the αvβ3 integrin receptor, activates the FAK/MEK/ERK signaling pathway, that, in turn activates the NF-kB protein complex, which is responsible for the stimulation of the expression of MMP-13, which promotes tumor cell migration (Tan et al. 2009). In a number of highly invasive tumors, MMP-14—also known as membrane type 1 matrix metalloproteinase (MT1-MMP)—plays several roles ranging from proliferation, invasion, angiogenesis, and tumor metastasis (Vos et al. 2021); in fibrosarcoma and breast carcinoma cell lines, for example, collagen degradation is closely related to the expression of MT1-MMP and can occur in two ways: a phagocytic collagen pathway driven solely by MT1-MMP; and collagen degradation triggered by the collagenolytic activity of MT1-MMP supported by the activation of MMP-2. Both collagen degradation pathways lead to a remodeling of the ECM that enables tumor cell invasion (Lee et al. 2007). Similarly, in malignant ovarian tumors, the collagenolytic activity of MT1-MMP is crucial for invading through the submesothelial matrix and allowing tumor cells to proliferate outside the primary site (Moss et al. 2009). MMP-14 has also been associated with the invasion and migration of esophageal squamous cell carcinoma (ESCC) cells and, indeed, its upregulation has been linked to the worst clinical outcomes in ESCC patients. It has been hypothesized that the metastasis of cancer cells is ensured by the proteolytic activity of MMP-14 (Chen et al. 2020).

ECM degradation and remodeling are also important in neoangiogenesis processes, and MMP-2, MMP-9, and MT1-MMP are among the main contributors in cancer new vessels formation (van Hinsbergh and Koolwijk 2008; Jabłońska-Trypuć et al. 2016). MMPs modulate VEGF expression by both direct and indirect mechanisms (Lederle et al. 2010; Deryugina and Quigley 2015). For instance, in an in vitro model of colorectal adenocarcinoma, MMP-9, and partially MMP-2, modify the expression of the Multimerin 2 (MMRN2), a matrix protein secreted by endothelial cells that, by binding factors such as VEGFA and other cytokines, has a negative control on the generation of new vessels; it was shown how these MMPs cleave MMRN2, thus facilitating the release of pro-angiogenic factors (i.e., VEGFA), which in turn allow endothelial cells to invade the surrounding tissue, a necessary step for neovascularization (Andreuzzi et al. 2017). Signaling via VEGFA in endothelial cells can also lead to new vessel formation through the overproduction of its receptor VEGFR1, regulated by MMP-1 (Mazor et al. 2013).

Finally, it should be taken into account that under physiological conditions, MMPs are regulated by various factors, including Tissue Inhibitors of Metalloproteinases (TIMPs); however, these control systems sometimes undergo alterations, thus contributing to tumor progression (Cabral-Pacheco et al. 2020). For example, in prostate cancer high levels of TIMP-1 have been linked to the accumulation of CAFs in the TME (tumor microenvironment) and the promotion of their proliferation and migration (Gong et al. 2013).

Roles like MMPs, but less investigated, were evaluated for other types of proteases, such as Cathepsin B. In an in vitro study on glioma, high levels of Cathepsin B and Tenascin-C were detected in comparison to normal brain tissue, as well as in association with the formation of new vessels within the tumor; although the mechanism has yet to be fully elucidated, the study showed that the relationship between these two factors is important for tumor progression and angiogenesis (Mai et al. 2002). In prostate cancer with bone metastases, it has been observed in vitro and ex vivo that Cathepsin B levels are increased in tumor cells and in the same cells when brought into contact with bone matrices (Podgorski et al. 2005).

What has been discussed so far are just some of the examples of mechanisms through which the ECM can be modified in physiological conditions, as well as in pathological processes including cancer. A further mechanism is represented by the transport by EVs of molecules that could contribute to ECM remodeling (Taraboletti et al. 2002; Nawaz et al. 2018; Thuault et al. 2022).

2.3 Extracellular Vesicles

EVs can be described as a class of spherical particles surrounded by a phospholipid bilayer, released by all cell types and organisms, and heterogeneous in shape, size, biogenesis, and function (Théry et al. 2018). These differences are reflected in the numerous subtypes of vesicles that have been described to date; however, they can generally be classified into three macro groups: exosomes (EXOs), microvesicles (MVs), and apoptotic bodies (ABs) (van Niel et al. 2018; Doyle and Wang 2019).

EXOs are characterized by a diameter between 50 and 150 nm. They are generated by inward invagination of the plasma membrane, starting with early endosomes and progressing to late endosomes; the membrane of late endosomes then undergoes further invagination, forming multivesicular bodies (MVBs) containing intraluminal vesicles (ILVs), that will ultimately fuse with the plasma membrane to release ILVs into the extracellular space as EXOs. The biogenesis of EXOs is regulated through two pathways: an ESCRT (endosomal sorting complex required for transport)-dependent pathway, in which sub-complexes of this system actively participate in membrane remodeling during the formation of ILVs; an ESCRT-independent pathway, in which ceramide formed by sphingomyelin is responsible for membrane shaping.

MVs reach a larger size than EXOs (up to 1000 nm) and generate directly from the plasma membrane outward budding (Fig. 2.1). Few is known about the molecular pathways that lead to the generation of MVs; however, it is currently known that it involves several changes that include the rearrangement of cytoskeleton proteins and membrane phospholipids (Akers et al. 2013; Juan and Fürthauer 2018; van Niel et al. 2018; Horbay et al. 2022).

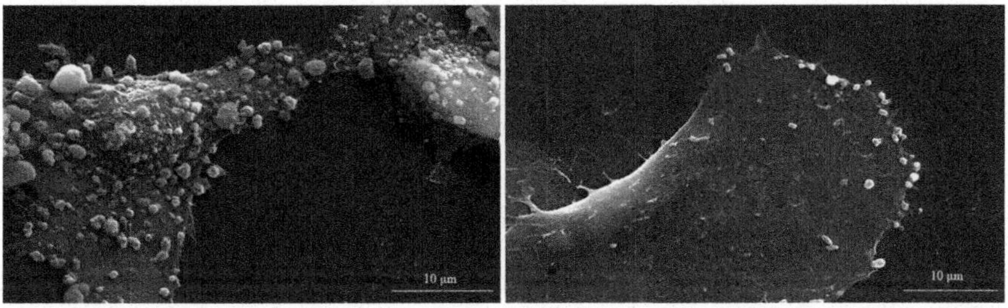

Fig. 2.1 Scanning Electron Microscopy (SEM) images of EVs. As described in literature, cancer cells are able to release a huge amount of MVs compared to normal ones (Muralidharan-Chari et al. 2010; Ståhl et al. 2019). The SEM images highlight this feature: indeed, human breast tumor cells MDA-MB-231 (on the left) exhibit a copious release of MVs from their surface, while the cell surface of normal human fibroblasts NHDF (on the right) is much less affected by this process. *Personal unpublished images*

Finally, apoptotic bodies, up to 2000 nm in size, are generated from apoptotic cells that lysate following a highly organized, caspase-mediated process (Zhou et al. 2022).

As comprehensively detailed in existing literature, the discovery of vesicles can be traced back to the mid-1990s, owing to research on coagulation. Since then, significant progress has been made in understanding their role, as well as their heterogeneity, largely due to advancements in isolation and characterization techniques (Chargaff and West 1946; Yáñez-Mó et al. 2015; Couch et al. 2021). Not surprisingly, their role as mediators in cell-cell communication in both physiological and pathological processes is well established (Fig. 2.2). Indeed, once released into the extracellular space, EVs can transfer molecules—such as various types of RNAs, DNA, lipids, and proteins derived from donor cells—to receiving cells.

2.3.1 Role of EVs in Health

Under physiological conditions EVs, particularly EXOs, are involved in maintaining cellular homeostasis, including acting as scavengers and eliminating what is harmful to the cell (Takeuchi et al. 2015; Vidal 2019); during pregnancy, specific factors found in EXOs play a crucial role in enabling communication and appropriate modification of the endometrial cavity that harbors the embryo (Greening et al. 2016); EVs derived from stem cell niches play a significant role in tissue regeneration by means of miRNAs or pro-angiogenic growth factors, such as VEGF, or by decreasing the levels of TNF-α and various cytokines, thus suppressing inflammation (Zhou et al. 2021; Gao et al. 2020). EVs can also modulate the pro- and anti-inflammatory response in innate immunity and inflammation and, in adaptive

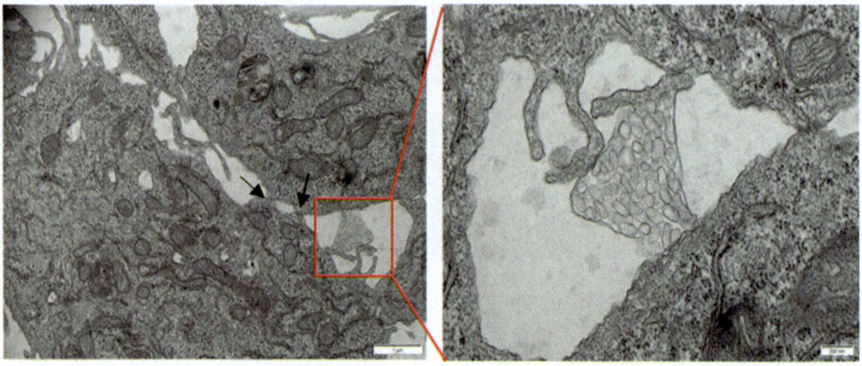

Fig. 2.2 Transmission Electron Microscopy on sections of an ovarian cancer cell grown in 3D. Migrasomes are a recent class of EVs whose biogenesis is linked to the migration process of the cell. Various functions have been assigned to migrasomes, such as homeostasis maintenance, embryonic development, cancer progression, as well as cell–cell communication (Yu and Yu 2022; Zhang et al. 2023a). The left figure (Scale bar: 1 μm) shows the connection between two cells fostered also by tight junctions (black arrows). Moreover, precisely a migrasome-like structure, identified by the red square, appears between the cells; this detail is shown at higher magnification in the figure on the right (Scale bar: 200 nm). *Personal unpublished images*

immunity, B-cell-derived EV's role in presenting antigens to T cells is well recognized (Greco et al. 2022; Buzas 2023).

2.3.2 *Role of EVs in Disease*

EVs also have a key involvement in a number of diseases ranging from cardiovascular diseases, inflammatory diseases, and metabolic diseases, to age-related diseases (Yin et al. 2021; Berumen Sánchez et al. 2021), and in cancer. Tumor-derived EVs (tEVs) influence all the hallmarks of cancer (which include alterations in cell proliferation, escape from apoptosis, promotion of angiogenesis, modulation of the immune response, and reprogramming of energy metabolism) leading to its initiation, progression, promotion, and resistance to treatment (Pan et al. 2022; Ren et al. 2019).

tEVs affect most of the different cell types found in the TME, including fibroblasts, endothelial and mesenchymal stem cells, and immune cells. These cells, in turn, through signaling molecules and EVs, contribute to tumor progression and spread (Giannandrea et al. 2022). Among all of them, fibroblasts, have an active role, since once they have acquired a pro-tumorigenic phenotype and functions, they modulate the tumor stroma in favor of tumor progression and spread. These fibroblasts are known as Cancer-Associated Fibroblasts (CAFs) and their activation is driven also by tEVs (Giusti et al. 2018, 2022; Shoucair et al. 2020). Likewise, tEVs are also liable for the phenotypic and functional change of macrophages in tumor-associated macrophage (TAM), which can lead to a remodeling of the tumor stroma in favor of an immunosuppressive microenvironment (Bellmunt et al. 2019; Rabe et al. 2021; Chen et al. 2023b).

However, EVs are able to interact and influence not only the TME and surrounding cells: in fact, through biological fluids, they can send and exchange signals at distant sites thus contributing to the metastatic process, from invasion to the formation of the premetastatic niche, to the dissemination of the metastasis in distant organs (Giannandrea et al. 2021; Dong et al. 2021; Giusti et al. 2023).

In addition to cell–cell communication, EVs facilitate communication between cells and the ECM; in tumors, through their interaction with the components of the ECM, EVs guide its remodeling in favor of a pro-tumor and pro-metastatic environment (Tkach and Théry 2016; Winkler et al. 2020).

2.4 ECM Remodeling Process and EVs Implication

EVs have been revealed as an important part of the extracellular space and the ECM is not only a diffusion path for them: they actively interact with matrix components, contributing to the remodeling process (Nawaz et al. 2018; Rilla et al. 2019).

In particular, both MMPs and cathepsins—described above—have been identified in the cargo of EVs (Giusti et al. 2008; Dankovich et al. 2021; Al Halawani et al. 2022; Thuault et al. 2022; Bellini et al. 2022; Patel et al. 2023).

Although the meshwork of the matrix is smaller than their size, EVs still get through the ECM meshwork mainly due to the aquaporins that mediate the deformability of the EVs, and the rigidity and relaxation of the matrix that act in combination to improve the transport of the EVs (Lenzini et al. 2020).

However, the role of EVs in matrix remodeling is carried out under both physiological and pathological conditions (Rilla et al. 2019).

2.4.1 Remodeling in Health

In physiological processes, the activity of EVs has been linked, for example, to the prevention of scar formation in wound healing processes. A study in mouse models has shown that administration of EVs derived from adipose stem cells reduces the differentiation of fibroblasts into myofibroblasts; increases the ratio of: collagen I to collagen III, TGF-β3 to TGF-β1, and MMP-1 to TIMP-3; through activation of the MAPK/ERK pathway, it increases the expression of MMP-3 in dermal fibroblasts. Taken together, these processes limit scar formation by promoting a normal healing process (Wang et al. 2017). Similarly, EVs derived from human amniotic epithelial cells (hAECs) showed the ability to promote wound healing without scar formation by promoting fibroblast proliferation and migration; in vivo, local injection into the wound site of hAECs derived-EVs also showed reorganization of collagen fibers that promoted rapid wound healing and decreased scarring. Although the underlying mechanisms are poorly understood, the authors hypothesized that these EVs carry microRNAs or proteins in their cargo that could be responsible for inhibiting differentiation into myofibroblasts and accelerating wound healing (Zhao et al. 2017). In vitro studies of astrocytic lesions have revealed that EVs from astrocytes cultured on different matrices (Laminin, Fibronectin, Tenascin C, Vitronectin) modulate the ability of complete wound healing in a positive manner, depending on the matrix source. In addition, it was found that EVs-related interleukin IL-1β may have a direct or indirect influence on wound healing, depending on the type of matrix used to culture the astrocytes. Despite this evidence, further investigations are needed to better understand the mechanism underlying the EVs–ECM–astrocytic wound repair relationship (Sutter et al. 2021).

EVs have been found also to be a key component in the mineralization of the bones ECM. Indeed, osteoblasts and chondrocytes release EVs into the matrix that initiate the formation of hydroxyapatite crystals and the conversion of the ECM into a mineralized architecture, both in endochondral calcification processes during development and during bone resorption in the adult. Apatite formation is mediated by the entry of calcium ions and phosphate into the vesicles via annexin channels and a Na-dependent phosphate carrier, respectively (Golub 2009; Veschi et al. 2020). Bone matrix and dentin mineralization are also driven by the transforming growth factor beta receptor II interacting protein 1 (TRIP-1): increased expression of this protein showed in vitro increased secretion of EVs from periosteoblasts, which were predominantly localized within the ECM where they probably play a role in the mineralization process. Overexpression of TRIP-1 was also correlated with increased expression of type I collagen and matrix fibronectin, which are crucial as they act as scaffolding in mineralized tissues (Chen and George 2018).

2.4.2 *Remodeling in Disease*

While EVs are implicated in ECM remodeling during normal cellular conditions, their role has been predominantly found and investigated in the occurrence of disease.

For instance, in chronic obstructive pulmonary disease (COPD) and bronchopulmonary dysplasia (BPD), EXOs enriched with the protease enzyme neutrophil elastase (NE), are released during polymorphonucleate leukocyte degranulation; NE drives the hydrolysis of collagen I—supported by the binding of Mac-1 integrin on the surface of the EXOs to the collagen—causing ECM degradation and the onset of typical disease symptoms (emphysema in COPD and pulmonary simplification in BPD) (Genschmer et al. 2019).

In venous malformations, the protein Rab27A seems to drive the secretion of EVs characterized by high levels of MMP-14 expression, linked to ECM degradation and the development of venous malformations (Chen et al. 2022).

However, the expression of MMPs, as already said, has been mainly investigated in cancer, where they are overregulated, thus leading to enhanced processes including cellular hyperproliferation and invasiveness, neovascularization, immune escape, and promotion of metastasis (Isaacson et al. 2017).

Indeed, EVs contain several enzymes in their cargo that act in a pro-tumorigenic manner, degrading the matrix and promoting the migration and proliferation of cancer cells. Among these molecules can be found many enzymes that act on specific matrix molecules, including lysyl oxidase type 2, heparinase, collagenase, and elastase (Patel et al. 2023); however, EVs-associated MMPs are the enzymes mainly involved in the active degradation and remodeling of the ECM (Shimoda and Khokha 2017).

In a mouse model in which breast cancer was induced, it was assessed how aspartate β-hydroxylase (ASPH) plays a key role in the promotion of metastasis: in particular, the ASPH-Notch axis mediates the production and secretion of tumor EXOs enriched with MMPs and ADAMs, which in turn act on matrix degradation and remodeling, promoting tumor cell invasiveness and the establishment of metastases (Lin et al. 2019).

Other studies performed in vitro and in vivo have shown that the Rab27B factor plays a key role in breast cancer by increasing proliferation, invasiveness, and development of peritoneal metastases; Rab27B is a member of the Rab family of proteins involved in vesicular trafficking. According to these studies, Rab27B has been identified in the cargo of tumor EVs and the signaling pathway mediated by this protein leads not only to the reorganization of the cytoskeleton and cell cycle progression and proliferation of tumor cells but also to the HSP90α-mediated MMP-2 activation, which acts by degrading the matrix and promoting tumor cell invasiveness (Hendrix et al. 2010).

MMP-2 is also responsible for matrix degradation in other tumors, including thyroid cancer: proteomic characterization of EVs derived from cocultures of fibroblasts and thyroid tumor cells showed that crosstalk between these two cell types provides a peculiar proteomic profile related to ECM degradation; this is due to the fact that EVs derived from tumor cells enhance the synthesis and activation of MMP-2 in the TME, promoting ECM degradation linked to thyroid tumor progression. MMP-2 activity in the TME was also potentiated by its increased expression in non-tumor cells (Bravo-Miana et al. 2022). Another in vitro study, performed on HPV-related cervical cancer cells revealed how EXOs secreted by these cells are enriched in MMP-2 and can increase the proliferation of HUVEC via the Akt signaling pathway, thus promoting tumor metastatic dissemination (Paolillo and Schinelli 2019).

In an in vitro study on nasopharyngeal cancer (NPC), elevated levels of MMP-13 carried by tumor EXOs (tEXOs) were found both in patient plasma samples and in conditioned cell culture media. By interacting with stromal cells, the potential of tEXOs to induce tumor proliferation and metastasis via MMP-13 was assessed. In particular, it was observed that human skin fibroblast cells treated with NPC tEXOs were able to upregulate MMP secretion in a time-dependent manner, suggesting a possible role in promoting tumor invasion by degrading several ECM components (You et al. 2015).

EVs, on the other hand, are able to actively regulate MMP expression as they can carry MMP regulators, such as the extracellular matrix metalloproteinase inducer (EMMPRIN) and TIMP, previously mentioned (Patel et al. 2023). An in vitro study has shown that EVs derived from medulloblastoma cells express high levels of EMMPRIN, which in turn correlates with increased MMP-2 expression; the latter, through ECM degradation, promotes migration and invasion of tumor cells, thereby promoting medulloblastoma metastasis (Jackson et al. 2023).

Normally, TIMP-1 acts in an inhibitory way towards MMPs, but some studies started to shed light on the evidence that it can modulate cancer progression independently by its inhibitory effect; indeed, in vitro and in vivo increased levels of TIMP-1 have been associated with activation of the PI3K/AKT/HIF-1/miR-210 pathway in lung adenocarcinoma cells. High levels of miR-210 were particularly found in the tEXOs and associated with pro-angiogenic activity (Cui et al. 2015).

In vivo and in vitro studies on glioblastoma have shown that mi-R21 acts by inhibiting genes associated with apoptosis, migration, and tumor invasiveness, as well as TIMP-3, the inactivation of which promotes the activity of various MMPs, mainly MMP-2 and MMP-9, degrading the ECM and allowing tumor cell invasion and motility (Gabriely et al. 2008).

A study performed on dermal fibroblasts obtained from TIMP-deficient mouse skin explants showed that loss of TIMP is enough to trigger a CAF-like state in fibroblasts. Moreover, the latter produce EXOs rich in ADAM10, responsible for the activation of tumor progression mediated by the Notch and RhoA signaling pathways when breast cancer cells are treated with these EXOs. TIMP-deficient fibroblasts also supply ECM proteins (including Tenascin C and periostin) contributing to the formation of pre-metastatic niches for tumor cells (Shimoda et al. 2014).

Finally, other EVs-related enzymes besides MMPs can cause matrix remodeling favoring a pro-tumorigenic and pro-invasive environment: in oral squamous cell carcinoma, CAFs release sEVs enriched in α-lysyl oxidase: this enzyme binds to matrix collagen via the surface integrin α2β1, triggering collagen crosslinking via the p-FAK/p-paxillin/YAP signaling pathway, which is necessary for the EMT (Liu et al. 2023).

It is known that ECM stiffness correlates with a worse tumor prognosis, due to its association with the pro-invasive behavior of tumor cells. Stiffness could also be related to EVs secretion and activity: for example, human hepatocellular carcinoma (HCC) cells, when cultured in vitro on a rigid matrix model, can release EXOs in higher quantities compared to the same cells cultured on a soft matrix. The process is driven by the Akt signaling pathway, leading to Rab8 activation and EXOs secretion. Furthermore, EXOs from cells grown on rigid ECM lead to activation of the Notch pathway, which correlates with tumor progression and poor prognosis of HCC (Wu et al. 2023). In breast cancer in vitro studies, ECM stiffness has been linked to elevated levels of Thrombospondin-1 transported by tEXOs; its simultaneous transport with integrins, fibronectins, and MMPs facilitates the ECM rigidity-dependent pro-invasive characteristics of breast cancer cells (Patwardhan et al. 2021).

2.5 Conclusion and Future Perspectives

The ECM is a highly dynamic structure that is heterogeneous in terms of molecular composition and physiologically undergoes remodeling. However, when normal cellular homeostasis is compromised, the ECM undergoes a series of changes that contribute to the development of diseases, including cancer. These changes may be due either to an overproduction (as occurs in the case of scarring) or to an excessive degradation of the molecules that make up the matrix (as occurs in the processes of tumor invasion). MMPs are among the main proponents of this role: their interaction with the matrix contributes to the remodeling of the ECM in its broadest meaning. Interestingly, MMPs can act in an adjuvant manner with other molecules, or exert their action through transport mediated by EVs, well known for their role in cell communication.

These three components, namely ECM, MMPs, and EVs, could be considered individually in terms of understanding their physiological and pathological roles, to evaluate possible therapeutic strategies.

Indeed, a better understanding of the structure and regulation of the ECM could lead to the development of new therapeutic approaches targeting different ECM components. To date, the decellularization of tissues and organs is the best strategy for studying and understanding ECM that comes as close as possible to in vivo conditions. However, the presence of possible "contaminants" and the lack of standardization of decellularization protocols are still limitations. At the same time, the introduction of biomaterials that mimic the structure of ECM has been found more useful in clinical applications. However, complications due to immune responses and rejection of human tissue make the use of these scaffolds of limited use. In conclusion, further investigation is required before these strategies can be considered promising tools in daily clinical practice (Aamodt and Grainger 2016; Fernández-Pérez and Ahearne 2019; Da et al. 2021; Valdoz et al. 2021; Zhang et al. 2022, 2023b; Chen et al. 2023a; Song et al. 2023).

MMPs are among the most important players in ECM remodeling and especially degradation; in cancer, they play a crucial role in the regulation of growth factors, and chemokines, as well as in the development of metastases and new vessel formation. They could be used, therefore, as possible targets for the development of new therapies, as well as potential biomarkers of disease. However, even in this case, there are limitations in clinical applications: among these, the complete understanding of the mechanisms of action by which MMPs operate on the matrix and the development of therapies targeting specific MMPs, which do not impact other factors, remain the major problems to be addressed at present before they can be considered as effective tools for eradicating disease in cancer (Shi et al. 2021b; Mustafa et al. 2022).

Finally, EVs could be used as carriers of molecules to inhibit the proteolytic activity of MMPs or, in a more generic way, all molecules responsible for the onset of disease because of ECM remodeling. However, the use of biologically derived EVs is also limited: on the one hand, the isolation of "pure" EVs is still a problem, and on the other hand, the administration of EVs could both induce immune responses and have an impact not only on the specific target but also on the surrounding environment. The use of engineered and/or artificial EVs is proving to be widely used, as it partially overcomes the issues discussed above; moreover, they can be produced in larger quantities and at lower costs. However, further investigations into the risk-benefit ratios arising from them will be necessary before they can be considered a safe tool for common clinical use (Murphy et al. 2019; Herrmann et al. 2021; Sajeesh et al. 2023; Chen et al. 2023a).

Summing up, however, and considering the close interdependence between these three components, these points need to encourage us to consider from another perspective, more integrated, the different aspects involving ECM, MMPs, and EVs in pathological conditions to identify optimized clinical strategies.

References

Aamodt JM, Grainger DW (2016) Extracellular matrix-based biomaterial scaffolds and the host response. Biomaterials 86:68–82. https://doi.org/10.1016/j.biomaterials.2016.02.003

Acerbi I, Cassereau L, Dean I, Shi Q, Au A, Park C, Chen YY, Liphardt J, Hwang ES, Weaver VM (2015) Human breast cancer invasion and aggression correlates with ECM stiffening and immune cell infiltration. Integr Biol 7:1120–1134. https://doi.org/10.1039/c5ib00040h

Acuna A, Drakopoulos MA, Leng Y, Goergen CJ, Calve S (2018) Three-dimensional visualization of extracellular matrix networks during murine development. Dev Biol 435:122–129. https://doi.org/10.1016/j.ydbio.2017.12.022

Akers JC, Gonda D, Kim R, Carter BS, Chen CC (2013) Biogenesis of extracellular vesicles (EV): exosomes, microvesicles, retrovirus-like vesicles, and apoptotic bodies. J Neurooncol 113:1–11. https://doi.org/10.1007/s11060-013-1084-8

Al Halawani A, Mithieux SM, Yeo GC, Hosseini-Beheshti E, Weiss AS (2022) Extracellular vesicles: interplay with the extracellular matrix and modulated cell responses. IJMS 23:3389. https://doi.org/10.3390/ijms23063389

Albrengues J, Shields MA, Ng D, Park CG, Ambrico A, Poindexter ME, Upadhyay P, Uyeminami DL, Pommier A, Küttner V, Bružas E, Maiorino L, Bautista C, Carmona EM, Gimotty PA, Fearon DT, Chang K, Lyons SK, Pinkerton KE, Trotman LC, Goldberg MS, Yeh JT-H, Egeblad M (2018) Neutrophil extracellular traps produced during inflammation awaken dormant cancer cells in mice. Science 361:eaao4227. https://doi.org/10.1126/science.aao4227

Andreuzzi E, Colladel R, Pellicani R, Tarticchio G, Cannizzaro R, Spessotto P, Bussolati B, Brossa A, De Paoli P, Canzonieri V, Iozzo RV, Colombatti A, Mongiat M (2017) The angiostatic molecule Multimerin 2 is processed by MMP-9 to allow sprouting angiogenesis. Matrix Biol 64:40–53. https://doi.org/10.1016/j.matbio.2017.04.002

Andrews JP, Marttala J, Macarak E, Rosenbloom J, Uitto J (2016) Keloids: the paradigm of skin fibrosis—pathomechanisms and treatment. Matrix Biol 51:37–46. https://doi.org/10.1016/j.matbio.2016.01.013

Aumailley M (2013) The laminin family. Cell Adhes Migr 7:48–55. https://doi.org/10.4161/cam.22826

Barkan D, Kleinman H, Simmons JL, Asmussen H, Kamaraju AK, Hoenorhoff MJ, Liu Z, Costes SV, Cho EH, Lockett S, Khanna C, Chambers AF, Green JE (2008) Inhibition of metastatic outgrowth from single dormant tumor cells by targeting the cytoskeleton. Cancer Res 68:6241–6250. https://doi.org/10.1158/0008-5472.CAN-07-6849

Baumann L, Bernstein EF, Weiss AS, Bates D, Humphrey S, Silberberg M, Daniels R (2021) Clinical relevance of elastin in the structure and function of skin. Aesthet Surg J Open Forum 3:ojab019. https://doi.org/10.1093/asjof/ojab019

Bellini S, Saraceno C, Benussi L, Geviti A, Longobardi A, Nicsanu R, Cimini S, Ricci M, Canafoglia L, Coppola C, Puoti G, Binetti G, Rossi G, Ghidoni R (2022) Plasma small extracellular vesicle Cathepsin D Dysregulation in GRN/C9orf72 and sporadic frontotemporal lobar degeneration. IJMS 23:10693. https://doi.org/10.3390/ijms231810693

Bellmunt ÀM, López-Puerto L, Lorente J, Closa D (2019) Involvement of extracellular vesicles in the macrophage-tumor cell communication in head and neck squamous cell carcinoma. PLoS One 14:e0224710. https://doi.org/10.1371/journal.pone.0224710

Bertero T, Oldham WM, Grasset EM, Bourget I, Boulter E, Pisano S, Hofman P, Bellvert F, Meneguzzi G, Bulavin DV, Estrach S, Feral CC, Chan SY, Bozec A, Gaggioli C (2019) Tumor-stroma mechanics coordinate amino acid availability to sustain tumor growth and malignancy. Cell Metab 29:124–140.e10. https://doi.org/10.1016/j.cmet.2018.09.012

Berumen Sánchez G, Bunn KE, Pua HH, Rafat M (2021) Extracellular vesicles: mediators of intercellular communication in tissue injury and disease. Cell Commun Signal 19:104. https://doi.org/10.1186/s12964-021-00787-y

Bravo-Miana RDC, Soler MF, Ceschin DG, Royo F, Negretti-Borga DM, Azkargorta M, Elortza F, Montesinos MDM, Pellizas CG, Falcón-Pérez JM, Donadio AC (2022) Extracellular vesicles from thyroid cancer harbor a functional machinery involved in extracellular matrix remodeling. Eur J Cell Biol 101:151254. https://doi.org/10.1016/j.ejcb.2022.151254

Buzas EI (2023) The roles of extracellular vesicles in the immune system. Nat Rev Immunol 23:236–250. https://doi.org/10.1038/s41577-022-00763-8

Cabral-Pacheco GA, Garza-Veloz I, Castruita-De la Rosa C, Ramirez-Acuña JM, Perez-Romero BA, Guerrero-Rodriguez JF, Martinez-Avila N, Martinez-Fierro ML (2020) The roles of matrix metalloproteinases and their inhibitors in human diseases. IJMS 21:9739. https://doi.org/10.3390/ijms21249739

Caley MP, Martins VLC, O'Toole EA (2015) Metalloproteinases and wound healing. Adv Wound Care 4:225–234. https://doi.org/10.1089/wound.2014.0581

Chargaff E, West R (1946) The biological significance of the thromboplastic protein of blood. J Biol Chem 166:189–197. https://doi.org/10.1016/S0021-9258(17)34997-9

Chen Y, George A (2018) TRIP-1 promotes the assembly of an ECM that contains extracellular vesicles and factors that modulate angiogenesis. Front Physiol 9:1092. https://doi.org/10.3389/fphys.2018.01092

Chen N, Zhang G, Fu J, Wu Q (2020) Matrix metalloproteinase-14 (MMP-14) downregulation inhibits esophageal squamous cell carcinoma cell migration, invasion, and proliferation. Thorac Cancer 11:3168–3174. https://doi.org/10.1111/1759-7714.13636

Chen G-H, Yang J-G, Xia H-F, Zhang L-Z, Chen Y-H, Wang K-M, Duan X, Wu L-Z, Zhao Y-F, Chen G (2022) Endothelial cells induce degradation of ECM through enhanced secretion of MMP14 carried on extracellular vesicles in venous malformation. Cell Tissue Res 389:517–530. https://doi.org/10.1007/s00441-022-03657-2

Chen C, Fu L, Luo Y, Zeng W, Qi X, Wei Y, Chen L, Zhao X, Li D, Tian M, Kang P (2023a) Engineered exosome-functionalized extracellular matrix-mimicking hydrogel for promoting bone repair in glucocorticoid-induced osteonecrosis of the femoral head. ACS Appl Mater Interfaces 15:28891–28906. https://doi.org/10.1021/acsami.3c01539

Chen QY, Gao B, Tong D, Huang C (2023b) Crosstalk between extracellular vesicles and tumor-associated macrophage in the tumor microenvironment. Cancer Lett 552:215979. https://doi.org/10.1016/j.canlet.2022.215979

Chute M, Aujla PK, Li Y, Jana S, Zhabyeyev P, Rasmuson J, Owen CA, Abraham T, Oudit GY, Kassiri Z (2022) ADAM15 is required for optimal collagen cross-linking and scar formation following myocardial infarction. Matrix Biol 105:127–143. https://doi.org/10.1016/j.matbio.2021.12.002

Couch Y, Buzàs EI, Di Vizio D, Gho YS, Harrison P, Hill AF, Lötvall J, Raposo G, Stahl PD, Théry C, Witwer KW, Carter DRF (2021) A brief history of nearly everything—the rise and rise of extracellular vesicles. J Extracell Vesicle 10. https://doi.org/10.1002/jev2.12144
Couchman JR, Pataki CA (2012) An introduction to proteoglycans and their localization. J Histochem Cytochem 60:885–897. https://doi.org/10.1369/0022155412464638
Cui H, Seubert B, Stahl E, Dietz H, Reuning U, Moreno-Leon L, Ilie M, Hofman P, Nagase H, Mari B, Krüger A (2015) Tissue inhibitor of metalloproteinases-1 induces a pro-tumourigenic increase of miR-210 in lung adenocarcinoma cells and their exosomes. Oncogene 34:3640–3650. https://doi.org/10.1038/onc.2014.300
Cui N, Hu M, Khalil RA (2017) Biochemical and biological attributes of matrix metalloproteinases. In: Progress in molecular biology and translational science. Elsevier, pp 1–73
Da L-C, Huang Y-Z, Xie H-Q, Zheng B-H, Huang Y-C, Du S-R (2021) Membranous extracellular matrix-based scaffolds for skin wound healing. Pharmaceutics 13:1796. https://doi.org/10.3390/pharmaceutics13111796
Dalton CJ, Lemmon CA (2021) Fibronectin: molecular structure, fibrillar structure and mechanochemical signaling. Cells 10:2443. https://doi.org/10.3390/cells10092443
Dankovich TM, Kaushik R, Olsthoorn LHM, Petersen GC, Giro PE, Kluever V, Agüi-Gonzalez P, Grewe K, Bao G, Beuermann S, Hadi HA, Doeren J, Klöppner S, Cooper BH, Dityatev A, Rizzoli SO (2021) Extracellular matrix remodeling through endocytosis and resurfacing of Tenascin-R. Nat Commun 12:7129. https://doi.org/10.1038/s41467-021-27462-7
de Almeida LGN, Thode H, Eslambolchi Y, Chopra S, Young D, Gill S, Devel L, Dufour A (2022) Matrix metalloproteinases: from molecular mechanisms to physiology, pathophysiology, and pharmacology. Pharmacol Rev 74:714–770. https://doi.org/10.1124/pharmrev.121.000349
Deng Z, Fear MW, Suk Choi Y, Wood FM, Allahham A, Mutsaers SE, Prêle CM (2020) The extracellular matrix and mechanotransduction in pulmonary fibrosis. Int J Biochem Cell Biol 126:105802. https://doi.org/10.1016/j.biocel.2020.105802
Deryugina EI, Quigley JP (2015) Tumor angiogenesis: MMP-mediated induction of intravasation- and metastasis-sustaining neovasculature. Matrix Biol 44–46:94–112. https://doi.org/10.1016/j.matbio.2015.04.004
Di Martino JS, Nobre AR, Mondal C, Taha I, Farias EF, Fertig EJ, Naba A, Aguirre-Ghiso JA, Bravo-Cordero JJ (2021) A tumor-derived type III collagen-rich ECM niche regulates tumor cell dormancy. Nat Cancer 3:90–107. https://doi.org/10.1038/s43018-021-00291-9
Dong Q, Liu X, Cheng K, Sheng J, Kong J, Liu T (2021) Pre-metastatic niche formation in different organs induced by tumor extracellular vesicles. Front Cell Dev Biol 9:733627. https://doi.org/10.3389/fcell.2021.733627
Doyle L, Wang M (2019) Overview of extracellular vesicles, their origin, composition, purpose, and methods for exosome isolation and analysis. Cells 8:727. https://doi.org/10.3390/cells8070727
Fernández-Pérez J, Ahearne M (2019) The impact of decellularization methods on extracellular matrix derived hydrogels. Sci Rep 9:14933. https://doi.org/10.1038/s41598-019-49575-2
Fonović M, Turk B (2014) Cysteine cathepsins and extracellular matrix degradation. Biochim Biophys Acta 1840:2560–2570. https://doi.org/10.1016/j.bbagen.2014.03.017
Gabriely G, Wurdinger T, Kesari S, Esau CC, Burchard J, Linsley PS, Krichevsky AM (2008) MicroRNA 21 promotes glioma invasion by targeting matrix metalloproteinase regulators. Mol Cell Biol 28:5369–5380. https://doi.org/10.1128/MCB.00479-08
Gao S, Chen T, Hao Y, Zhang F, Tang X, Wang D, Wei Z, Qi J (2020) Exosomal miR-135a derived from human amnion mesenchymal stem cells promotes cutaneous wound healing in rats and fibroblast migration by directly inhibiting LATS2 expression. Stem Cell Res Ther 11:56. https://doi.org/10.1186/s13287-020-1570-9
Genschmer KR, Russell DW, Lal C, Szul T, Bratcher PE, Noerager BD, Abdul Roda M, Xu X, Rezonzew G, Viera L, Dobosh BS, Margaroli C, Abdalla TH, King RW, McNicholas CM, Wells JM, Dransfield MT, Tirouvanziam R, Gaggar A, Blalock JE (2019) Activated PMN exosomes:

pathogenic entities causing matrix destruction and disease in the lung. Cell 176:113–126.e15. https://doi.org/10.1016/j.cell.2018.12.002

Giannandrea D, Citro V, Lesma E, Bignotto M, Platonova N, Chiaramonte R (2021) Restoring tissue homeostasis at metastatic sites: a focus on extracellular vesicles in bone metastasis. Front Oncol 11:644109. https://doi.org/10.3389/fonc.2021.644109

Giannandrea D, Platonova N, Colombo M, Mazzola M, Citro V, Adami R, Maltoni F, Ancona S, Dolo V, Giusti I, Basile A, Pistocchi A, Cantone L, Bollati V, Casati L, Calzavara E, Turrini M, Lesma E, Chiaramonte R (2022) Extracellular vesicles mediate the communication between multiple myeloma and bone marrow microenvironment in a NOTCH dependent way. Haematologica. https://doi.org/10.3324/haematol.2021.279716

Giusti I, D'Ascenzo S, Millimaggi D, Taraboletti G, Carta G, Franceschini N, Pavan A, Dolo V (2008) Cathepsin B mediates the pH-dependent proinvasive activity of tumor-shed microvesicles. Neoplasia 10:481–488. https://doi.org/10.1593/neo.08178

Giusti I, Francesco MD, Ascenzo SD, Palmerini MG, Macchiarelli G, Carta G, Dolo V (2018) Ovarian cancer-derived extracellular vesicles affect normal human fibroblast behavior. Cancer Biol Ther 19(8):722–734. https://doi.org/10.1080/15384047.2018.1451286

Giusti I, Di Francesco M, Poppa G, Esposito L, D'Ascenzo S, Dolo V (2022) Tumor-derived extracellular vesicles activate Normal human fibroblasts to a cancer-associated fibroblast-like phenotype, sustaining a pro-tumorigenic microenvironment. Front Oncol 12:839880. https://doi.org/10.3389/fonc.2022.839880

Giusti I, Poppa G, Di Fazio G, D'Ascenzo S, Dolo V (2023) Metastatic dissemination: role of tumor-derived extracellular vesicles and their use as clinical biomarkers. IJMS 24:9590. https://doi.org/10.3390/ijms24119590

Golub EE (2009) Role of matrix vesicles in biomineralization. Biochim Biophys Acta 1790:1592–1598. https://doi.org/10.1016/j.bbagen.2009.09.006

Gong Y, Scott E, Lu R, Xu Y, Oh WK, Yu Q (2013) TIMP-1 promotes accumulation of cancer associated fibroblasts and cancer progression. PLoS One 8:e77366. https://doi.org/10.1371/journal.pone.0077366

Greco MF, Rizzuto AS, Zarà M, Cafora M, Favero C, Solazzo G, Giusti I, Adorni MP, Zimetti F, Dolo V, Banfi C, Ferri N, Sirtori CR, Corsini A, Barbieri SS, Pistocchi A, Bollati V, Macchi C, Ruscica M (2022) PCSK9 confers inflammatory properties to extracellular vesicles released by vascular smooth muscle cells. IJMS 23:13065. https://doi.org/10.3390/ijms232113065

Greening DW, Nguyen HPT, Elgass K, Simpson RJ, Salamonsen LA (2016) Human endometrial exosomes contain hormone-specific cargo modulating trophoblast adhesive capacity: insights into endometrial-embryo interactions. Biol Reprod 94. https://doi.org/10.1095/biolreprod.115.134890

Guldager Kring Rasmussen D, Karsdal MA (2016) Laminins. In: Biochemistry of collagens, laminins and elastin. Elsevier, pp 163–196

Hebert JD, Myers SA, Naba A, Abbruzzese G, Lamar JM, Carr SA, Hynes RO (2020) Proteomic profiling of the ECM of xenograft breast cancer metastases in different organs reveals distinct metastatic niches. Cancer Res 80:1475–1485. https://doi.org/10.1158/0008-5472.CAN-19-2961

Hendrix A, Maynard D, Pauwels P, Braems G, Denys H, Van den Broecke R, Lambert J, Van Belle S, Cocquyt V, Gespach C, Bracke M, Seabra MC, Gahl WA, De Wever O, Westbroek W (2010) Effect of the secretory small GTPase Rab27B on breast cancer growth, invasion, and metastasis. JNCI J Natl Cancer Inst 102:866–880. https://doi.org/10.1093/jnci/djq153

Herrmann IK, Wood MJA, Fuhrmann G (2021) Extracellular vesicles as a next-generation drug delivery platform. Nat Nanotechnol 16:748–759. https://doi.org/10.1038/s41565-021-00931-2

Hohenester E (2019) Structural biology of laminins. Essays Biochem 63:285–295. https://doi.org/10.1042/EBC20180075

Horbay R, Hamraghani A, Ermini L, Holcik S, Beug ST, Yeganeh B (2022) Role of ceramides and lysosomes in extracellular vesicle biogenesis, cargo sorting and release. IJMS 23:15317. https://doi.org/10.3390/ijms232315317

Huang J, Heng S, Zhang W, Liu Y, Xia T, Ji C, Zhang L (2022) Dermal extracellular matrix molecules in skin development, homeostasis, wound regeneration and diseases. Semin Cell Dev Biol 128:137–144. https://doi.org/10.1016/j.semcdb.2022.02.027

Hynes RO, Naba A (2012) Overview of the matrisome—an inventory of extracellular matrix constituents and functions. Cold Spring Harb Perspect Biol 4:a004903. https://doi.org/10.1101/cshperspect.a004903

Ikeda M, Naitoh M, Kubota H, Ishiko T, Yoshikawa K, Yamawaki S, Kurokawa M, Utani A, Nakamura T, Nagata K, Suzuki S (2009) Elastic fiber assembly is disrupted by excessive accumulation of chondroitin sulfate in the human dermal fibrotic disease, keloid. Biochem Biophys Res Commun 390:1221–1228. https://doi.org/10.1016/j.bbrc.2009.10.125

Iozzo RV, Schaefer L (2015) Proteoglycan form and function: a comprehensive nomenclature of proteoglycans. Matrix Biol 42:11–55. https://doi.org/10.1016/j.matbio.2015.02.003

Isaacson KJ, Martin Jensen M, Subrahmanyam NB, Ghandehari H (2017) Matrix-metalloproteinases as targets for controlled delivery in cancer: an analysis of upregulation and expression. J Control Release 259:62–75. https://doi.org/10.1016/j.jconrel.2017.01.034

Jabłońska-Trypuć A, Matejczyk M, Rosochacki S (2016) Matrix metalloproteinases (MMPs), the main extracellular matrix (ECM) enzymes in collagen degradation, as a target for anticancer drugs. J Enzyme Inhib Med Chem 31:177–183. https://doi.org/10.3109/14756366.2016.1161620

Jackson HK, Mitoko C, Linke F, Macarthur D, Kerr ID, Coyle B (2023) Extracellular vesicles potentiate medulloblastoma metastasis in an EMMPRIN and MMP-2 dependent manner. Cancers 15:2601. https://doi.org/10.3390/cancers15092601

Jariwala N, Ozols M, Bell M, Bradley E, Gilmore A, Debelle L, Sherratt MJ (2022) Matrikines as mediators of tissue remodelling. Adv Drug Deliv Rev 185:114240. https://doi.org/10.1016/j.addr.2022.114240

Jones MG, Andriotis OG, Roberts JJ, Lunn K, Tear VJ, Cao L, Ask K, Smart DE, Bonfanti A, Johnson P, Alzetani A, Conforti F, Doherty R, Lai CY, Johnson B, Bourdakos KN, Fletcher SV, Marshall BG, Jogai S, Brereton CJ, Chee SJ, Ottensmeier CH, Sime P, Gauldie J, Kolb M, Mahajan S, Fabre A, Bhaskar A, Jarolimek W, Richeldi L, O'Reilly KM, Monk PD, Thurner PJ, Davies DE (2018) Nanoscale dysregulation of collagen structure-function disrupts mechano-homeostasis and mediates pulmonary fibrosis. eLife 7:e36354. https://doi.org/10.7554/eLife.36354

Juan T, Fürthauer M (2018) Biogenesis and function of ESCRT-dependent extracellular vesicles. Semin Cell Dev Biol 74:66–77. https://doi.org/10.1016/j.semcdb.2017.08.022

Kadler KE, Baldock C, Bella J, Boot-Handford RP (2007) Collagens at a glance. J Cell Sci 120:1955–1958. https://doi.org/10.1242/jcs.03453

Kai F, Drain AP, Weaver VM (2019) The extracellular matrix modulates the metastatic journey. Dev Cell 49:332–346. https://doi.org/10.1016/j.devcel.2019.03.026

Karamanos NK, Theocharis AD, Neill T, Iozzo RV (2019) Matrix modeling and remodeling: a biological interplay regulating tissue homeostasis and diseases. Matrix Biol 75–76:1–11. https://doi.org/10.1016/j.matbio.2018.08.007

Karamanos NK, Theocharis AD, Piperigkou Z, Manou D, Passi A, Skandalis SS, Vynios DH, Orian-Rousseau V, Ricard-Blum S, Schmelzer CEH, Duca L, Durbeej M, Afratis NA, Troeberg L, Franchi M, Masola V, Onisto M (2021) A guide to the composition and functions of the extracellular matrix. FEBS J 288:6850–6912. https://doi.org/10.1111/febs.15776

Kleiser S, Nyström A (2020) Interplay between cell-surface receptors and extracellular matrix in skin. Biomolecules 10:1170. https://doi.org/10.3390/biom10081170

Lederle W, Hartenstein B, Meides A, Kunzelmann H, Werb Z, Angel P, Mueller MM (2010) MMP13 as a stromal mediator in controlling persistent angiogenesis in skin carcinoma. Carcinogenesis 31:1175–1184. https://doi.org/10.1093/carcin/bgp248

Lee H, Sodek KL, Hwang Q, Brown TJ, Ringuette M, Sodek J (2007) Phagocytosis of collagen by fibroblasts and invasive cancer cells is mediated by MT1-MMP. Biochem Soc Trans 35:704–706. https://doi.org/10.1042/BST0350704

Lenzini S, Bargi R, Chung G, Shin J-W (2020) Matrix mechanics and water permeation regulate extracellular vesicle transport. Nat Nanotechnol 15:217–223. https://doi.org/10.1038/s41565-020-0636-2

Li L, Zhao Q, Kong W (2018) Extracellular matrix remodeling and cardiac fibrosis. Matrix Biol 68–69:490–506. https://doi.org/10.1016/j.matbio.2018.01.013

Li X, Guo L, Yang X, Wang J, Hou Y, Zhu S, Du J, Feng J, Xie Y, Zhuang L, He X, Liu Y (2020) TGF-β1-induced connexin43 promotes scar formation *via* the Erk/MMP-1/collagen III pathway. J Oral Rehabil 47:99–106. https://doi.org/10.1111/joor.12829

Lin Q, Chen X, Meng F, Ogawa K, Li M, Song R, Zhang S, Zhang Z, Kong X, Xu Q, He F, Bai X, Sun B, Hung M-C, Liu L, Wands J, Dong X (2019) ASPH-notch axis guided exosomal delivery of prometastatic secretome renders breast cancer multi-organ metastasis. Mol Cancer 18:156. https://doi.org/10.1186/s12943-019-1077-0

Liu Y, Cao X (2016) Characteristics and significance of the pre-metastatic niche. Cancer Cell 30:668–681. https://doi.org/10.1016/j.ccell.2016.09.011

Liu M, de Juan L, Abad B, Cheng K (2021) Cardiac fibrosis: myofibroblast-mediated pathological regulation and drug delivery strategies. Adv Drug Deliv Rev 173:504–519. https://doi.org/10.1016/j.addr.2021.03.021

Liu X, Li J, Yang X, Li X, Kong J, Qi D, Zhang F, Sun B, Liu Y, Liu T (2023) Carcinoma-associated fibroblast-derived lysyl oxidase-rich extracellular vesicles mediate collagen crosslinking and promote epithelial-mesenchymal transition via p-FAK/p-paxillin/YAP signaling. Int J Oral Sci 15:32. https://doi.org/10.1038/s41368-023-00236-1

Loeser RF (2009) Aging and osteoarthritis: the role of chondrocyte senescence and aging changes in the cartilage matrix. Osteoarthr Cartil 17:971–979. https://doi.org/10.1016/j.joca.2009.03.002

Lotz MK, Caramés B (2011) Autophagy and cartilage homeostasis mechanisms in joint health, aging and OA. Nat Rev Rheumatol 7:579–587. https://doi.org/10.1038/nrrheum.2011.109

Mai J, Sameni M, Mikkelsen T, Sloane BF (2002) Degradation of extracellular matrix protein Tenascin-C by Cathepsin B: an interaction involved in the progression of gliomas. Biol Chem 383. https://doi.org/10.1515/BC.2002.159

Marozzi M, Parnigoni A, Negri A, Viola M, Vigetti D, Passi A, Karousou E, Rizzi F (2021) Inflammation, extracellular matrix remodeling, and proteostasis in tumor microenvironment. IJMS 22:8102. https://doi.org/10.3390/ijms22158102

Mazor R, Alsaigh T, Shaked H, Altshuler AE, Pocock ES, Kistler EB, Karin M, Schmid-Schönbein GW (2013) Matrix metalloproteinase-1-mediated up-regulation of vascular endothelial growth factor-2 in endothelial cells. J Biol Chem 288:598–607. https://doi.org/10.1074/jbc.M112.417451

Mittal R, Patel AP, Debs LH, Nguyen D, Patel K, Grati M, Mittal J, Yan D, Chapagain P, Liu XZ (2016) Intricate functions of matrix metalloproteinases in physiological and pathological conditions: role of MMPs in health and disease. J Cell Physiol 231:2599–2621. https://doi.org/10.1002/jcp.25430

Moretti L, Stalfort J, Barker TH, Abebayehu D (2022) The interplay of fibroblasts, the extracellular matrix, and inflammation in scar formation. J Biol Chem 298:101530. https://doi.org/10.1016/j.jbc.2021.101530

Moss NM, Barbolina MV, Liu Y, Sun L, Munshi HG, Stack MS (2009) Ovarian cancer cell detachment and multicellular aggregate formation are regulated by membrane type 1 matrix metalloproteinase: a potential role in I.p. Metastatic dissemination. Cancer Res 69:7121–7129. https://doi.org/10.1158/0008-5472.CAN-08-4151

Muralidharan-Chari V, Clancy JW, Sedgwick A, D'Souza-Schorey C (2010) Microvesicles: mediators of extracellular communication during cancer progression. J Cell Sci 123:1603–1611. https://doi.org/10.1242/jcs.064386

Murphy DE, De Jong OG, Brouwer M, Wood MJ, Lavieu G, Schiffelers RM, Vader P (2019) Extracellular vesicle-based therapeutics: natural versus engineered targeting and trafficking. Exp Mol Med 51:32. https://doi.org/10.1038/s12276-019-0223-5

Mustafa S, Koran S, AlOmair L (2022) Insights into the role of matrix metalloproteinases in cancer and its various therapeutic aspects: a review. Front Mol Biosci 9:896099. https://doi.org/10.3389/fmolb.2022.896099

Naba A, Clauser KR, Ding H, Whittaker CA, Carr SA, Hynes RO (2016) The extracellular matrix: tools and insights for the "omics" era. Matrix Biol 49:10–24. https://doi.org/10.1016/j.matbio.2015.06.003

Najafi M, Farhood B, Mortezaee K (2019) Extracellular matrix (ECM) stiffness and degradation as cancer drivers. J Cell Biochem 120:2782–2790. https://doi.org/10.1002/jcb.27681

Nawaz M, Shah N, Zanetti B, Maugeri M, Silvestre R, Fatima F, Neder L, Valadi H (2018) Extracellular vesicles and matrix Remodeling enzymes: the emerging roles in extracellular matrix remodeling, progression of diseases and tissue repair. Cells 7:167. https://doi.org/10.3390/cells7100167

Pan J, Sheng S, Ye L, Xu X, Ma Y, Feng X, Qiu L, Fan Z, Wang Y, Xia X, Zheng JC (2022) Extracellular vesicles derived from glioblastoma promote proliferation and migration of neural progenitor cells via PI3K-Akt pathway. Cell Commun Signal 20:7. https://doi.org/10.1186/s12964-021-00760-9

Paolillo S, Schinelli S (2019) Extracellular matrix alterations in metastatic processes. IJMS 20(19):4947. https://doi.org/10.3390/ijms20194947

Parks WC, Wilson CL, López-Boado YS (2004) Matrix metalloproteinases as modulators of inflammation and innate immunity. Nat Rev Immunol 4:617–629. https://doi.org/10.1038/nri1418

Patel NJ, Ashraf A, Chung EJ (2023) Extracellular vesicles as regulators of the extracellular matrix. Bioengineering 10:136. https://doi.org/10.3390/bioengineering10020136

Patwardhan S, Mahadik P, Shetty O, Sen S (2021) ECM stiffness-tuned exosomes drive breast cancer motility through thrombospondin-1. Biomaterials 279:121185. https://doi.org/10.1016/j.biomaterials.2021.121185

Pickup MW, Mouw JK, Weaver VM (2014) The extracellular matrix modulates the hallmarks of cancer. EMBO Rep 15:1243–1253. https://doi.org/10.15252/embr.201439246

Podgorski I, Linebaugh BE, Sameni M, Jedeszko C, Bhagat S, Cher ML, Sloane BF (2005) Bone microenvironment modulates expression and activity of cathepsin B in prostate cancer. Neoplasia 7:207–223. https://doi.org/10.1593/neo.04349

Rabe DC, Walker ND, Rustandy FD, Wallace J, Lee J, Stott SL, Rosner MR (2021) Tumor extracellular vesicles regulate macrophage-driven metastasis through CCL5. Cancers 13:3459. https://doi.org/10.3390/cancers13143459

Raskov H, Gaggar S, Tajik A, Orhan A, Gögenur I (2023) The matrix reloaded—the role of the extracellular matrix in cancer. Cancers 15:2057. https://doi.org/10.3390/cancers15072057

Rath M, Pitiot A, Kirr M, Fröhlich W, Plosnita B, Schliep S, Bauerschmitz J, Baur AS, Ostalecki C (2021) Multi-antigen imaging reveals inflammatory DC, ADAM17 and neprilysin as effectors in keloid formation. IJMS 22:9417. https://doi.org/10.3390/ijms22179417

Ren W, Hou J, Yang C, Wang H, Wu S, Wu Y, Zhao X, Lu C (2019) Extracellular vesicles secreted by hypoxia pre-challenged mesenchymal stem cells promote non-small cell lung cancer cell growth and mobility as well as macrophage M2 polarization via miR-21-5p delivery. J Exp Clin Cancer Res 38:62. https://doi.org/10.1186/s13046-019-1027-0

Ricard-Blum S, Perez S (2022) Glycosaminoglycan interaction networks and databases. Curr Opin Struct Biol 74:102355. https://doi.org/10.1016/j.sbi.2022.102355

Ricard-Blum S, Ruggiero F (2005) The collagen superfamily: from the extracellular matrix to the cell membrane. Pathol Biol 53:430–442. https://doi.org/10.1016/j.patbio.2004.12.024

Rilla K, Mustonen A-M, Arasu UT, Härkönen K, Matilainen J, Nieminen P (2019) Extracellular vesicles are integral and functional components of the extracellular matrix. Matrix Biol 75–76:201–219. https://doi.org/10.1016/j.matbio.2017.10.003

Rousselle P, Montmasson M, Garnier C (2019) Extracellular matrix contribution to skin wound re-epithelialization. Matrix Biol 75–76:12–26. https://doi.org/10.1016/j.matbio.2018.01.002

Roy R, Morad G, Jedinak A, Moses MA (2020) Metalloproteinases and their roles in human cancer. Anat Rec 303:1557–1572. https://doi.org/10.1002/ar.24188

Sajeesh S, Camardo A, Dahal S, Ramamurthi A (2023) Surface-functionalized stem cell-derived extracellular vesicles for vascular elastic matrix regenerative repair. Mol Pharm 20:2801–2813. https://doi.org/10.1021/acs.molpharmaceut.2c00769

Sariano PA, Mizenko RR, Shirure VS, Brandt AK, Nguyen BB, Nesiri C, Shergill BS, Brostoff T, Rocke DM, Borowsky AD, Carney RP, George SC (2023) Convection and extracellular matrix binding control interstitial transport of extracellular vesicles. J Extracell Vesicle 12:e12323. https://doi.org/10.1002/jev2.12323

Shi D, Sheng A, Chi L (2021a) Glycosaminoglycan-protein interactions and their roles in human disease. Front Mol Biosci 8:639666. https://doi.org/10.3389/fmolb.2021.639666

Shi Y, Ma X, Fang G, Tian X, Ge C (2021b) Matrix metalloproteinase inhibitors (MMPIs) as attractive therapeutic targets: recent progress and current challenges. NanoImpact 21:100293. https://doi.org/10.1016/j.impact.2021.100293

Shimoda M, Khokha R (2017) Metalloproteinases in extracellular vesicles. Biochim Biophys Acta Mol Cell Res 1864:1989–2000. https://doi.org/10.1016/j.bbamcr.2017.05.027

Shimoda M, Principe S, Jackson HW, Luga V, Fang H, Molyneux SD, Shao YW, Aiken A, Waterhouse PD, Karamboulas C, Hess FM, Ohtsuka T, Okada Y, Ailles L, Ludwig A, Wrana JL, Kislinger T, Khokha R (2014) Loss of the Timp gene family is sufficient for the acquisition of the CAF-like cell state. Nat Cell Biol 16:889–901. https://doi.org/10.1038/ncb3021

Shoucair I, Weber Mello F, Jabalee J, Maleki S, Garnis C (2020) The role of cancer-associated fibroblasts and extracellular vesicles in tumorigenesis. IJMS 21:6837. https://doi.org/10.3390/ijms21186837

Sidgwick GP, Bayat A (2012) Extracellular matrix molecules implicated in hypertrophic and keloid scarring: extracellular matrix molecules in raised scarring. J Eur Acad Dermatol Venereol 26:141–152. https://doi.org/10.1111/j.1468-3083.2011.04200.x

Soans KG, Norden C (2021) Shining a light on extracellular matrix dynamics in vivo. Semin Cell Dev Biol 120:85–93. https://doi.org/10.1016/j.semcdb.2021.05.008

Song Y, You Y, Xu X, Lu J, Huang X, Zhang J, Zhu L, Hu J, Wu X, Xu X, Tan W, Du Y (2023) Adipose-derived mesenchymal stem cell-derived exosomes biopotentiated extracellular matrix hydrogels accelerate diabetic wound healing and skin regeneration. Adv Sci 10:2304023. https://doi.org/10.1002/advs.202304023

Ståhl A, Johansson K, Mossberg M, Kahn R, Karpman D (2019) Exosomes and microvesicles in normal physiology, pathophysiology, and renal diseases. Pediatr Nephrol 34:11–30. https://doi.org/10.1007/s00467-017-3816-z

Sutter PA, Rouillard ME, Alshawi SA, Crocker SJ (2021) Extracellular matrix influences astrocytic extracellular vesicle function in wound repair. Brain Res 1763:147462. https://doi.org/10.1016/j.brainres.2021.147462

Takeuchi T, Suzuki M, Fujikake N, Popiel HA, Kikuchi H, Futaki S, Wada K, Nagai Y (2015) Intercellular chaperone transmission via exosomes contributes to maintenance of protein homeostasis at the organismal level. Proc Natl Acad Sci USA 112. https://doi.org/10.1073/pnas.1412651112

Tan T-W, Lai C-H, Huang C-Y, Yang W-H, Chen H-T, Hsu H-C, Fong Y-C, Tang C-H (2009) CTGF enhances migration and MMP-13 up-regulation via αvβ3 integrin, FAK, ERK, and NF-κB-dependent pathway in human chondrosarcoma cells. J Cell Biochem 107:345–356. https://doi.org/10.1002/jcb.22132

Taraboletti G, D'Ascenzo S, Borsotti P, Giavazzi R, Pavan A, Dolo V (2002) Shedding of the matrix metalloproteinases MMP-2, MMP-9, and MT1-MMP as membrane vesicle-associated components by endothelial cells. Am J Pathol 160:673–680. https://doi.org/10.1016/S0002-9440(10)64887-0

Theocharis A, Gialeli C, Hascall V, Karamanos NK (2012) 1.1 extracellular matrix: a functional scaffold. In: Karamanos N (ed) Extracellular matrix: pathobiology and signaling. De Gruyter, pp 3–20

Theocharis AD, Manou D, Karamanos NK (2019) The extracellular matrix as a multitasking player in disease. FEBS J 286:2830–2869. https://doi.org/10.1111/febs.14818

Théry C, Witwer KW, Aikawa E, Alcaraz MJ, Anderson JD, Andriantsitohaina R, Antoniou A, Arab T, Archer F, Atkin-Smith GK, Ayre DC, Bach J-M, Bachurski D, Baharvand H, Balaj L, Baldacchino S, Bauer NN, Baxter AA, Bebawy M, Beckham C, Bedina Zavec A, Benmoussa A, Berardi AC, Bergese P, Bielska E, Blenkiron C, Bobis-Wozowicz S, Boilard E, Boireau W, Bongiovanni A, Borràs FE, Bosch S, Boulanger CM, Breakefield X, Breglio AM, Brennan MÁ, Brigstock DR, Brisson A, Broekman ML, Bromberg JF, Bryl-Górecka P, Buch S, Buck AH, Burger D, Busatto S, Buschmann D, Bussolati B, Buzás EI, Byrd JB, Camussi G, Carter DR, Caruso S, Chamley LW, Chang Y-T, Chen C, Chen S, Cheng L, Chin AR, Clayton A, Clerici SP, Cocks A, Cocucci E, Coffey RJ, Cordeiro-da-Silva A, Couch Y, Coumans FA, Coyle B, Crescitelli R, Criado MF, D'Souza-Schorey C, Das S, Datta Chaudhuri A, de Candia P, De Santana EF, De Wever O, del Portillo HA, Demaret T, Deville S, Devitt A, Dhondt B, Di Vizio D, Dieterich LC, Dolo V, Dominguez Rubio AP, Dominici M, Dourado MR, Driedonks TA, Duarte FV, Duncan HM, Eichenberger RM, Ekström K, EL Andaloussi S, Elie-Caille C, Erdbrügger U, Falcón-Pérez JM, Fatima F, Fish JE, Flores-Bellver M, Försönits A, Frelet-Barrand A, Fricke F, Fuhrmann G, Gabrielsson S, Gámez-Valero A, Gardiner C, Gärtner K, Gaudin R, Gho YS, Giebel B, Gilbert C, Gimona M, Giusti I, Goberdhan DC, Görgens A, Gorski SM, Greening DW, Gross JC, Gualerzi A, Gupta GN, Gustafson D, Handberg A, Haraszti RA, Harrison P, Hegyesi H, Hendrix A, Hill AF, Hochberg FH, Hoffmann KF, Holder B, Holthofer H, Hosseinkhani B, Hu G, Huang Y, Huber V, Hunt S, Ibrahim AG-E, Ikezu T, Inal JM, Isin M, Ivanova A, Jackson HK, Jacobsen S, Jay SM, Jayachandran M, Jenster G, Jiang L, Johnson SM, Jones JC, Jong A, Jovanovic-Talisman T, Jung S, Kalluri R, Kano S, Kaur S, Kawamura Y, Keller ET, Khamari D, Khomyakova E, Khvorova A, Kierulf P, Kim KP, Kislinger T, Klingeborn M, Klinke DJ, Kornek M, Kosanović MM, Kovács ÁF, Krämer-Albers E-M, Krasemann S, Krause M, Kurochkin IV, Kusuma GD, Kuypers S, Laitinen S, Langevin SM, Languino LR, Lannigan J, Lässer C, Laurent LC, Lavieu G, Lázaro-Ibáñez E, Le Lay S, Lee M-S, Lee YXF, Lemos DS, Lenassi M, Leszczynska A, Li IT, Liao K, Libregts SF, Ligeti E, Lim R, Lim SK, Linē A, Linnemannstöns K, Llorente A, Lombard CA, Lorenowicz MJ, Lörincz ÁM, Lötvall J, Lovett J, Lowry MC, Loyer X, Lu Q, Lukomska B, Lunavat TR, Maas SL, Malhi H, Marcilla A, Mariani J, Mariscal J, Martens-Uzunova ES, Martin-Jaular L, Martinez MC, Martins VR, Mathieu M, Mathivanan S, Maugeri M, McGinnis LK, McVey MJ, Meckes DG, Meehan KL, Mertens I, Minciacchi VR, Möller A, Møller Jørgensen M, Morales-Kastresana A, Morhayim J, Mullier F, Muraca M, Musante L, Mussack V, Muth DC, Myburgh KH, Najrana T, Nawaz M, Nazarenko I, Nejsum P, Neri C, Neri T, Nieuwland R, Nimrichter L, Nolan JP, Nolte-'t Hoen EN, Noren Hooten N, O'Driscoll L, O'Grady T, O'Loghlen A, Ochiya T, Olivier M, Ortiz A, Ortiz LA, Osteikoetxea X, Østergaard O, Ostrowski M, Park J, Pegtel DM, Peinado H, Perut F, Pfaffl MW, Phinney DG, Pieters BC, Pink RC, Pisetsky DS, Pogge von Strandmann E, Polakovicova I, Poon IK, Powell BH, Prada I, Pulliam L, Quesenberry P, Radeghieri A, Raffai RL, Raimondo S, Rak J, Ramirez MI, Raposo G, Rayyan MS, Regev-Rudzki N, Ricklefs FL, Robbins PD, Roberts DD, Rodrigues SC, Rohde E, Rome S, Rouschop KM, Rughetti A, Russell AE, Saá P, Sahoo S, Salas-Huenuleo E, Sánchez C, Saugstad JA, Saul MJ, Schiffelers RM, Schneider R, Schøyen TH, Scott A, Shahaj E, Sharma S, Shatnyeva O, Shekari F, Shelke GV, Shetty AK, Shiba K, Siljander PR-M, Silva AM, Skowronek A, Snyder OL, Soares RP, Sódar BW, Soekmadji C, Sotillo J, Stahl PD, Stoorvogel W, Stott SL, Strasser EF, Swift S, Tahara H, Tewari M, Timms K, Tiwari S, Tixeira R, Tkach M, Toh WS, Tomasini R, Torrecilhas AC, Tosar JP, Toxavidis V, Urbanelli L, Vader P, van Balkom BW, van der Grein SG, Van Deun J, van Herwijnen MJ, Van Keuren-Jensen K, van Niel G, van Royen ME, van Wijnen AJ, Vasconcelos MH, Vechetti IJ, Veit TD, Vella LJ, Velot É, Verweij FJ, Vestad B, Viñas JL, Visnovitz T, Vukman KV, Wahlgren J, Watson DC, Wauben MH, Weaver A, Webber JP, Weber V, Wehman AM, Weiss DJ, Welsh JA, Wendt S, Wheelock AM, Wiener Z, Witte L, Wolfram J, Xagorari A, Xander P, Xu J, Yan X, Yáñez-Mó M, Yin H, Yuana Y, Zappulli V, Zarubova J, Žėkas V, Zhang J, Zhao Z, Zheng L, Zheutlin AR, Zickler AM, Zimmermann P, Zivkovic AM, Zocco D, Zuba-Surma EK (2018) Minimal information for studies of extracellular vesicles 2018 (MISEV2018): a position statement of the International Society for Extracellular Vesicles

and update of the MISEV2014 guidelines. J Extracell Vesicles 7:1535750. https://doi.org/10.1080/20013078.2018.1535750
Thuault S, Ghossoub R, David G, Zimmermann P (2022) A journey on extracellular vesicles for matrix metalloproteinases: a mechanistic perspective. Front Cell Dev Biol 10:886381. https://doi.org/10.3389/fcell.2022.886381
Tkach M, Théry C (2016) Communication by extracellular vesicles: where we are and where we need to go. Cell 164:1226–1232. https://doi.org/10.1016/j.cell.2016.01.043
To WS, Midwood KS (2011) Plasma and cellular fibronectin: distinct and independent functions during tissue repair. Fibrogenesis Tissue Repair 4:21. https://doi.org/10.1186/1755-1536-4-21
Tracy LE, Minasian RA, Caterson EJ (2016) Extracellular matrix and dermal fibroblast function in the healing wound. Adv Wound Care 5:119–136. https://doi.org/10.1089/wound.2014.0561
Valdoz JC, Johnson BC, Jacobs DJ, Franks NA, Dodson EL, Sanders C, Cribbs CG, Van Ry PM (2021) The ECM: to scaffold, or not to scaffold, that is the question. IJMS 22:12690. https://doi.org/10.3390/ijms222312690
van Hinsbergh VWM, Koolwijk P (2008) Endothelial sprouting and angiogenesis: matrix metalloproteinases in the lead. Cardiovasc Res 78:203–212. https://doi.org/10.1093/cvr/cvm102
van Niel G, D'Angelo G, Raposo G (2018) Shedding light on the cell biology of extracellular vesicles. Nat Rev Mol Cell Biol 19:213–228. https://doi.org/10.1038/nrm.2017.125
van Niel G, Carter DRF, Clayton A, Lambert DW, Raposo G, Vader P (2022) Challenges and directions in studying cell–cell communication by extracellular vesicles. Nat Rev Mol Cell Biol 23:369–382. https://doi.org/10.1038/s41580-022-00460-3
Veschi EA, Bolean M, Strzelecka-Kiliszek A, Bandorowicz-Pikula J, Pikula S, Granjon T, Mebarek S, Magne D, Ramos AP, Rosato N, Millán JL, Buchet R, Bottini M, Ciancaglini P (2020) Localization of annexin A6 in matrix vesicles during physiological mineralization. IJMS 21:1367. https://doi.org/10.3390/ijms21041367
Vidal M (2019) Exosomes: revisiting their role as "garbage bags". Traffic 20:815–828. https://doi.org/10.1111/tra.12687
Vizovišek M, Fonović M, Turk B (2019) Cysteine cathepsins in extracellular matrix remodeling: extracellular matrix degradation and beyond. Matrix Biol 75–76:141–159. https://doi.org/10.1016/j.matbio.2018.01.024
Vos MC, van der Wurff AAM, van Kuppevelt TH, Massuger LFAG (2021) The role of MMP-14 in ovarian cancer: a systematic review. J Ovarian Res 14:101. https://doi.org/10.1186/s13048-021-00852-7
Wang L, Hu L, Zhou X, Xiong Z, Zhang C, Shehada HMA, Hu B, Song J, Chen L (2017) Exosomes secreted by human adipose mesenchymal stem cells promote scarless cutaneous repair by regulating extracellular matrix remodelling. Sci Rep 7:13321. https://doi.org/10.1038/s41598-017-12919-x
Wang W-J, Yang W, Ouyang Z-H, Xue J-B, Li X-L, Zhang J, He W-S, Chen W-K, Yan Y-G, Wang C (2018) MiR-21 promotes ECM degradation through inhibiting autophagy via the PTEN/akt/mTOR signaling pathway in human degenerated NP cells. Biomed Pharmacother 99:725–734. https://doi.org/10.1016/j.biopha.2018.01.154
Winkler J, Abisoye-Ogunniyan A, Metcalf KJ, Werb Z (2020) Concepts of extracellular matrix remodelling in tumour progression and metastasis. Nat Commun 11:5120. https://doi.org/10.1038/s41467-020-18794-x
Wiseman BS, Sternlicht MD, Lund LR, Alexander CM, Mott J, Bissell MJ, Soloway P, Itohara S, Werb Z (2003) Site-specific inductive and inhibitory activities of MMP-2 and MMP-3 orchestrate mammary gland branching morphogenesis. J Cell Biol 162:1123–1133. https://doi.org/10.1083/jcb.200302090
Wu H, Du Q, Dai Q, Ge J, Cheng X (2018) Cysteine protease cathepsins in atherosclerotic cardiovascular diseases. J Atheroscler Thromb 25:111–123. https://doi.org/10.5551/jat.RV17016
Wu B, Liu D-A, Guan L, Myint PK, Chin L, Dang H, Xu Y, Ren J, Li T, Yu Z, Jabban S, Mills GB, Nukpezah J, Chen YH, Furth EE, Gimotty PA, Wells RG, Weaver VM, Radhakrishnan R,

Wang XW, Guo W (2023) Stiff matrix induces exosome secretion to promote tumour growth. Nat Cell Biol 25:415–424. https://doi.org/10.1038/s41556-023-01092-1

Xie Y, Zinkle A, Chen L, Mohammadi M (2020) Fibroblast growth factor signalling in osteoarthritis and cartilage repair. Nat Rev Rheumatol 16:547–564. https://doi.org/10.1038/s41584-020-0469-2

Yáñez-Mó M, Siljander PR-M, Andreu Z, Bedina Zavec A, Borràs FE, Buzas EI, Buzas K, Casal E, Cappello F, Carvalho J, Colás E, Cordeiro-da Silva A, Fais S, Falcon-Perez JM, Ghobrial IM, Giebel B, Gimona M, Graner M, Gursel I, Gursel M, Heegaard NHH, Hendrix A, Kierulf P, Kokubun K, Kosanovic M, Kralj-Iglic V, Krämer-Albers E-M, Laitinen S, Lässer C, Lener T, Ligeti E, Linē A, Lipps G, Llorente A, Lötvall J, Manček-Keber M, Marcilla A, Mittelbrunn M, Nazarenko I, Nolte-'t Hoen ENM, Nyman TA, O'Driscoll L, Olivan M, Oliveira C, Pállinger É, del Portillo HA, Reventós J, Rigau M, Rohde E, Sammar M, Sánchez-Madrid F, Santarém N, Schallmoser K, Stampe Ostenfeld M, Stoorvogel W, Stukelj R, Van der Grein SG, Helena Vasconcelos M, MHM W, De Wever O (2015) Biological properties of extracellular vesicles and their physiological functions. J Extracell Vesicles 4:27066. https://doi.org/10.3402/jev.v4.27066

Yao X, Cheng F, Yu W, Rao T, Li W, Zhao S, Zhou X, Ning J (2019) Cathepsin S regulates renal fibrosis in mouse models of mild and severe hydronephrosis. Mol Med Rep. https://doi.org/10.3892/mmr.2019.10230

Yin Y, Chen H, Wang Y, Zhang L, Wang X (2021) Roles of extracellular vesicles in the aging microenvironment and age-related diseases. J Extracell Vesicles 10. https://doi.org/10.1002/jev2.12154

You Y, Shan Y, Chen J, Yue H, You B, Shi S, Li X, Cao X (2015) Matrix metalloproteinase 13-containing exosomes promote nasopharyngeal carcinoma metastasis. Cancer Sci 106:1669–1677. https://doi.org/10.1111/cas.12818

Yu S, Yu L (2022) Migrasome biogenesis and functions. FEBS J 289:7246–7254. https://doi.org/10.1111/febs.16183

Zappe A, Miller RL, Struwe WB, Pagel K (2022) State-of-the-art glycosaminoglycan characterization. Mass Spectrom Rev 41:1040–1071. https://doi.org/10.1002/mas.21737

Zhang W, Liu Y, Zhang H (2021) Extracellular matrix: an important regulator of cell functions and skeletal muscle development. Cell Biosci 11:65. https://doi.org/10.1186/s13578-021-00579-4

Zhang X, Chen X, Hong H, Hu R, Liu J, Liu C (2022) Decellularized extracellular matrix scaffolds: recent trends and emerging strategies in tissue engineering. Bioact Mater 10:15–31. https://doi.org/10.1016/j.bioactmat.2021.09.014

Zhang X, Yao L, Meng Y, Li B, Yang Y, Gao F (2023a) Migrasome: a new functional extracellular vesicle. Cell Death Discov 9:381. https://doi.org/10.1038/s41420-023-01673-x

Zhang Y, Qi G, Yan Y, Wang C, Wang Z, Jiang C, Jiang Z, Ma T, Zhang C, Yan Z (2023b) Exosomes derived from bone marrow mesenchymal stem cells pretreated with decellularized extracellular matrix enhance the alleviation of osteoarthritis through miR-3473b/phosphatase and tensin homolog axis. J Gene Med 25:e3510. https://doi.org/10.1002/jgm.3510

Zhao B, Zhang Y, Han S, Zhang W, Zhou Q, Guan H, Liu J, Shi J, Su L, Hu D (2017) Exosomes derived from human amniotic epithelial cells accelerate wound healing and inhibit scar formation. J Mol Hist 48:121–132. https://doi.org/10.1007/s10735-017-9711-x

Zhou Y, Zhao B, Zhang X-L, Lu Y, Lu S-T, Cheng J, Fu Y, Lin L, Zhang N-Y, Li P-X, Zhang J, Zhang J (2021) Combined topical and systemic administration with human adipose-derived mesenchymal stem cells (hADSC) and hADSC-derived exosomes markedly promoted cutaneous wound healing and regeneration. Stem Cell Res Ther 12:257. https://doi.org/10.1186/s13287-021-02287-9

Zhou M, Li Y-J, Tang Y-C, Hao X-Y, Xu W-J, Xiang D-X, Wu J-Y (2022) Apoptotic bodies for advanced drug delivery and therapy. J Control Release 351:394–406. https://doi.org/10.1016/j.jconrel.2022.09.045

Zito F (2012) Role of extracellular matrix in regulating embryonic epithelial-mesenchymal transition. Biomol Concepts 3:333–344. https://doi.org/10.1515/bmc-2011-0065

Chapter 3
Extracellular Matrix Modulation by Cancer-Derived Extracellular Vesicles: Impact on Cancer Malignancy

Nao Nishida-Aoki and Takahiro Ochiya

Abstract The tumor microenvironment (TME) is an intricate network that influences cancer progression, comprising cancer cells, stromal cells, immune cells, and the extracellular matrix (ECM). The ECM, a scaffold of macroproteins like collagens and proteoglycans, is critical for maintaining tissue structure and regulating cellular functions. It undergoes significant remodeling in cancer, affecting cell proliferation, invasion, and drug response. Recent studies highlight the role of extracellular vesicles (EVs), lipid bilayer vesicles carrying bioactive molecules, secreted by cancer cells in modulating the ECM. Cancer EVs modulate the ECM by stimulating cancer-associated fibroblasts (CAFs) and other stromal cells to produce ECM components and by directly altering ECM structure through enzymes they carry or by modulating cancer cell–ECM interactions. The altered ECM facilitates cancer progression by establishing pre-metastatic niches, enhancing tumor invasion, and affecting drug efficacy. This chapter summarizes the roles of cancer EVs in remodeling ECM in tumors, highlighting how EVs influence ECM secretion by stromal cells, directly modify ECM structures, and impact tumor traits and clinical outcomes. Understanding the interaction between EVs and the ECM in the TME is crucial for developing novel diagnostic and therapeutic strategies targeting EV-mediated ECM remodeling to disrupt the metastatic cascade and improve cancer treatment efficacy.

N. Nishida-Aoki (✉)
Waseda Institute for Advanced Study, Waseda University, Shinjuku-ku, Tokyo, Japan
e-mail: nnishida@aoni.waseda.jp

T. Ochiya
Department of Molecular and Cellular Medicine, Center for Future Medical Research, Institute of Medical Science, Tokyo Medical University, Shinjuku-ku, Tokyo, Japan
e-mail: tochiya@tokyo-med.ac.jp

K. Rilla (ed.), *Extracellular Vesicles as Matrix Messengers*, Biology of Extracellular Matrix 15, https://doi.org/10.1007/978-3-031-68811-9_3

3.1 Introduction

The tumor microenvironment (TME) is a complex ecosystem comprising not only cancer cells but also stromal cells like cancer-associated fibroblasts (CAFs), immune cells, and extracellular matrix (ECM) components. This dynamic milieu plays a critical role in tumor initiation, progression, and metastasis (Anderson and Simon 2020). ECM provides structural support to tissues and regulates cellular behavior. Within tumor, ECM is largely modulated by both cancer and the surrounding cells to become stiffer and increased in amount, with altered components and crosslinks in primary tumor and metastasis (Winkler et al. 2020). The modulated ECM has a large impact on cancer traits and modulates the hallmarks of cancer; cell proliferation, anti-apoptosis, tumor invasion, angiogenesis, and drug response (Pickup et al. 2014). Indeed, stiffer ECM is directly connected to cancer cell malignancy (Levental et al. 2009). Further, the altered ECM modulates the functions of the surrounding non-cancerous cells, including immune cells, by physical barriers for infiltration and functional changes (Sutherland et al. 2023). Therefore, ECM remodeling alters cancer cell drug efficacy in multiple ways and is therefore expected as therapeutic targets (Henke et al. 2020; Sleeboom et al. 2024).

ECM consists of ~300 unique macroproteins such as collagens, proteoglycans, and glycoproteins (Naba et al. 2016). The ECM exists in two primary forms: the interstitial matrix and the basement membrane. The interstitial matrix, comprising mainly collagens and elastin, maintains tissue structural integrity, modulates processes like cell differentiation and migration, and plays a significant role in cancer by altering cell signaling and tumor progression. The basement membrane, composed of collagen IV and laminins, is denser and more stable, playing a crucial role in cell polarity, developmental processes, and tissue homeostasis. These ECM structures are modulated by 700 proteins, affecting the ECM topology and properties (Naba et al. 2016; Winkler et al. 2020). The mechanisms of ECM structural changes involve ECM deposition, chemical modification, proteolytic degradation, and physical remodeling. Remodeling the basement membrane is essential for cancer cell invasion (Winkler et al. 2020).

Recent research has shed light on the pivotal role of extracellular vesicles (EVs) in orchestrating ECM modulation. EVs are lipid bilayer vesicles that serve as carriers of bioactive molecules, such as proteins, nucleic acids, lipids, and others, enabling them to function as mediators of cell-to-cell communication. Importantly, EVs can transfer their cargo to recipient cells, thereby modulating various cellular properties (Yáñez-Mó et al. 2015). Cancer cells secrete large amounts of EVs and utilize them to modulate the surrounding cells and ECM structures to establish TME, stimulating tumor progression and metastasis (Tkach and Théry 2016). In the initiation of cancer at primary or metastatic environment, cancer EVs require to modulate ECM to establish TME. As the tumor grows, ECM reorganization is necessary to invade surrounding tissues and stimulate angiogenesis. Moreover, EVs delivered to distal organs modulate ECM composition to form pre-metastatic niches to facilitate the adhesion, survival, and proliferation of metastatic cancer cells. Cancer EVs increase ECM deposition, and alter components and structures through multiple mechanisms; indirect modulation through effects on the stromal cells, especially CAFs, and direct enzymatic and nonenzymatic modulation of ECM (Fig. 3.1).

This chapter delves into the emerging understanding of how cancer EVs regulate ECM dynamics in the context of progression and tumor malignancy; how cancer EVs stimulate ECM secretion of stromal cells, directly modulate ECM structure, and their impact on tumor traits and clinical impacts. Understanding the interplay between EVs and the ECM within the TME is essential for elucidating the mechanisms driving cancer progression and metastasis. This knowledge holds significant implications for the development of novel diagnostic and therapeutic strategies for cancer. By targeting EV-mediated ECM remodeling, researchers aim to disrupt the metastatic cascade and improve patient outcomes.

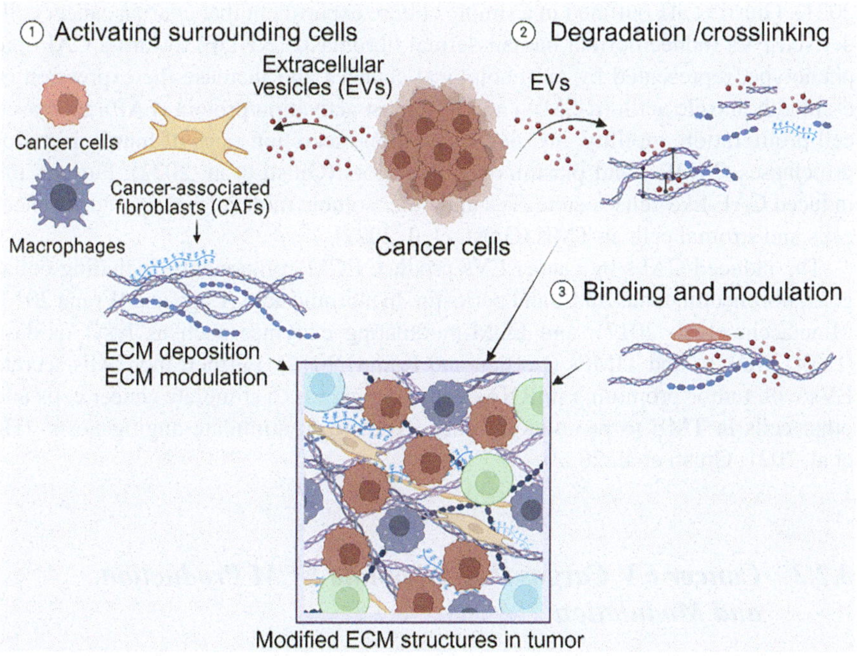

Fig. 3.1 Cancer EVs can modulate ECM structures in multiple ways. Cancer EVs can activate the surrounding fibroblasts, other cancer cells, or macrophages to deposit different compositions of ECM and ECM-modulating enzymes. Cancer EVs load ECM-modulating enzymes such as MMPs on their surface and possess enzymatic activities on ECM. Further, some EVs with ECM-binding receptors (e.g., integrins) bind directly on ECM and modulate their properties, affecting interactions with the cells

3.2 Cancer EV Modulating Tumor ECM by Stimulating Stromal Cells

3.2.1 Cancer EVs Stimulate Cancer-Associated Fibroblasts (CAF)

Within TME, ECM structures are modulated by various cell types; cancer, cancer-associated fibroblasts (CAFs), endothelial cells, tumor-associated macrophages (TAMs) and other immune cells, and pericytes. Among them, CAFs are considered the most responsible cells for establishing altered ECM in TME. CAFs originate from various sources, including mesenchymal stem cells (MSCs), bone marrow cells, pericytes, endothelial cells (ECs), and resident fibroblasts, which introduces heterogeneity of CAFs (LeBleu and Neilson 2020). Recent studies have illuminated the transformative role of EVs derived from tumor cells in converting normal fibroblasts into CAFs (Webber et al. 2010) (Fig. 3.1). Fibroblast induction by cancer EVs is reported in many cancer types (Paggetti et al. 2015; Bhome et al. 2017; He et al. 2021). Giusti et al., outlined in a simple culture experiment that ovarian cancer cell-derived EVs induce normal human dermal fibroblasts (NHDF) toward a CAF-like phenotype, represented by morphological changes and increase the expression of α-smooth muscle actin (α-SMA) and fibroblast activation protein (FAP), increased cell proliferation, motility, invasive capabilities, secretion rates of matrix metalloproteinases (MMPs) and plasminogen activators (Giusti et al. 2022). Further, the induced CAF-like cells secrete EVs and other soluble molecules that induce cancer cells and stromal cells in TME (Giusti et al. 2022).

The induced CAFs by cancer EVs produce ECM components, including collagens, fibronectin, tenascin C, and periostin, hyaluronic acid (Luga and Wrana 2013; Minciacchi et al. 2017), and ECM-modulating enzymes such as lysyl oxidase (LOX), MMPs, and TIMPs (LeBleu and Kalluri 2018). Further, the CAFs secrete EVs with tumor-promoting miRNAs and other cargos to stimulate cancer cells and other cells in TME to pro-tumoral phenotypes (e.g., stimulate angiogenesis) (He et al. 2021; Giusti et al. 2022).

3.2.2 Cancer EV Cargos that Stimulate ECM Production and Modulation

Various cargos of cancer EVs can modulate TME. Of them, microRNAs (miRNAs) contribute largely to cancer EV's capability of altering the recipient cells (Li et al. 2022). Cancer EVs carry varieties of bioactive miRNAs to modulate the recipient cells, which are reviewed elsewhere in many cancer types (Shoucair et al. 2020; He et al. 2021; Samuels et al. 2023). Some of the identified miRNAs involved in CAFs are shared among many cancers; miR-21 is identified in EVs from hepatocellular carcinoma, ovarian cancer, breast cancer, prostate, and multiple myeloma. While varieties of miRNAs are identified in different cancer cell types; miR-9, miR-146a, and miR-125b were identified in breast cancer, miR-1247-3p in hepatocarcinoma (Fang et al. 2018), and miR-142-3p in lung cancer (Lawson et al. 2019), suggesting both shared and cancer-type-specific miRNAs are loaded into cancer EVs to induce CAFs. The miRNAs transferred through cancer EVs are released in the recipient cells and become functional to alter gene expression.

Vu et al. showed in mouse breast cancer that the transfer of miR-125b via EVs induces the transformation of NF into CAFs, contributing to tumor progression. The authors confirmed that cancer EVs from both exogenously administered and released orthotopically were efficiently incorporated into mouse lung fibroblasts in mouse models using flow cytometry and immunohistochemistry (Vu et al. 2019). Ren et al. demonstrated that breast cancer cell-derived EVs carrying miR-370-3p transform NF into CAFs by suppressing CYLD and activating NF-κB signaling. This process leads to enhanced migration, invasion, and epithelial–mesenchymal transition of breast cancer cells, facilitating lung metastasis in vivo (Ren et al. 2021). In liver cancer, extravesicular miR-1247-3p activates CAFs by targeting B4GALT3 and activating β1-integrin/NF-κB signaling (Fang et al. 2018). This leads to the release of pro-inflammatory cytokines like IL-6 and IL-8, which promote lung metastasis. In colorectal cancer (CRC), cells release EVs enriched with miR-10b, which can initiate transformation of normal cells into CAFs (Dai et al. 2018). This process involves the downregulation of PIK3CA expression and reduced activity of the PI3K/AKT/mTOR signaling pathway, which reduced cell proliferation but upregulated TGF-β and SM α-actins. High levels of miR-27a in EVs from gastric cancer (GC) cells can reprogram fibroblasts into cancer-associated fibroblasts (CAFs), activate oncogenes, and enhance the proliferation, motility, and metastasis of cancer cells (Wang et al. 2018).

Although miRNAs are the most well-studied EV cargos to stimulate the stromal cells for ECM production and modulation, some components other than miRNAs are reported as cancer EV cargos inducing CAF polarization. In earlier work, Dr. Crayton's group showed that cancer EVs from mesothelioma cancer cell lines as well as prostate (DU145 and PC3) and breast cancer simulate fibroblast in myofibroblast through TGF-β represented at the surface of cancer EVs by transmembrane proteoglycan β-glycan (Webber et al. 2010). The EV–TGF-β complex was fully functional in eliciting SMAD3-related signaling pathways. TGF-β loading and their contribution to CAF differentiation are later shown in other cancers, like prostate (Webber et al. 2014), bladder cancer, (Goulet et al. 2018). However, among cell lines tested in Webber et al., prostate (LNCAP) and colorectal cancer did not incorporate TGF-β. Although TGF-β is a master regulator of fibroblast differentiation, other components in EVs are definitely contributing to CAF polarization. This is also supported by a report of oral cancer (Arebro et al. 2023). EVs from SCC-9 and SCC-25 cell lines associated with TGF-β and stimulated fibroblasts to pro-inflammatory CAF, but the characteristics of EV-induced CAFs were not equivalent to TGF-β-activated CAFs. In Lawson et al. that miR-142-3p in EVs, known to suppress tumors in lung adenocarcinoma cells (LAC), induces CAF phenotype without relying on TGFβ signaling (Lawson et al. 2019). Therefore, loading components of EVs to induce fibroblasts to CAFs are different among cells, and several cargos will have synergistic effects.

TGF-β could be also loaded as an mRNA format into cancer EVs (Urciuoli et al. 2018). The authors showed in human osteosarcoma EVs uptaking to murine fibroblast cells that human mRNAs of TNF-α, IL-6, and TGF-β were identified, and further induced murine MMP-9 and TNF-α expression to have CAF-like phenotypes. Also, other proteins, mRNAs, and lncRNAs were identified as EV components which involved in CAF differentiation (Shoucair et al. 2020).

3.2.3 Cells Other than CAFs that Are Involved in ECM Modulation

Besides CAFs, cancer and other stromal cells can deposit and modulate ECM. Cancer cells themselves produce collagen, fibronectin, and laminin, and also express ECM degradation and modifying enzymes, such as LOX (Henke et al. 2020; Winkler et al. 2020).

Also, immune cells, including macrophages in TME, are involved in ECM modulation. Macrophages in TME are induced by cancer EVs and other secretory factors to polarize into tumor-associated macrophages (TAM) (Baig et al. 2020), which share phenotypes with alternatively activated macrophages, expressing fibronectin (Gratchev et al. 2001). TAMs secrete cytokines, growth factors, to activate CAFs, and deposit ECM components (e.g., fibronectin, laminin, osteopontin, and versican), modulate collagen-degrading enzymes, and uptaking collagens for degradation (Sutherland et al. 2023). In pancreatic cancer, which is known for desmoplastic cancer with more than 70% of the total volume being stroma (Ferrara et al. 2021), macrophages deposit a distinct subset of collagen from fibroblast to TME (Zhu et al. 2017). In colorectal cancer, collagen-producing macrophages were more prevalent than fibroblasts and contributed to TME remodeling (Afik et al. 2016). The monocytes differentiating to TAMs produce and assemble collagens I, VI, and XIV, to modulate ECM structures in TME. Deligne and Midwood summarized in their review the mutual relationships of TAM and ECM in breast cancer (Deligne and Midwood 2021). In breast cancer, TAM can reorganize collagen fibers and produce ECM (collagen and osteopontin). ECM is also involved in the localization and infiltration of TAMs. In breast cancer cells, tenacin-C is recognized by TAM to be colocalized, and blocking tenacin-C reduces infiltration of TAMs into tumors. These examples show that ECM components and structures are regulated in various ways by multiple cells, and the altered ECM further modulates these cellular phenotypes.

3.2.4 ECM Modulation in Bone Metastasis

Cancer EVs also affect bone ECM to establish metastatic lesions and promote cancer progression. Bone consists of minerals and mostly collagens and other ECM components (Fontcuberta-Rigo et al. 2023), with osteocytes, osteoblasts, and osteoclasts incorporated. Bone is maintained by balanced cycles of bone resorption and bone formation, regulated by osteoclast and osteoblast, respectively (Salhotra et al. 2020). Bone-metastatic cancer, like prostate cancer and breast cancer, or bone-derived cancer, secrete EVs to affect these cells to compromise homeostasis of bone metabolism to cause either osteolysis or osteogenesis (Probert et al. 2019; Satcher and Zhang 2021). There are many reviews that summarize detailed mechanisms of actions of cancer EVs on bone metastasis and resulting bone defects (Rossi et al. 2018; Tamura et al. 2020; Giannandrea et al. 2021).

3.3 Direct ECM Remodeling by Cancer-Derived EVs

3.3.1 Direct ECM Remodeling by Enzymatic Activities of Cancer-Derived EVs

In addition to modulating the surrounding cells to alter ECM structures, cancer EVs can directly modify ECM through enzymatic activities (Fig. 3.1). Cancer EV contains various ECM-modifying enzymes, including matrix metalloproteinases (MMPs), tissue inhibitors of metalloproteinases (TIMPs), a disintegrin and metalloproteinases (ADAMs), and ADAMs with thrombospondin motifs (ADAMTSs), heparanases, hyaluronidases, extracellular matrix metalloproteinase inducer (EMMPRIN)/CD147, and lysyl oxidase (LOX) (Nawaz et al. 2018; Li et al. 2019), and therefore EVs themselves have ECM degradation enzymatic activities. The expression of these ECM-modulating enzymes is associated with tumor malignancy and progression, and is considered to be a biomarker of the prognostic and diagnosis (Roy et al. 2020). As cellular expression increases, the amount of enzymes loaded to the cancer EVs is generally proportionally increased.

Shimoda et al. provided a comprehensive review of metalloproteinase on extracellular vesicles in a journal (Shimoda and Khokha 2017) and in a book (Shimoda 2019). Metalloproteinases include matrix metalloproteinases (MMPs), ADAMs, and ADAMTSs, and in humans, 23 MMPs, 20 ADAMs, and 19 ADAMTSs are identified (Kelwick et al. 2015; Cui et al. 2017; Souza et al. 2020). Recent proteomic analyses identified MMPs, ADAMs, and ADAMTSs from isolated EVs (Shimoda and Khokha 2013). At the time of the review in 2017, 11 MMPs, 6 ADAMs, and 7 ADAMTSs in EVs derived from various cell types have been identified (Shimoda and Khokha 2017).

Metalloproteinases are highly expressed in tumors and loaded in proportion to their EV, and associated with cancer malignancy. Tauro et al. explicitly showed it with EVs from Madin-Darby canine kidney (MDCK) cells with or without oncogenic H-Ras. The EVs from the oncogenic cells were enriched with MMP-1, MMP-14, MMP-10, ADAM-10, and ADAMTS1, and integrins (ITGB1, ITGA3, and ITGA) compared to the non-oncogenic cells of the same origin (Tauro et al. 2013). Mu et al. showed in EVs from rat adenocarcinoma cell lines with or without HA receptor CD44v. EVs from the rat pancreatic adenocarcinoma cell line (BSp73ASML wt) contain active proteases such as uPAR, MMP3, ADAM17, ADAMTS1, ADAMTS8, and CD13, while EVs from CD44vKD cells show lower abundance of uPAR, MMP9, MMP13, and ADAM17, but higher levels of HAdase. ASML EVs and ASML-CD44v Kd EVs exhibit differential binding affinities to ECM, with the latter showing less binding to HA, collagens, and FN. WT EVs are more efficient in degrading Coll I, coll IV, LN111, LN332, and FN compared to CD44vKD EVs (Mu et al. 2013). These results suggest that malignant tumors secrete EVs enriched with ECM degradation activities in cargos and binding efficiency.

3.3.1.1 Matrix Metalloproteinases

MMPs are a family of zinc-dependent endopeptidases responsible for the cleavage of extracellular matrix (ECM) components, involved in physiological processes including tissue remodeling, wound healing, embryonic development, and angiogenesis. In cancer, MMP facilitates tumor progression by degrading extracellular matrix components, enabling cancer cells to invade adjacent tissues and metastasize to distant sites. Additionally, MMPs contribute to the remodeling of the TME, promoting angiogenesis, immune evasion, and creating conditions conducive to tumor growth and spread (Kessenbrock et al. 2010). Most of the MMPs are secreted in soluble form, while six of them are membrane-type MMP (MT-MMPs); MT1-, MT2-, MT3-, and MT5-MMP function as transmembrane proteases, while MT4- and MT6-MMP are tethered via a GPI-anchor (Bonnans et al. 2014; Itoh 2015). MMPs cover most ECM proteins as substrates.

Several types of MMPs have been identified in cancer-derived EVs, including MMP-2, MMP-9, and MMP-14 (MT-MMP1), among others (Shimoda and Khokha 2013). MMP-2 and MMP-9, also known as gelatinases, are particularly notable for their ability to degrade type IV collagen, a major component of basement membranes (Zeng et al. 1999). This degradation is crucial for tumor invasion and metastasis, as it allows cancer cells to breach the basement membrane and infiltrate surrounding tissues. Nawaz et al. provided detailed mechanisms of MMP loading and functions on ECM in their review (Nawaz et al. 2018). Although the loading mechanisms of MMPs and other proteins are largely unknown, cortactin may be involved in stabilizing MMPs in MVBs and their secretion to EVs, and a specific set of Rabs are involved in the surface presentation.

MMP-14 (MT1-MMP) is a critical transmembrane MMP that not only directly degrades ECM components but also activates other MMPs, such as MMP-2 and MMP-13 (Itoh 2015). MMP-14/MT1-MMP can degrade fibrillar collagens (types I, II, and III), fibronectin, vitronectin and laminins, fibrin/fibrinogen, parlecan, and aggrecan (Itoh 2022). MMP-14 also works as sheddase by cleaving membrane proteins, syndecan-1, ICAM-1, and CD44, into soluble molecules (reviewed in Thuault et al. 2022). MMP-14 is delivered to invadopodia, cellular structures that facilitate tumor invasion, and focal adhesion to stimulate cellular invasion and motility (Itoh 2022).

MMPs are closely related to cancer, stimulating their invasion and penetration through basal membranes (Niland et al. 2022). MMPs on cancer EVs also possess similar functions to cellular MMPs, but some additional functions. Hakulinen et al. showed in earlier studies that MMP-14/MT1-MMP is released in extracellular space via EVs using human fibrosarcoma (HT-1080) and melanoma (G361) cells (Hakulinen et al. 2008). They found both full-length (60 kDa) and processed (43 kDa) forms in the EVs. The EVs had enzymatic activities on pro-MMP-2, type 1 collagen, and gelatin suggesting the MMP-14/MT1-MMP on EVs are functionally active. Thuault et al. reviewed detailed molecular mechanisms of MMP-14/MT1-MMP loading to EVs, trafficking, and their functions at recipient cells (Thuault et al. 2022). Delivery of MMP-14/MT1-MMP to invadopodia share the same molecular machinery with MVB fusion to plasma membrane, such as SNAREs. In addition, Hoshino et al. revealed that EV secretion is necessary for invadopodia formation, and EV-loaded MMP-14/MT1-MMP are likely to regulate invadopodia maturation and ECM degradation. Invadopodia formation further regulates EV secretion, indicating a mutial relationship between EVs and invadopodia (Hoshino et al. 2013). Recently, Wenzel et al. showed intercellular transfer of MT-1-MMP via EVs to noninvasive MT1-MMP negative cells to induce invasiveness. TKS4 and TKS5, adaptor proteins for invadopodia formation, promote ADAM-mediated cleavage of MMP-14/MT1-MMP in endosome lumen in donor cells, and the secreted EVs dock onto the recipient cells to induce ECM degradation (Wenzel et al. 2024).

The modification of ECM by EVs stimulates remodeling of ECM in TME, cancer cell proliferation, invasion, and metastasis, and also affects angiogenesis and immune cell reactions.

3.3.1.2 ADAMS and ADAMTS

To date, 21 ADAMs and 19 ADAMTSs proteins have been identified in humans (Przemyslaw et al. 2013). ADAMs are transmembrane proteins involved in proteolytic processing of surface proteins and growth factors, while ADAMTSs are secretory proteins exist at ECM and primarily process and cleavage ECM components such as proteoglycans and procollagens. Several ADAMs and ADAMTSs are identified in cancer EVs. Lischnig et al. compared protein cargos in small and large EVs from metastatic human breast cancer cells. Among nine ADAMs and ADAMTSs identified in their assay, ADAM-8, 9, 10, 17, 19 and ADAMTS-1, 13, 15 were enriched in small EVs of human breast cancer cell lines compared to large EVs. Especially, ADAMTS1 was enriched in highly metastatic cancer small EVs (sEVs) (Lischnig et al. 2022). ADAM10 has been identified and enriched in sEVs from other cancers, such as neuroblastoma (Keerthikumar et al. 2015), human colon cancer (Xu et al. 2015), and human cervical and ovarian cancer cells, as well as non-cancerous dendritic cells and HEK293 (Kowal et al. 2016) ADAM10 mediate shedding of various membrane proteins including APP (precursor of amyloid peptide Ab) and Notch receptors (Przemyslaw et al. 2013). ADAM17 sheddases have wide varieties of substrates, and involve in releasing cleaving ectodomains of various transmembrane proteins, and are implicated in cancer (Gooz 2010). Growth et al. showed ADAM10 and ADAM17 are expressed on the surface of lung epithelial cancer cell line (A549), and PMA induction downregulated mature ADAM17 amount on cell surface by releasing ADAM17 through EV secretion without affecting ADAM10 (Groth et al. 2016). These results suggest selective loading of ADAM17 to EVs. This ADAM17 is a mature form and could degrade its substrate, TGFα and amphiregulin, and enhance their shedding. Gutiérrez-López et al. suggested that the sheddase activity of ADAM17 is regulated by tetraspanin CD9, which is commonly found in EVs (Gutiérrez-López et al. 2011). Although the study focus is on the cell surface, the regulation might extend to EV surface.

ADAMTS is a soluble proteases with broad substrate and is involved in ECM formation and remodeling (Rose et al. 2021). Modulation of ECM protein cleavage by ADAMTS is required for the formation, remodeling, and physiological adaptation of ECM. In cancer, ADAMTSs are rather silenced but in colon cancer ADAMTS12 were highly expressed in stromal cells (Rose et al. 2021). ADAMTSs are identified on EVs, but there is no direct proof for the EV-dependent roles of ADAMTSs in cancer. Regarding MMPs and ADAMs, ADAMTSs loaded on EVs are also likely to function as enzymes for ECM modulation.

3.3.1.3 Tissue Inhibitors of Metalloproteinases

MMP inhibitors, TIMPs are found in EVs as well. TIMP alterations are identified in all cancers but at different levels among cancer types, and are suggested to regulate protein loading to EVs together with proteases (especially ADAM10) (Jackson et al. 2016). Rao et al. showed that TIMP-1 in EVs is a noninvasive independent prognostic marker in clinical cohorts and potential therapeutic target in colorectal liver metastases (Rao et al. 2022). TIMP-1 protein exists in EVs using colorectal cell lines (HCT 116, HT29, and SW620). Extravesicular TIMP-1 upregulated TIMP-1 in recipient fibroblast and stimulated ECM contraction. An interaction between TIMP-1EVs and heatshock protein HSP90AA may attenuate its function in ECM remodeling. These findings suggest TIMP-1 EV remodels ECM in the liver microenvironment to promote colorectal cancer progression and can be a potential therapeutic target.

3.3.1.4 Lysyl Oxidase (LOX)

Li et al. showed that Lysyl oxidase-like 4 (LOXL4) is upregulated in hepatocellular carcinoma (HCC), and LOXL4 was loaded onto HCC-derived EVs. The EV-loaded LOXL4 induced migration of other HCC cells, and when transferred to endothelial cells, promoted angiogenesis (Li et al. 2019). Besides cancer, Liu et al. showed that sEVs from CAFs of oral squamous cell carcinoma are rich in an active form of LOX. CAF sEV-associated integrin α2β1 mediated the binding to collagen I, and the vesicle-loaded LOX crosslink collagen to promote epithelial–mesenchymal transition (EMT) of OSCC through FAK/paxillin/YAP pathway. The active LOX seemed to be associated with the surface of EVs as FN/POSTN/BMP-1/LOX complex (Liu et al. 2023).

3.3.1.5 Neuraminidase

Paolini et al. suggested that the NEU3 sialidase was associated on the surface of EVs from HeLa cells with inducible murine NEU3 (Paolini et al. 2017). In microglial cells, LPS stimulation increases Neu1 loading on EVs and can degrade microglial cell surface polysialic acid (Sumida et al. 2015), suggesting cancer EV-loaded neuraminidases will also have enzymatic activities to modulate glycocalyx moiety of cells and ECM.

3.3.1.6 Heparanase (HPSE)

Both cancer cells and EV surfaces are covered with heparan sulfate. Dr. Sanderson's group showed several roles of heparanases (HPSE) on cancer EVs in myeloma cells (Sanderson et al. 2019). EVs from HPSE-high myeloma load a high amount of HPSE at the surface, and they have enzymatic activities on degrading heparan sulfate in ECM. The degradation liberates growth factors bound to heparan sulfate, aiding in processes like cancer cell invasion, angiogenesis, and metastasis. Further, endogenous expression or exposure to exogenous HPSE stimulate EV secretion by myeloma cells and even alter EV cargos by higher levels of syndecan-1, VEGF, and hepatocyte growth factor, stimulating cancer cells and endothelial cell motility (Thompson et al. 2013). Chemotherapy treatment on myeloma cells induces HPSE loading at the surface of to their EVs, and activates ERK signaling pathway, and shedding of syndecan-1 proteoglycan in the recipient myeloma cells or activate macrophages to migrate and secrete TNF-α, a myeloma growth factor (Bandari et al. 2018).

3.3.2 Loading Formats and Locations of ECM Degradation Enzymes

Basically, EV-loaded metalloproteinases are assumed to be associated with the surface of EVs and to work as ECM degrading enzymes, based on the enzymatic activities of cancer EVs. Transmembrane-type metalloproteinases (e.g., MMP-14/MT1-MMP and ADAMs) should be integrated into EV membrane; EVs are essential carriers of transporting transmembrane-type ECM degrading enzymes to distant sites and beneficial to concentrate enzymes on the surface through associating proteins like integrin (Shimoda and Khokha 2017). In addition, EV surfaces have recently been shown to be enriched with associated proteins and other molecules, and this outer layer of associated molecules is termed EV "corona" (Buzás et al. 2018; Tóth et al. 2021; Buzas 2022). Therefore, soluble forms of ECM-modulating enzymes are very likely to be associated at the surface through non-covalent binding with membrane or interaction with surface receptors. To distinguish the surface and internal EV cargo, usually digesting EVs surface with proteinase K, trypsin, and Lys-C to degrade the surface, or labeling the surface molecules with biotinylating is used (Buzas 2022).

ADAMs are primarily membrane-bound enzymes that include a transmembrane domain, which makes ADAMs be loaded to EVs as membrane-anchored form (Seifert et al. 2021). In contrast, ADAMTSs lack a transmembrane domain, making them secretory proteins, and possess thrombospondin motifs which enable them to attach to ECM to function as enzymes (Rose et al. 2021), and therefore are likelye to be loaded onto EV surface through non-covalent interaction.

However, some reports suggest these enzymes are loaded as mRNA forms inside of EVs. Yokoi et al. showed that advanced ovarian cancer-derived EVs incorporate MMP1 in mRNA form. The MMP1 mRNA-loaded EVs elevated MMP1 protein amount in mesothelial cells, suggesting that the transported mRNA was translated to MMP1 protein in the receiver cells (Yokoi et al. 2017).

TIMP-1 also works not just in protein forms to inhibit the function of MMP-1. Dias et al. found that in clear cell renal cell carcinoma (ccRCC), higher levels of TIMP-1 mRNA in plasma EVs are associated with localized but larger tumors and metastatic conditions (Dias et al. 2020). High TIMP-1 levels correlated with poorer survival in metastatic cases. Notably, only the 786-O ccRCC cell line showed detectable TIMP-1 mRNA in EVs, while TIMP-2 mRNA was detectable in all cell lines at a similar level, indicating selective mRNA loading among cancer subtypes. While the study did not explore TIMP-1 mRNA's mechanisms of action, it suggests the mRNA could be translated into functional protein in recipient cells. Additionally, Cui et al. demonstrated that increased TIMP-1 in lung adenocarcinoma cells boosts miR-210 expression through CD63/PI3K/AKT/HIF-1-dependent pathway, and miR-210 loaded to EVs promoted angiogenesis, thus highlighting TIMP-1's role in modifying the TME via EVs (Cui et al. 2014). Therefore, overall, cancer cells utilize ECM modification enzymes in multiform to become tumorigenic.

3.3.3 Cancer EVs Can Decorate ECM by Direct Binding

A subpopulation of EVs is prone to bound to ECM, and a portion of EVs are even considered to be a component of ECM (Huleihel et al. 2016; Rilla et al. 2019; Debnath et al. 2023). Decellularized tissues retain ECM-bound EVs, and they have been analyzed for their component and functions. Cancer EVs display ECM-bound receptors, such as integrins to collagen, and CD44 to hyaluronan (Buzás et al. 2018).

Indeed, Palmulli et al. directly observed in in vitro experiments that human melanoma cell-derived EVs bound to collagen I fiber, and different melanoma strains (MNT-1, A375, and WM1716) with different binding capacities (Palmulli et al. 2023). CD9-positive EVs were more efficient to bind to CD63-positive EVs.

The interactions with ECM retain EVs around proximal stromal cells, and the ECM-bound EVs seem to have specific functions compared to liquid-phase EVs (Fig. 3.1). As Mu et al. showed and as discussed above, tumor EVs interact with host ECM, especially collagens and laminins, work as ECM-modulating enzymes (Mu et al. 2013). Dr. Badylak and colleagues examined the characteristics of ECM-bound EVs in their studies (Huleihel et al. 2016). They isolated matrix-bound vesicles through partial enzymatic digestion of porcine dermal, urinary bladder, and small intestinal matrices. The isolated EVs contained miRNA capable of activating macrophages and inducing neuroblastoma differentiation. Subsequently, the team explored ECM-bound EVs using mouse fibroblasts (3T3) and identified distinct miRNAs and lipid components compared to liquid-phase EVs (Hussey et al. 2020). Their findings suggest that ECM-unbound EVs may serve as carriers for selective cargo destined for delivery to distant organs. In a recent publication, Tassinari et al. analyzed matrix-entrapped EVs of colorectal cancer and its adjacent mucosa, to show EVs from the adjacent mucosa of stage IV cancer included high immune surface markers compared to adjacent tissue of lower-stage cancer (Tassinari et al. 2024).

The interaction between extravesicular heparanase and cell surface heparan sulfate is vital for EVs docking with target cells (Christianson et al. 2013; Purushothaman et al. 2016). Sanderson's group showed that heparan sulfate on the surface of myeloma EVs captures fibronectin, and heparan sulfate at the surface of the recipient cells facilitates interactions with and incorporation of fibronectin-bound EVs (Purushothaman et al. 2016). The involvement of heparan sulfate on the surface of the cells in EV internalization and functions in the recipient cells was also shown by Christianson et al. (Christianson et al. 2013).

The directional deposition of cancer EVs involves cancer cell invasions. Dr. Weaver's research team discovered that cancer cells strategically deposit their EVs onto the ECM in the direction of invasion, thereby promoting cell motility, utilizing intravital imaging of chick embryos (Sung et al. 2015). These EVs utilize fibronectin on their surface to facilitate matrix attachment through integrins. EV secretion is induced at invadopodia, invasive structures with actin, where CD63 and Rab27a-positive EVs are located. The secreted EVs mediate invadopodia biogenesis, stability, and activities and stimulate cancer cell invasion (Hoshino et al. 2013).

Emerging studies showed that the cells also deposit migrasomes, a ring-like organelle, together with a tubular network called retraction fibers at the rear site when they migrate (Ma et al. 2014). Migrasomes contain many smaller vesicles and cytosolic contents, and they are released as migrasomes breaks. They seem to be used for trajectory for the following cell migration. Although there are no detailed molecular analyses on cancer migrasomes, contribution of migrasomes to tumor progressions are suggested. Pan-cancer analyses suggested migrasomes seemed to be related to tumor development and immune escape (Qin et al. 2022). Cheng et al.

demonstrated that nanoparticles blocking recognition and incorporation of migrasomes reduced breast cancer cell migration, and partial metastasis suppression effect in vivo (Cheng et al. 2022). Baschieri et al. showed that migrating CAFs deposit migrasomes to guide cancer cell migration (Baschieri et al. 2023). The migrasomes further decorate ECM in tumors to promote cancer cell migration and metastasis.

Moreover, the unique surface receptor achieves specific binding to ECM in distant organs. Hoshino et al. showed that integrins at the surface of cancer EVs determine organ tropisms through binding to organ-specific ECM; Integrins $\alpha 6\beta 4$ and $\alpha 6\beta 1$ associate with laminin to develop lung metastasis, while integrin $\alpha v\beta 5$ was associated with fibronectins for liver metastasis (Hoshino et al. 2015). By binding to specific ECMs, cancer EVs can colocalize with their target cell types.

Altogether, these reports suggest that a subset of EVs appears to be bound to the ECM, remaining close to the donor cells and exhibiting distinct functions from bound-free EVs. Investigating these matrix-bound vesicles promises to reveal novel insights into the roles and transfer mechanisms of EVs.

3.4 Outcomes of Modulated ECM by Cancer EVs in Tumor Tissues

ECM modulation has large impacts on tumor tissues in several ways. Stiff ECM is directly connected to tumor aggressiveness, as a hard scaffold induces mechanotransduction via integrins and other ECM-binding receptors (Levental et al. 2009; Mouw et al. 2014; Cox 2021). Additionally, degradation of ECM releases growth factors and cytokines entrapped in ECM (Henke et al. 2020), and the degraded ECM itself further serves as growth factors (matrikines) to stimulate cancer growth (Boyd and Thomas 2017). The ECM remodeling helps vascularize inside of the tumor (Quintero-Fabián et al. 2019), together with direct activation of endothelial cells by cancer EVs (Nawaz et al. 2018).

In cancer treatment, stiffer ECM prevents penetration of therapeutic drugs and provides resistance to radiotherapy, lowering therapeutic efficacies (Henke et al. 2020). These effects are especially prominent for desmoplastic tumors, e.g., pancreatic cancer rich in hyaluronan. The induced cellular signaling pathways stimulate cancer cell proliferation and survival, invasion, migration, EMTs, and metastatic capabilities.

Of note, immune cell responses within tumors significantly influence cancer cell behavior, often being suppressed within the TME. The ECM plays a pivotal role in regulating immune responses in cancer. Recent studies highlight a mutually dependent relationship between immune cells and the ECM, not only in cancer but also in other contexts (reviewed in Sutherland et al. 2023). Stiffer ECM composition within tumors prevents immune cell invasion inside of the tumors and impacts activation and function within the TME (Henke et al. 2020). Therefore, in addition to cancer

EVs direcly modulate immune cell reactions, cancer EVs can further modulate immune cells by altering ECM structures.

In addition to altering cancer cell properties and behaviors as discussed above, Halawani et al. reviewed that the ECM stiffness affects EV release efficiency (Halawani et al. 2022); Although non-cancerous examples, fibroblasts under stress release more plasma membrane-derived small EVs into a collagen matrix (Lee et al. 1993), and mesenchymal stem cells on soft hydrogels with lower integrin-ligand interaction emit more EVs than those on stiff substrates (Lenzini et al. 2021), highlighting that physical properties like viscoelasticity and interaction with ECM influence EV release. The similar effect of mechanical stresses on EV secretion can apply to cancer cells to alter their EV-mediated modification of the cells.

These examples highlight how ECM modulation by cancer EVs affects cancer properties and therapeutic reponsiveness, and the modulated ECM further affect EV production and release in various physiological and pathological contexts. Understanding mutual relationship of ECM and EVs interactions can support development of effective cancer therapies.

3.5 Clinical Implications

Understanding the role of EVs in ECM remodeling has significant implications for applying cancer EVs for cancer diagnosis, prognosis, and therapy (Henke et al. 2020). Cancer EVs with unique components from cancer cells are expected as clinical biomarkers for cancer in liquid biopsy, which is less invasive and helps detect cancer at an early stage (Hinestrosa et al. 2022; Matsuzaki et al. 2023). Currently, approximately 50 clinical trials are registered for early detection of cancer using EVs in clinicaltrials.gov. ECM degradation enzymes like MMPs are enriched in cancer EVs, and they are also considered clinical biomarkers (Nawaz et al. 2018).

ECM-targeted therapies are being explored to overcome the immunosuppressive effects of ECM in cancer. As discussed above, ECM acts as a barrier for therapeutic drugs and immune cells induce drug resistance and malignancy of cancer cells. Therefore, modulating cancer ECM by directly targeting ECM (collagens, HA), targeting ECM-modulating enzymes (MMPs, ADAMs, and LOX), cell surface receptors recognizing ECM (integrins, hedgehog, FAK), and ECM-producing CAFs themselves are under clinical trials (Sleeboom et al. 2024). EV-based ECM modulating will become an additional possible strategy for treating cancer by enhancing conventional drug efficacy. Clinical trials targeting the specific ECM-associating cancer EVs are yet to be performed. However, preclinical and clinical strategies to target cancer EVs to inhibit their protumoral functions of cancer proliferation, invasion, and metastasis have been under development (Kosaka et al. 2013; Vader et al. 2014; Nishida-Aoki et al. 2017). In clinicaltrials.gov, approximately ~10 have been registered for cancer EV-targeting therapy. Approaches vary from disrupting EV biogenesis, cargo loading, uptake, removing EVs from circulation, or utilizing EVs as therapeutic agents, which are expected to attenuate the pro-tumorigenic effects of cancer EVs. The effect of these therapies will be applicable to their roles on modulating ECM and enhance conventional therapeutic efficacy.

3.6 Conclusion and Future Perspectives

EVs play a pivotal role in modulating the ECM within the TME, thereby promoting cancer progression and metastasis. Their ability to shuttle bioactive molecules to CAFs and other stromal cells to modulate ECM production, and directly modify ECM structures by their enzymatic activities underscores their significance as key mediators of intercellular communication in cancer. By deciphering the intricate mechanisms underlying EV-mediated ECM remodeling, researchers aim to uncover novel therapeutic avenues for combating metastatic disease and improving patient outcomes.

Acknowledgments I would like to thank Ochiya lab members for their direct and indirect support. Additionally, I would like to express my heartfelt thanks to my family for their continuous support.

NN-A research activities are supported by the following fundings: JSPS KAKENHI Grant-in-Aid for Young Scientists (Start-up) (22 K20805), Grant-in-Aid for Scientific Research (C) (23 K06680), the Noguchi Institute research fund, Waseda University personal research funding, Waseda University Grants for Special Research Project, and collaborative research fund from Asfreya. Inc. The figure was prepared using BioRender.

Compliance with Ethical Standards Does not apply.

References

Afik R, Zigmond E, Vugman M et al (2016) Tumor macrophages are pivotal constructors of tumor collagenous matrix. J Exp Med 213:2315–2331. https://doi.org/10.1084/JEM.20151193

Al Halawani A, Mithieux SM, Yeo GC et al (2022) Extracellular vesicles: interplay with the extracellular matrix and modulated cell responses. Int J Mol Sci 23:3389

Anderson NM, Simon MC (2020) Tumor microenvironment. Curr Biol 30:R921. https://doi.org/10.1016/J.CUB.2020.06.081

Arebro J, Towle R, Lee CM et al (2023) Extracellular vesicles promote activation of pro-inflammatory cancer-associated fibroblasts in oral cancer. Front Cell Dev Biol 11:1240159. https://doi.org/10.3389/FCELL.2023.1240159

Baig MS, Roy A, Rajpoot S et al (2020) Tumor-derived exosomes in the regulation of macrophage polarization. Inflamm Res 69(5):435–451. https://doi.org/10.1007/S00011-020-01318-0

Bandari SK, Purushothaman A, Ramani VC et al (2018) Chemotherapy induces secretion of exosomes loaded with heparanase that degrades extracellular matrix and impacts tumor and host cell behavior. Matrix Biol 65:104–118. https://doi.org/10.1016/J.MATBIO.2017.09.001

Baschieri F, Illand A, Barbazan J et al (2023) Fibroblasts generate topographical cues that steer cancer cell migration. Sci Adv 9(10):1126

Bhome R, Goh RW, Bullock MD et al (2017) Exosomal microRNAs derived from colorectal cancer-associated fibroblasts: role in driving cancer progression. Aging 9. https://doi.org/10.18632/aging.101355

Bonnans C, Chou J, Werb Z (2014) Remodelling the extracellular matrix in development and disease. Nat Rev Mol Cell Biol 15:786. https://doi.org/10.1038/NRM3904

Boyd DF, Thomas PG (2017) Towards integrating extracellular matrix and immunological pathways. Cytokine 98:79–86. https://doi.org/10.1016/j.cyto.2017.03.004

Buzas EI (2022) Opportunities and challenges in studying the extracellular vesicle corona. Nat Cell Biol 24(9):1322–1325. https://doi.org/10.1038/s41556-022-00983-z

Buzás EI, Tóth E, Sódar BW, Szabó-Taylor K (2018) Molecular interactions at the surface of extracellular vesicles. Semin Immunopathol 40(5):453–464. https://doi.org/10.1007/S00281-018-0682-0

Cheng Y, Ren J, Fan S et al (2022) Nanoparticulates reduce tumor cell migration through affinity interactions with extracellular migrasomes and retraction fibers. Nanoscale Horiz 7:779–789. https://doi.org/10.1039/D2NH00067A

Christianson HC, Svensson KJ, Van Kuppevelt TH et al (2013) Cancer cell exosomes depend on cell-surface heparan sulfate proteoglycans for their internalization and functional activity. Proc Natl Acad Sci USA 110:17380–17385. https://doi.org/10.1073/PNAS.1304266110

Cox TR (2021) The matrix in cancer. Nat Rev Cancer 21(4):217–238. https://doi.org/10.1038/s41568-020-00329-7

Cui H, Seubert B, Stahl E et al (2014) Tissue inhibitor of metalloproteinases-1 induces a protumourigenic increase of miR-210 in lung adenocarcinoma cells and their exosomes. Oncogene 34(28):3640–3650. https://doi.org/10.1038/onc.2014.300

Cui N, Hu M, Khalil RA (2017) Biochemical and biological attributes of matrix metalloproteinases. Prog Mol Biol Transl Sci 147:1. https://doi.org/10.1016/BS.PMBTS.2017.02.005

Dai G, Yao X, Zhang Y et al (2018) Colorectal cancer cell-derived exosomes containing miR-10b regulate fibroblast cells via the PI3K/Akt pathway. Bull Cancer 105:336–349. https://doi.org/10.1016/J.BULCAN.2017.12.009

Debnath K, Las Heras K, Rivera A et al (2023) Extracellular vesicle–matrix interactions. Nat Rev Mater 8:390–402. https://doi.org/10.1038/s41578-023-00551-3

Deligne C, Midwood KS (2021) Macrophages and extracellular matrix in breast cancer: partners in crime or protective allies? Front Oncol 11:1. https://doi.org/10.3389/FONC.2021.620773

Dias F, Teixeira AL, Nogueira I et al (2020) Plasma extracellular vesicle-derived TIMP-1 mRNA as a prognostic biomarker in clear cell renal cell carcinoma: a pilot study. Int J Mol Sci 21:4624. https://doi.org/10.3390/IJMS21134624

Fang T, Lv H, Lv G et al (2018) Tumor-derived exosomal miR-1247-3p induces cancer-associated fibroblast activation to foster lung metastasis of liver cancer. Nat Commun 9. https://doi.org/10.1038/S41467-017-02583-0

Ferrara B, Pignatelli C, Cossutta M et al (2021) The extracellular matrix in pancreatic cancer: description of a complex network and promising therapeutic options. Cancers (Basel) 13. https://doi.org/10.3390/CANCERS13174442

Fontcuberta-Rigo M, Nakamura M, Puigbò P (2023) Phylobone: a comprehensive database of bone extracellular matrix proteins in human and model organisms. Bone Res 11:1–9. https://doi.org/10.1038/s41413-023-00281-w

Giannandrea D, Citro V, Lesma E et al (2021) Restoring tissue homeostasis at metastatic sites: a focus on extracellular vesicles in bone metastasis. Front Oncol 11:644109. https://doi.org/10.3389/FONC.2021.644109

Giusti I, Di Francesco M, Poppa G et al (2022) Tumor-derived extracellular vesicles activate normal human fibroblasts to a cancer-associated fibroblast-like phenotype, sustaining a protumorigenic microenvironment. Front Oncol 12:1. https://doi.org/10.3389/FONC.2022.839880

Gooz M (2010) ADAM-17: the enzyme that does it all. Crit Rev Biochem Mol Biol 45:146. https://doi.org/10.3109/10409231003628015

Goulet CR, Bernard G, Tremblay S et al (2018) Exosomes induce fibroblast differentiation into cancer-associated fibroblasts through TGFb signaling. Mol Cancer Res 16:1196–1204. https://doi.org/10.1158/1541-7786.MCR-17-0784/81263

Gratchev A, Guillot P, Hakiy N et al (2001) Alternatively activated macrophages differentially express fibronectin and its splice variants and the extracellular matrix protein betaIG-H3. Scand J Immunol 53:386–392. https://doi.org/10.1046/J.1365-3083.2001.00885.X

Groth E, Pruessmeyer J, Babendreyer A, et al (2016) Stimulated release and functional activity of surface expressed metalloproteinase ADAM17 in exosomes Biochim Biophys Acta 1863:2795–2808. https://doi.org/10.1016/J.BBAMCR.2016.09.002

Gutiérrez-López MD, Gilsanz A, Yàñez-Mó M et al (2011) The sheddase activity of ADAM17/TACE is regulated by the tetraspanin CD9. Cell Mol Life Sci 68:3275–3292. https://doi.org/10.1007/S00018-011-0639-0

Hakulinen J, Sankkila L, Sugiyama N et al (2008) Secretion of active membrane type 1 matrix metalloproteinase (MMP-14) into extracellular space in microvesicular exosomes. J Cell Biochem 105:1211–1218. https://doi.org/10.1002/JCB.21923

He C, Wang L, Li L, Zhu G (2021) Extracellular vesicle-orchestrated crosstalk between cancer-associated fibroblasts and tumors. Transl Oncol 14:101231. https://doi.org/10.1016/J.TRANON.2021.101231

Henke E, Nandigama R, Ergün S (2020) Extracellular matrix in the tumor microenvironment and its impact on cancer therapy. Front Mol Biosci 6:470149. https://doi.org/10.3389/FMOLB.2019.00160

Hinestrosa JP, Kurzrock R, Lewis JM et al (2022) Early-stage multi-cancer detection using an extracellular vesicle protein-based blood test. Commun Med 2:1–9. https://doi.org/10.1038/s43856-022-00088-6

Hoshino D, Kirkbride KC, Costello K et al (2013) Exosome secretion is enhanced by invadopodia and drives invasive behavior. Cell Rep 5:1159–1168

Hoshino A, Costa-Silva B, Shen T-L et al (2015) Tumour exosome integrins determine organotropic metastasis. Nature 527:329–335. https://doi.org/10.1038/nature15756

Huleihel L, Hussey GS, Naranjo JD et al (2016) Matrix-bound nanovesicles within ECM bioscaffolds. Sci Adv 2:e1600502

Hussey GS, Molina CP, Cramer MC et al (2020) Lipidomics and RNA sequencing reveal a novel subpopulation of nanovesicle within extracellular matrix biomaterials. Sci Adv 6. https://doi.org/10.1126/SCIADV.AAY4361

Itoh Y (2015) Membrane-type matrix metalloproteinases: their functions and regulations. Matrix Biol 44–46:207–223. https://doi.org/10.1016/J.MATBIO.2015.03.004

Itoh Y (2022) Proteolytic modulation of tumor microenvironment signals during cancer progression. Front Oncol 12. https://doi.org/10.3389/FONC.2022.935231

Jackson HW, Defamie V, Waterhouse P, Khokha R (2016) TIMPs: versatile extracellular regulators in cancer. Nat Rev Cancer 17(1):38–53. https://doi.org/10.1038/nrc.2016.115

Keerthikumar S, Gangoda L, Liem M et al (2015) Proteogenomic analysis reveals exosomes are more oncogenic than ectosomes. Oncotarget 6:15375–15396. https://doi.org/10.18632/ONCOTARGET.3801

Kelwick R, Desanlis I, Wheeler GN, Edwards DR (2015) The ADAMTS (a Disintegrin and metalloproteinase with thrombospondin motifs) family. Genome Biol 16:1–16. https://doi.org/10.1186/S13059-015-0676-3

Kessenbrock K, Plaks V, Werb Z (2010) Matrix metalloproteinases: regulators of the tumor microenvironment. Cell 141:52–67. https://doi.org/10.1016/J.CELL.2010.03.015

Kosaka N, Takeshita F, Yoshioka Y et al (2013) Exosomal tumor-suppressive microRNAs as novel cancer therapy: "exocure" is another choice for cancer treatment. Adv Drug Deliv Rev 65:376–382

Kowal J, Arras G, Colombo M et al (2016) Proteomic comparison defines novel markers to characterize heterogeneous populations of extracellular vesicle subtypes. Proc Natl Acad Sci USA 113:E968–E977. https://doi.org/10.1073/PNAS.1521230113/SUPPL_FILE/PNAS.1521230113.SD03.XLS

Lawson J, Dickman C, Towle R et al (2019) Extracellular vesicle secretion of miR-142-3p from lung adenocarcinoma cells induces tumor promoting changes in the stroma through cell-cell communication. Mol Carcinog 58:376–387. https://doi.org/10.1002/MC.22935

LeBleu VS, Kalluri R (2018) A peek into cancer-associated fibroblasts: origins, functions and translational impact. Dis Model Mech 11. https://doi.org/10.1242/DMM.029447

LeBleu VS, Neilson EG (2020) Origin and functional heterogeneity of fibroblasts. FASEB J 34:3519–3536. https://doi.org/10.1096/FJ.201903188R

Lee TL, Lin YC, Mochitate K, Grinnell F (1993) Stress-relaxation of fibroblasts in collagen matrices triggers ectocytosis of plasma membrane vesicles containing actin, annexins II and VI, and beta 1 integrin receptors. J Cell Sci 105:167–177

Lenzini S, Debnath K, Joshi JC et al (2021) Cell–matrix interactions regulate functional extracellular vesicle secretion from mesenchymal stromal cells. ACS Nano 15:17439–17452

Levental KR, Yu H, Kass L et al (2009) Matrix crosslinking forces tumor progression by enhancing integrin signaling. Cell 139. https://doi.org/10.1016/j.cell.2009.10.027

Li R, Wang Y, Zhang X et al (2019) Exosome-mediated secretion of LOXL4 promotes hepatocellular carcinoma cell invasion and metastasis. Mol Cancer 18:1–19. https://doi.org/10.1186/S12943-019-0948-8

Li C, Zhou T, Chen J et al (2022) The role of exosomal miRNAs in cancer. J Transl Med 20:1–15. https://doi.org/10.1186/S12967-021-03215-4

Lischnig A, Bergqvist M, Ochiya T, Lässer C (2022) Quantitative proteomics identifies proteins enriched in large and small extracellular vesicles. Mol Cell Proteomics 21:100273. https://doi.org/10.1016/J.MCPRO.2022.100273

Liu X, Li J, Yang X et al (2023) Carcinoma-associated fibroblast-derived lysyl oxidase-rich extracellular vesicles mediate collagen crosslinking and promote epithelial-mesenchymal transition via p-FAK/p-paxillin/YAP signaling. Int J Oral Sci 15:1–15. https://doi.org/10.1038/s41368-023-00236-1

Luga V, Wrana JL (2013) Tumor-stroma interaction: Revealing fibroblast-secreted exosomes as potent regulators of Wnt-planar cell polarity signaling in cancer metastasis. Cancer Res 73:6843–6847. https://doi.org/10.1158/0008-5472.CAN-13-1791

Ma L, Li Y, Peng J et al (2014) Discovery of the migrasome, an organelle mediating release of cytoplasmic contents during cell migration. Cell Res 25(1):24–38. https://doi.org/10.1038/cr.2014.135

Matsuzaki J, Kato K, Oono K et al (2023) Prediction of tissue-of-origin of early stage cancers using serum miRNomes. JNCI Cancer Spectr 7. https://doi.org/10.1093/JNCICS/PKAC080

Minciacchi VR, Spinelli C, Reis-Sobreiro M, et al (2017) MYC Mediates Large Oncosome-Induced Fibroblast Reprogramming in Prostate Cancer. Cancer Res 77:2306–2317. https://doi.org/10.1158/0008-5472.CAN-16-2942

Mouw JK, Yui Y, Damiano L et al (2014) Tissue mechanics modulate microRNA-dependent PTEN expression to regulate malignant progression. Nat Med 20:360–367. https://doi.org/10.1038/nm.3497

Mu W, Rana S, Zöller M (2013) Host matrix modulation by tumor exosomes promotes motility and invasiveness. Neoplasia 15. https://doi.org/10.1593/neo.13786

Naba A, Clauser KR, Ding H et al (2016) The extracellular matrix: tools and insights for the "omics" era. Matrix Biol 49:10–24. https://doi.org/10.1016/J.MATBIO.2015.06.003

Nawaz M, Shah N, Zanetti BR et al (2018) Extracellular vesicles and matrix remodeling enzymes: the emerging roles in extracellular matrix remodeling, progression of diseases and tissue repair. Cells 7:167. https://doi.org/10.3390/CELLS7100167

Niland S, Riscanevo AX, Eble JA (2022) Matrix metalloproteinases shape the tumor microenvironment in cancer progression. Int J Mol Sci 23. https://doi.org/10.3390/IJMS23010146

Nishida-Aoki N, Tominaga N, Takeshita F et al (2017) Disruption of circulating extracellular vesicles as a novel therapeutic strategy against cancer metastasis. Mol Ther 25. https://doi.org/10.1016/j.ymthe.2016.10.009

Paggetti J, Haderk F, Seiffert M et al (2015) Exosomes released by chronic lymphocytic leukemia cells induce the transition of stromal cells into cancer-associated fibroblasts. Blood 126:1106–1117. https://doi.org/10.1182/BLOOD-2014-12-618025

Palmulli R, Bresteau E, Raposo G et al (2023) In vitro interaction of melanoma-derived extracellular vesicles with collagen. Int J Mol Sci 24. https://doi.org/10.3390/IJMS24043703

Paolini L, Orizio F, Busatto S et al (2017) Exosomes secreted by hela cells shuttle on their surface the plasma membrane-associated sialidase NEU3. Biochemistry 56:6401–6408. https://doi.org/10.1021/ACS.BIOCHEM.7B00665

Pickup MW, Mouw JK, Weaver VM (2014) The extracellular matrix modulates the hallmarks of cancer. EMBO Rep 15:1243. https://doi.org/10.15252/EMBR.201439246

Probert C, Dottorini T, Speakman A et al (2019) Communication of prostate cancer cells with bone cells via extracellular vesicle RNA; a potential mechanism of metastasis. Oncogene 38. https://doi.org/10.1038/s41388-018-0540-5

Przemyslaw L, Oguslaw HA, Elzbieta S, Malgorzata SM (2013) ADAM and ADAMTS family proteins and their role in the colorectal cancer etiopathogenesis. BMB Rep 46:139. https://doi.org/10.5483/BMBREP.2013.46.3.176

Purushothaman A, Bandari SK, Liu J et al (2016) Fibronectin on the surface of myeloma cell-derived exosomes mediates exosome-cell interactions. J Biol Chem 291:1652. https://doi.org/10.1074/JBC.M115.686295

Qin Y, Yang J, Liang C et al (2022) Pan-cancer analysis identifies migrasome-related genes as a potential immunotherapeutic target: a bulk omics research and single cell sequencing validation. Front Immunol 13. https://doi.org/10.3389/FIMMU.2022.994828/FULL

Quintero-Fabián S, Arreola R, Becerril-Villanueva E et al (2019) Role of matrix metalloproteinases in angiogenesis and cancer. Front Oncol 9. https://doi.org/10.3389/FONC.2019.01370

Rao VS, Gu Q, Tzschentke S et al (2022) Extravesicular TIMP-1 is a noninvasive independent prognostic marker and potential therapeutic target in colorectal liver metastases. Oncogene 41:1809. https://doi.org/10.1038/S41388-022-02218-9

Ren Z, Lv M, Yu Q et al (2021) MicroRNA-370-3p shuttled by breast cancer cell-derived extracellular vesicles induces fibroblast activation through the CYLD/Nf-κB axis to promote breast cancer progression. FASEB J 35:e21383. https://doi.org/10.1096/FJ.202001430RR

Rilla K, Mustonen A-M, Arasu UT et al (2019) Extracellular vesicles are integral and functional components of the extracellular matrix. Matrix Biol 75:201–219

Rose KWJ, Taye N, Karoulias SZ, Hubmacher D (2021) Regulation of ADAMTS proteases. Front Mol Biosci 8:701959. https://doi.org/10.3389/FMOLB.2021.701959

Rossi M, Battafarano G, D'Agostini M, Del Fattore A (2018) The role of extracellular vesicles in bone metastasis. Int J Mol Sci 19:1136. https://doi.org/10.3390/IJMS19041136

Roy R, Morad G, Jedinak A, Moses MA (2020) Metalloproteinases and their roles in human cancer. Anat Rec 303:1557–1572. https://doi.org/10.1002/AR.24188

Salhotra A, Shah HN, Levi B, Longaker MT (2020) Mechanisms of bone development and repair. Nat Rev Mol Cell Biol 21(11):696–711. https://doi.org/10.1038/s41580-020-00279-w

Samuels M, Jones W, Towler B et al (2023) The role of non-coding RNAs in extracellular vesicles in breast cancer and their diagnostic implications. Oncogene 42(41):3017–3034. https://doi.org/10.1038/s41388-023-02827-y

Sanderson RD, Bandari SK, Vlodavsky I (2019) Proteases and glycosidases on the surface of exosomes: newly discovered mechanisms for extracellular remodeling. Matrix Biol 75–76:160–169. https://doi.org/10.1016/J.MATBIO.2017.10.007

Satcher RL, Zhang XHF (2021) Evolving cancer–niche interactions and therapeutic targets during bone metastasis. Nat Rev Cancer 22(2):85–101. https://doi.org/10.1038/s41568-021-00406-5

Seifert A, Düsterhöft S, Wozniak J et al (2021) The metalloproteinase ADAM10 requires its activity to sustain surface expression. Cell Mol Life Sci 78:715–732. https://doi.org/10.1007/S00018-020-03507

Shimoda M (2019) Extracellular vesicle-associated MMPs: a modulator of the tissue microenvironment. Adv Clin Chem 88:35–66. https://doi.org/10.1016/BS.ACC.2018.10.006

Shimoda M, Khokha R (2013) Proteolytic factors in exosomes. Proteomics 13:1624–1636. https://doi.org/10.1002/PMIC.201200458

Shimoda M, Khokha R (2017) Metalloproteinases in extracellular vesicles. Biochim Biophys Acta Mol Cell Res 1864:1989–2000. https://doi.org/10.1016/J.BBAMCR.2017.05.027

Shoucair I, Mello FW, Jabalee J et al (2020) The role of cancer-associated fibroblasts and extracellular vesicles in tumorigenesis. Int J Mol Sci 21:1–37. https://doi.org/10.3390/IJMS21186837

Sleeboom JJF, van Tienderen GS, Schenke-Layland K et al (2024) The extracellular matrix as hallmark of cancer and metastasis: from biomechanics to therapeutic targets. Sci Transl Med 16:3840. https://doi.org/10.1126/SCITRANSLMED.ADG3840/SUPPL_FILE/SCITRANSLMED.ADG3840_SM.PDF

Souza JSM, Lisboa ABP, Santos TM et al (2020) The evolution of ADAM gene family in eukaryotes. Genomics 112:3108–3116. https://doi.org/10.1016/J.YGENO.2020.05.010

Sumida M, Hane M, Yabe U et al (2015) Rapid trimming of cell surface polysialic acid (PolySia) by exovesicular sialidase triggers release of preexisting surface neurotrophin. J Biol Chem 290:13202–13214. https://doi.org/10.1074/jbc.M115.638759

Sung BH, Ketova T, Hoshino D et al (2015) Directional cell movement through tissues is controlled by exosome secretion. Nat Commun 6. https://doi.org/10.1038/ncomms8164

Sutherland TE, Dyer DP, Allen JE (2023) The extracellular matrix and the immune system: a mutually dependent relationship. Science 379(6633):eabp8964

Tamura T, Yoshioka Y, Sakamoto S et al (2020) Extracellular vesicles in bone metastasis: key players in the tumor microenvironment and promising therapeutic targets. Int J Mol Sci 21:6680. https://doi.org/10.3390/IJMS21186680

Tassinari S, D'Angelo E, Caicci F et al (2024) Profile of matrix-entrapped extracellular vesicles of microenvironmental and infiltrating cell origin in decellularized colorectal cancer and adjacent mucosa. J Extracell Biol 3:e144. https://doi.org/10.1002/JEX2.144

Tauro BJ, Mathias RA, Greening DW et al (2013) Oncogenic H-ras reprograms Madin-Darby canine kidney (MDCK) cell-derived exosomal proteins following epithelial-mesenchymal transition. Mol Cell Proteomics 12:2148. https://doi.org/10.1074/MCP.M112.027086

Thompson CA, Purushothaman A, Ramani VC et al (2013) Heparanase regulates secretion, composition, and function of tumor cell-derived exosomes. J Biol Chem 288:10093–10099. https://doi.org/10.1074/jbc.C112.444562

Thuault S, Ghossoub R, David G, Zimmermann P (2022) A journey on extracellular vesicles for matrix metalloproteinases: a mechanistic perspective. Front Cell Dev Biol 10:886381. https://doi.org/10.3389/FCELL.2022.886381/BIBTEX

Tkach M, Théry C (2016) Communication by extracellular vesicles: where we are and where we need to go. Cell 164:1226–1232. https://doi.org/10.1016/J.CELL.2016.01.043

Tóth E, Turiák L, Visnovitz T et al (2021) Formation of a protein corona on the surface of extracellular vesicles in blood plasma. J Extracell Vesicles 10. https://doi.org/10.1002/JEV2.12140

Urciuoli E, Giorda E, Scarsella M et al (2018) Osteosarcoma-derived extracellular vesicles induce a tumor-like phenotype in normal recipient cells. J Cell Physiol 233:6158–6172. https://doi.org/10.1002/JCP.26464

Vader P, Breakefield XO, Wood MJA (2014) Extracellular vesicles: emerging targets for cancer therapy. Trends Mol Med. https://doi.org/10.1016/j.molmed.2014.03.002

Vu LT, Peng B, Zhang DX et al (2019) Tumor-secreted extracellular vesicles promote the activation of cancer-associated fibroblasts via the transfer of microRNA-125b. J Extracell Vesicles 8:1599680. https://doi.org/10.1080/20013078.2019.1599680

Wang J, Guan X, Zhang Y et al (2018) Exosomal miR-27a derived from gastric cancer cells regulates the transformation of fibroblasts into cancer-associated fibroblasts. Cell Physiol Biochem 49:869–883. https://doi.org/10.1159/000493218

Webber J, Steadman R, Mason MD et al (2010) Cancer exosomes trigger fibroblast to myofibroblast differentiation. Cancer Res 70:9621–9630. https://doi.org/10.1158/0008-5472.CAN-10-1722/656495

Webber JP, Spary LK, Sanders AJ et al (2014) Differentiation of tumour-promoting stromal myofibroblasts by cancer exosomes. Oncogene 34(3):290–302. https://doi.org/10.1038/onc.2013.560

Wenzel EM, Pedersen NM, Elfmark LA et al (2024) Intercellular transfer of cancer cell invasiveness via endosome-mediated protease shedding. Nat Commun 15:1–22. https://doi.org/10.1038/s41467-024-45558-8

Winkler J, Abisoye-Ogunniyan A, Metcalf KJ, Werb Z (2020) Concepts of extracellular matrix remodelling in tumour progression and metastasis. Nat Commun 11:1–19. https://doi.org/10.1038/s41467-020-18794-x

Xu R, Greening DW, Rai A et al (2015) Highly-purified exosomes and shed microvesicles isolated from the human colon cancer cell line LIM1863 by sequential centrifugal ultrafiltration are biochemically and functionally distinct. Methods 87:11–25. https://doi.org/10.1016/J.YMETH.2015.04.008

Yáñez-Mó M, Siljander PR-M, Andreu Z et al (2015) Biological properties of extracellular vesicles and their physiological functions. J Extracell Vesicles 4:27066. https://doi.org/10.3402/jev.v4.27066

Yokoi A, Yoshioka Y, Yamamoto Y et al (2017) Malignant extracellular vesicles carrying MMP1 mRNA facilitate peritoneal dissemination in ovarian cancer. Nat Commun 8:1–15. https://doi.org/10.1038/ncomms14470

Zeng ZS, Cohen AM, Guillem JG (1999) Loss of basement membrane type IV collagen is associated with increased expression of metalloproteinases 2 and 9 (MMP-2 and MMP-9) during human colorectal tumorigenesis. Carcinogenesis 20:749–755. https://doi.org/10.1093/CARCIN/20.5.749

Zhu Y, Herndon JM, Sojka DK et al (2017) Tissue-resident macrophages in pancreatic ductal adenocarcinoma originate from embryonic hematopoiesis and promote tumor progression. Immunity 47:323–338.e6. https://doi.org/10.1016/j.immuni.2017.07.014

Chapter 4
Role of Hyaluronan in Interactions of Tumor-Derived Extracellular Vesicles with Cells of Monocyte Origin

Rafał Szatanek, Izabela Siemińska, Marzena Lenart, Jarek Baran, and Monika Baj-Krzyworzeka

Abstract Extracellular vesicles (EVs) as a new form of intercellular communication have preoccupied researchers for many years now. In the field of cancer biology, specifically the role of tumor-derived EVs (TEVs) in tumor progression and their interactions with immunological cells have been a highly investigated research topic. In this context, an emphasis has been put on the cargo material transported via TEVs from the parental cells to the target "recipients" in different organs and tissues. Here we focused on hyaluronic acid (HA) as an important player in these interactions, with relevance to myeloid cells of monocyte origin, known as having both anti- and pro-tumorigenic effects. Despite the general information regarding the HA structure, its forms, and cellular receptors, based on the recent studies, we provide evidence on the role of the tumor tissue HA and HA-TEVs complexes in modulation of monocytes, macrophages, and myeloid-derived suppressor cells activities during cancer development and progression. We also present the possible forms of HA-targeted therapies as potentially promising approach in future treatment of cancer.

Rafał Szatanek and Izabela Siemińska contributed equally.

R. Szatanek · M. Lenart · J. Baran · M. Baj-Krzyworzeka (✉)
Department of Clinical Immunology, Medical College, Jagiellonian University, Kraków, Poland
e-mail: rafal.szatanek@uj.edu.pl; m.lenart@uj.edu.pl; jarek.baran@uj.edu.pl; monika.baj-krzyworzeka@uj.edu.pl

I. Siemińska
Institute of Veterinary Sciences, University Center of Veterinary Medicine JU-AU, University of Agriculture in Krakow, Kraków, Poland
e-mail: izabela.sieminska@urk.edu.pl

K. Rilla (ed.), *Extracellular Vesicles as Matrix Messengers*, Biology of Extracellular Matrix 15, https://doi.org/10.1007/978-3-031-68811-9_4

4.1 Introduction

Cancer is a major public health burden and economic challenge worldwide. In the European Union cancer accounts for a quarter of all deaths and has recently overtaken cardiovascular disease as the number one cause of death for people aged 45–64 years in an increasing number of the member states. According to the European Cancer Information System, the number of cancer incidences systematically increases, and by 2040 it is estimated to reach 3.25 million new cases a year (https://ecis.jrc.ec.europa.eu). This will be almost an 18.5% increase in comparison to 2.74 million patients diagnosed in 2022. Late diagnosis, a lack of curative treatment for many cancer types and demographic changes are primary reasons for such a perspective. Although the projected estimates are staggering, recent advances in technology, most notably biomarker discovery and artificial intelligence machine learning, provide new perspectives for treatment therapies and better cancer diagnosis. The application of these novel approaches is geared toward more personalized therapies relying on thorough analysis of the tumor itself, as well as the tumor microenvironment to identify specific agent(s) (i.e., receptors, signaling, and immune checkpoint molecules), which can be later used as potential treatment targets.

4.2 Tumor-Derived Extracellular Vesicles (TEVs)

For over a decade now, extracellular vesicles (EVs), which are being released by all types of cells, including tumor cells, have been the subject of extensive studies by scientists around the world. The term "tumor-derived extracellular vesicles" (TEVs) specifically refers to a population of EVs which are being released by tumor cells. As with other types of EVs, TEVs are considered to represent a mode of communication that allows transfer of bioactive cargo between cells. Although several types of EVs and different mechanisms for their release have been described (Marostica et al. 2021), the majority of EVs, including TEVs, are formed through either outward budding and fission of the plasma membrane or within multivesicular bodies (MVBs), which later fuse with cell surface and are then released to the surroundings. TEVs had been shown to contain surface, cytosolic, and nuclear proteins as well as nucleic acids such as double-stranded DNA, mRNA, and noncoding RNA (miRNA, tRNA, rRNA), which upon their transfer had substantially modified the targeted cells. For example, the addition of TEVs obtained from pancreatic tumor cells to NIH/3T3 fibroblasts caused molecular changes in their DNA, which led to tumor formation in mice (Stefanius et al. 2019). Breast cancer-derived EVs caused the activation of STAT3 signaling and an increase in macrophage survival together with the release of pro-tumorigenic cytokines, including IL-6 (Ham et al. 2018). In the glioblastoma model, the delivery of miR-387a to the tumor microenvironment (TME) by TEVs led to the downregulation of granzyme B expression and consequently the impairment of the infiltrating NK cell cytolytic functions (Briand et al. 2020). Besides the tumorigenic capabilities, TEVs have been also implicated in the modification of extracellular matrix (ECM) and premetastatic niche formation, further propagating tumor progression. In a melanoma cell model, TEVs lacking MT1-MMP, a major driver of TEV-mediated invasiveness, resulted in a significant reduction in invasive capacity and a loss of directionality in amoeboid-type tumor cells (Clancy et al. 2015). Additionally, it was shown that TEVs containing migration inhibitory factors released by pancreatic tumor cells primed liver as the metastatic site for this type of cancer (Peinado et al. 2012).

Tumor cells can also be engaged in auto- and paracrine signaling via TEVs. Several studies suggest the ability of tumor cells to employ a signaling mechanism using TEVs in order to avoid an immune response (Al-Nedawi et al. 2008; Gamperl et al. 2016; Szatanek et al. 2020). In all of the cases, TEVs released by one type of tumor cells were able to transfer proteins/mRNA (i.e., TF, EGFRvIII, and CXCR4) to another tumor cells in the same tumor resulting in the modification of the behavior of the latter (i.e., signal transduction, chemotactic potential, and gene expression) and subsequently further tumor progression.

EVs, including TEVs, are also subjected to the formation of protein corona (PC) at their surface (Sódar et al. 2016; Németh et al. 2017; Tóth et al. 2021). Although the concept of PC was at first explored in terms of artificial nanoparticles (NPs) and their biomedical application, it has been recently also applied to EVs. The overall idea of PC formation involves adsorption of biomolecules, mainly proteins, present in the surrounding biological environment to the surface of NPs/EVs. Analogically to NPs, the formation of PC around EVs has been a subject of recent studies focusing on PC's composition and impact on EV distribution in biological fluids (Kowal et al. 2016; Nakai et al. 2016; Tóth et al. 2021). Indeed, the presence of corona proteins (e.g., ApoA1, ApoB, ApoC3, ApoE, complement factors 3 and 4B, fibrinogen α-chain, and immunoglobulin heavy constant γ2 and γ4 chains) in the PC surrounding EVs isolated from blood plasma have been confirmed, although their specific impact on EVs functionality needs to be further investigated (Tóth et al. 2021). In addition to proteins and lipids, EVs carry glycans on their surface and carbohydrate-binding proteins located inside (Saito et al. 2018). The corona of EVs was described to be especially enriched by proteoglycans containing heparan sulfate glycosaminoglycan and hyaluronan (Buzás et al. 2018).

4.3 Hyaluronan

Hyaluronan, also referred to as hyaluronic acid (HA), is a ubiquitously distributed natural polysaccharide, which is particularly abundant in the ECM of adult soft connective tissue and the embryonic tissue in vertebrates (Abatangelo et al. 2020). HA is composed of D-glucuronic acid and N-acetyl-D-glucosamine repeating disaccharide units, which are linked via β-1,4- or β-1,3-glycosidic bonds (Schanté et al. 2011). Due to its carboxyl groups, HA is negatively charged and highly hydrophilic

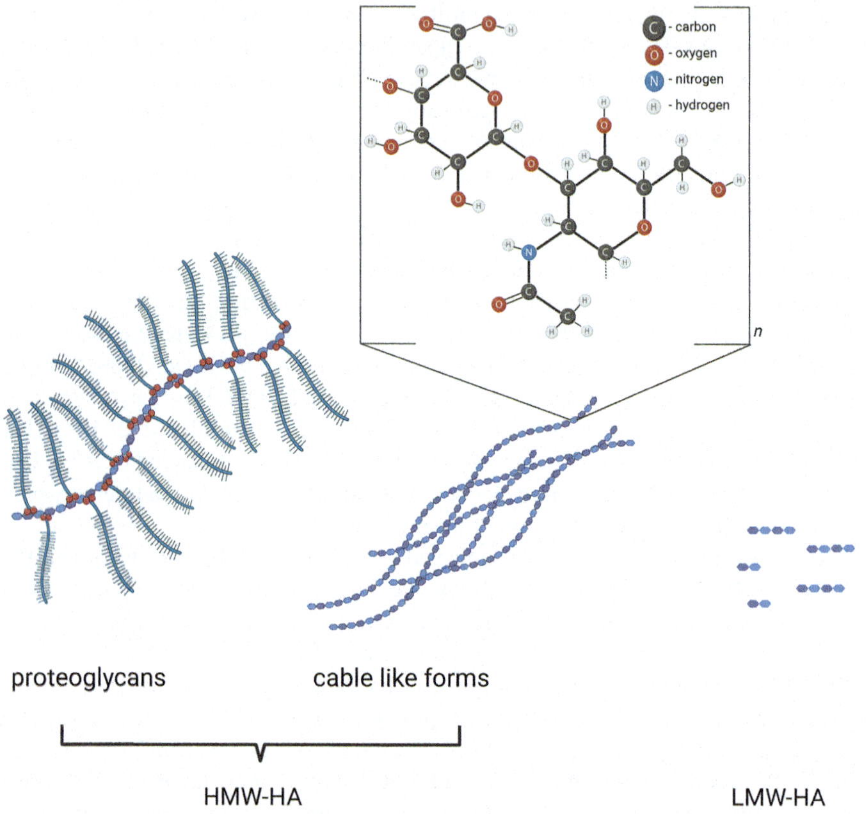

Fig. 4.1 Chemical structure of HA and its possible biological forms. HA can be classified according to their molecular weight. Low molecular weight (LMW)-HA is composed of several individual HA molecules; its molecular weight does not exceed 1×10^6 Da. High molecular weight (HMW)-HA (>2×10^6 Da) can form "cable like structures" that are composed of multiple HA molecules. Additionally, these long chains of HA can also interact with proteoglycans to form more sophisticated complexes

(Abatangelo et al. 2020). A graphical representation of the chemical structure and different forms of HA is presented in Fig. 4.1. HA hydrates ECM and is responsible for maintaining tissue homeostasis and resistance to forces of compression (Abatangelo et al. 2020). Furthermore, HA has the ability to form coats around most cells, and, thus, interact with its binding proteins as a signaling molecule (Abatangelo et al. 2020).

Synthesis of HA occurs on the inner side of the plasma membrane and is mediated by three transmembrane enzymes—HA synthase 1 (HAS1), HAS2, and HAS3 (Weigel et al. 1997; Itano and Kimata 2002). Following the HA synthesis, the HAS enzymes, through a pore in their structures, translocate HA to the extracellular space together with the elongation of the polymeric chain (Abatangelo et al. 2020). The expression level of the three HAS enzymes differs depending on the pathophysiological conditions and during morphogenesis. The HAS enzymes also generate HA of different molecular weights (MW) with HAS3 producing molecules with low MW (LMW, 1×10^5–1×10^6 Da), whereas HAS1 and HAS2 with large MW (HMW, $>2 \times 10^6$ Da) (Abatangelo et al. 2020).

HA present in the tissue is degraded outside the cell by hyaluronidases (Hyals), reactive oxygen species, superoxide, nitric oxide, and peroxynitrite (Heldin et al. 2019). Out of this group, the Hyal family has been characterized the most. So far, there have been six Hyals identified in humans: Hyal-1-4, PH-20, and Hyalp-1 (Heldin et al. 2019). Much of the activity of Hyal-3, Hyal-4, and Hyalp1 is still unknown, however, Hyal-1 and Hyal-2 are fairly characterized. These enzymes are widely expressed in somatic tissue and seem to cooperate during the process of HA degradation. Hyal-2 binds and internalizes HA engulfed in vesicles, which is followed by its cleavage into 20 kDa fragments. These fragments are then further digested into tetrasaccharides by Hyal-1 (Tammi et al. 2001).

Apart from Hyals, two other proteins, cell migration-inducing and hyaluronan-binding protein (CEMIP/KIAA1199) and transmembrane protein 2 (TMEM2), are involved in the degradation of HA outside the cell. CEMIP by itself does not have hyaluronidase activity, but its secretion to the outside of the cells seems to be necessary for the HA degradation process (Domanegg et al. 2022). HA degradation occurs mainly in lysosomes, where it is carried out mostly by Hyals (Hyal-1 and -2) under acidic conditions (pH ~5) (Yamaguchi et al. 2019). Although an accelerated turnover of HA/HA fragment formation has been observed in most cancers (Tammi et al. 2019), it does not correspond with Hyal-1 and -2 expressions, which varies depending on the tumor type. Elevated levels of Hyal-1 have been observed in both bladder and prostate cancers whereas increased expression of Hyal-2 has been reported in dysplastic nevi and melanoma (Tammi et al. 2019). On the other hand, in ovarian, pancreatic, and endometrial cancers, both Hyal-1 and -2 expression levels were shown to be reduced (Tammi et al. 2019).

Multiple reports suggest that MW of HA has a profound effect on its activity. HA's role has been, above all, associated with cell movement, both during embryonic development and in tumor progression (Abatangelo et al. 2020). It has been observed that HMW-HA was linked to cancer resistance, whereas LMW-HA favored tumor progression. The exact mechanism behind this phenomenon is still largely unknown, however, it has been postulated that HMW-HA has an ability to cluster HA receptors at the cell surface, followed by subsequent activation of intracellular signaling pathways (Yang et al. 2012). Additionally, it was confirmed that HMW-HA was able to trigger anti-inflammatory response by altering the recruitment of inflammatory cells, whereas LMW-HA caused a pro-inflammatory reaction resulting in angiogenesis and tissue remodeling in the wound-healing process (Day and de la Motte 2005; Stern et al. 2006). Moreover, LMW-HA has been also reported to be present in the blood under severed inflammatory conditions (Tammi et al. 2019).

There seems to be an intriguing association between TEVs and HA. Aaltonen et al. observed that TEV shedding occurs in parallel with HA synthesis in rapidly proliferating cells such as tumor cells (Aaltonen et al. 2022). The authors suggested that this phenomenon might be linked to the overexpression of HAS or the processing/digestion of glycocalyx produced by tumor cells (Aaltonen et al. 2022). Indeed, other groups also reported that significant amounts of HA were present in EVs that were being secreted by mesothelial cells (Koistinen et al. 2015), colorectal (Paul et al. 2020; Babula et al. 2023), gastric (Härkönen et al. 2019), pancreatic (Lenart et al. 2017; Purushothaman et al. 2023), and chondrosarcoma cancer cell lines (Rilla et al. 2013). A similar association can be observed when comparing both the HA and EV levels between healthy individuals and cancer patients. Based on the data from multiple reports, higher concentrations of HA and EVs were detected in cancer patients than in healthy subjects (Ponting et al. 1992; Hernández et al. 1995; Xing et al. 2008; Momen-Heravi and Bala 2018; Choi et al. 2020; Vinik et al. 2020). Moreover, an increase in HA concentration seems to be also dependent on cancer stage with patients with more advanced tumors showing higher levels of this polymer (Xing et al. 2008).

It is believed that the amount of HA transported by TEVs reflects the HA secretion intensity of the parental cells. For example, HCT116 colon cancer cells secrete more HA compared to SW116 colon adenocarcinoma cells, and this difference is also evident in the HA content transported by their respective TEVs (Babula et al. 2023). The size of HA carried by TEVs is of low and intermediate molecular weight as defined by single-molecule force microscopy (Paul et al. 2022). Interestingly, the actual density of HA molecules presence seems to be significantly higher on TEVs, as exemplified by colon cancer cells, than on EVs from normal cells (Paul et al. 2020). This may suggest the ability of EVs/TEVs to synthesize HA themselves, which may be supported by the HAS presence on their membrane (Rilla et al. 2013; Härkönen et al. 2019; Rilla 2021), however, currently there is no legitimate evidence to validate this claim.

4.3.1 Hyaluronan Receptors

HA is a major component of the ECM and its abundance is especially evident in the TME where it plays a pro-tumorigenic role. Hyaladherins compose a group of cell surface proteins, i.e., CD44, RHAMM (CD168), TLR2, TLR4, and LYVE-1, which are responsible for HA binding and subsequent downstream signaling (Fig. 4.2) (Abatangelo et al. 2020). Two other receptors, HARE and layilin, also belong to hyaladherins but are not expressed on cells of monocytic origin and thus were not included in Fig. 4.2. CD44 is considered to be one of the main HA surface receptors and it will be the main focus of this review.

4.3.1.1 CD44

CD44 molecule was identified in the early 1980s as a glycoprotein present in human and murine mesenchymal cells (Hughes et al. 1981; Stamenkovic et al. 1989). CD44 is a non-kinase transmembrane glycoprotein residing at the cell surface of

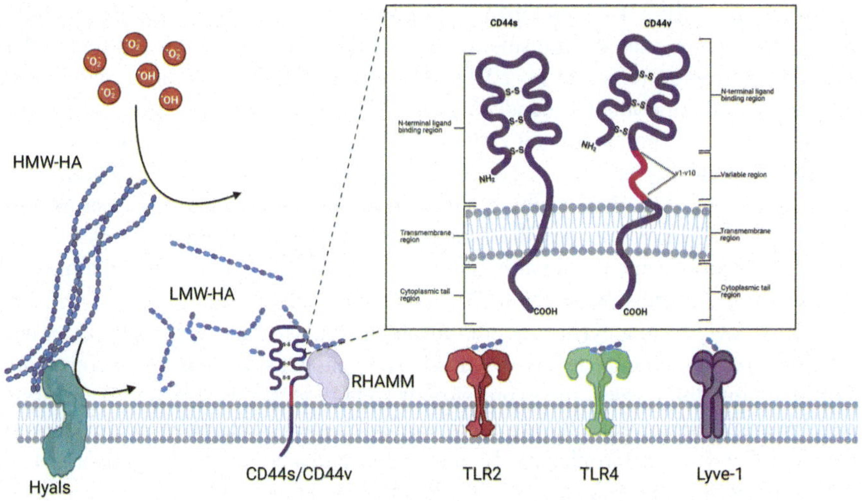

Fig. 4.2 HA receptors and molecules on cells of monocytic origin. Presence of hyaluronidases (Hyals) in the cell membrane as well as reactive oxygen species (i.e., superoxide and hydroxide) results in the degradation of HA and its interactions with appropriate receptors. CD44 (both standard and variant forms), RHAMM, TLR2, TLR4, and Lyve-1, collectively referred to as hyaladherins, are responsible for binding of HA and subsequent downstream signaling in cells of monocytic origin

leukocytes and epithelial cells (Hassn Mesrati et al. 2021). It was defined as a single polypeptide chain encoded by a conserved gene located on chromosome 11 in humans (Goodfellow et al. 1982). CD44 is also referred to as In (Lu)-related protein p80, Pgp-1/Ly-24, ECMRIII, HUTCH-1, Hermes antigen, and HA receptor (Hassn Mesrati et al. 2021). Currently, CD44 is considered to be the main surface receptor for HA and a member of the cell adhesion molecules (CAMs) family, which are regarded as major players in cellular communication and adhesion between cells and ECM (Goodfellow et al. 1982). Additionally, CD44 was also observed to be involved in lymphopoiesis and myelopoiesis, leukocyte activation, angiogenesis, and cytokine release as well as metastasis, epithelial–mesenchymal transition (EMT), cellular proliferation, migration, and invasion (Hassn Mesrati et al. 2021).

There are two known forms of CD44, a standard (CD44s) form and variant (CD44v) isoforms, which are formed as a consequence of alternative splicing of the CD44 gene. The CD44 gene is composed of 20 exons, where exons 1–5 and 16–20 are considered to encode the N- and C-terminal domains, respectively. These domains are homologous and are shared by all CD44 family members (Loh et al. 2015). CD44s comprise 10 constant exons with no variants and are considered to be the smallest and the most expressed CD44 isoform (Rall and Rustgi 1995). On the other hand, CD44v isoforms can have single or multiple insertions or excisions of alternatively spliced exons between the N- and C-terminal domains (Mishra et al. 2019).

Structurally, the CD44 protein is composed of four regions: the extracellular region, the stem region, the transmembrane region, and the C-terminal intracellular/cytoplasmic region (Chen et al. 2020). The extracellular fragment is made up of seven extracellular domains encoded by the 1–5, 6, and 7 constant exons, which also includes the N-terminal domains (ligand-binding region) (Hassn Mesrati et al. 2021). The stem region, regarded as the alternative splicing fragment, is defined by insertion of one or multiple variant exons between exons 5 and 6 (Hassn Mesrati et al. 2021). Exon 8 encodes the transmembrane region whereas exons 10 or 9 encode the cytoplasmic region although exon 9 is routinely spliced out in almost all CD44 cDNA isoforms (Misra et al. 2015; Mishra et al. 2019).

Lymphocytes, monocytes, smooth muscle, fibroblasts, and various types of epithelia express significant levels of CD44. Whereas CD44s isoform is commonly expressed in adult tissues, the expression of CD44v isoforms seems to be reserved for selected types of cells during their specific stage of activation, maturation, or development (i.e., macrophages, lymphocytes, and keratinocytes) (Mishra et al. 2019). Expression of HA receptors on myeloid cells of monocytic origin is summarized in Table 4.1. Currently, three activation states of CD44 have been defined

Table 4.1 HA receptors on myeloid cells of monocytic origin

Cell population		Receptor	Molecule involved in HA processing	References
Monocytes		CD44		Webb et al. (1990), Stamenkovic et al. (1991), Culty et al. (1992), Zembala et al. (1994), Weiss et al. (1998), Lee-Sayer et al. (2015), Babula et al. (2023)
		TLR2		Yang et al. (1998), Ropert (2019)
		TLR4		Adib-Conquy and Cavaillon (2002), Sokolowska et al. (2014)
			Hyal	Girard et al. (2002)
Macrophages	TAMs	LYVE-1		Anstee et al. (2023)
			Hyal1 and Hyal2	Witschen et al. (2023)
	Macrophages	RHAMM		Foley et al. (2012)
		CD44		Yan et al. (2022)
		TLR2		Mosser and Zhang (2008)
		Lyve-1		Schyns et al. (2019)
MDSCs			Hyal	Kusmartsev et al. (2022)
		RHAMM		Willemen et al. (2016)
		TLR2		Fu et al. (2021)
		CD44	Hyal2	Dominguez-Gutierrez et al. (2021)
Dendritic cells		TLR2 and TLR4		Kaisho and Akira (2000)

among cell lines and normal cell populations (including monocytes and other myeloid cells) (Lesley and Hyman 1998). The first one concerns CD44 in its active form, which enables HA binding constitutively. The second category includes CD44 in an inducible state, which needs to be activated before HA binding. The last involves CD44 in its inactive form, which cannot bind HA even after its activation (Lesley and Hyman 1998).

Overexpression of CD44 has been implicated with tumorigenesis although its role in this process is still unclear. Elevated CD44 mRNA levels have been observed in lymphoma, breast, colon, and endometrial cancers (Fujita et al. 1994). Other cancer types, such as gallbladder, prostate, ovarian, oral squamous cell carcinoma, and gastric cancer, also showed overexpression of this protein, which correlated with increased aggressiveness and poor prognosis in these tumors (Yu and Stamenkovic 1999). There have been contradictory reports concerning CD44 isoforms and cancer progression. For example, it has been suggested that CD44s, and not CD44v, promote breast cancer initiation by activating the PDGFRβ/STAT3 cascade-signaling pathway (Zhang et al. 2019). Another group showed that shifting the CD44 isoform from variant to standard resulted in the activation of the AKT signaling pathway and epithelial–mesenchymal transition in breast cancer cells leading to an increase in their metastatic potential (Brown et al. 2011). On the other hand, it was found that CD44v8-10, and not CD44 expression, defined cancer stem cells in colon cancer (Tanabe et al. 1993). Furthermore, CD44v5-6 and CD44v8-9, but not CD44s, mediated tumor cell proliferation and poor prognosis by activating the KRAS/MAPK signaling pathway (Zhao et al. 2013).

CD44 is one of the many proteins being transported by TEVs. It was shown that TEVs of epithelial ovarian cancer origin were found to be enriched with CD44 and were able to transfer this receptor to human peritoneal mesothelial cells, which led to cancer progression (Nakamura et al. 2017). In another study, TEVs isolated from plasma of head and neck squamous cell carcinoma patients expressed elevated levels of CD44v3, which also correlated with lymph node metastasis in these individuals (Theodoraki et al. 2020). Huang et al. reported that TEVs released by gastric cancer cells contained CD44, which was transferred to bone marrow mesenchymal stem cells. This transfer induced activation of the STAT3 and NFkβ signaling pathways resulting in IL-8 secretion in these cells (Huang et al. 2023).

4.3.1.2 HA–CD44 Interactions

The formation of HA–CD44 complex can result in the activation of downstream signaling cascades involved in proliferation, differentiation, cell motility, and the degradation of HA itself (Sherman et al. 1994; Dicker et al. 2014). The intensity of these signals seems to be governed by the affinity of the CD44 receptor to HA, which, in turn, depends both on the concentration and MW of the molecule, by glycosylation of extracellular domains and/or by phosphorylation of the serine residues (Skelton et al. 1998; Lesley and Hyman 1998). Furthermore, it has been also observed that HMW-HA can cause CD44 clustering, which facilitates further interactions with ECM molecules, growth factors, matrix metalloproteinases, and cytokines (Lesley and Hyman 1998; Yang et al. 2012).

According to the study where the MDA-MB-231 breast cancer cell line was used, CD44 expression seems to increase upon HA binding in a gradient-dependent manner, suggesting a feedback response in these cells (influence of HA). In the same study, the authors also showed that HA density had a positive impact on motility of the MDA-MB-231 cells. Concurrently, blocking of CD44 in this case resulted in a decrease of the migratory abilities of these cells suggesting that the HA-CD44 complex plays an important role in this process. Moreover, HA–CD44 interactions activated stemness-related signaling pathways resulting in tumor recurrence in oral squamous cell carcinoma. Recently, an under-appreciated aspect has come to light where EVs have the capacity to capture components of the extracellular milieu onto their own surface. Such "corona" forms a "halo" around the vesicles (Tóth et al. 2021), allowing for the interactions with specific receptors on the target cells. In addition, an innate "corona" may also coat EVs originating from multivesicular endosomes, where factors within endosomes can bind to intraluminal vesicles prior to their eventual extracellular secretion (Yerneni et al. 2022). In this context, it cannot be excluded that the HA–CD44 complexes can be established during "corona" formation on EVs.

In CRC, patients that display high HA staining have a lower survival rate suggesting the engagement of HA/CD44 in the cancer-initiating cells homing (Ropponen et al. 2001). In the pancreatic cancer rat model that cannot grow locally, TEVs have been shown to be associated with premetastatic niche formation. In this model, cancer cells with the knockdown CD44v6 gene were unable to promote tumor growth whereas TEVs, isolated from wild-type tumor matrix, initiated tumorigenesis. These TEVs preferentially bound HA, laminin, and collagen and contained high content of proteases that efficiently degraded ECM liberating growth factors and chemokines that promoted cell migration and proliferation (Klingbeil and Isacke 2011).

These and other examples may suggest some form of interdependency between TEVs and the HA–CD44 complex. There have been reports suggesting that HA is carried by TEVs jointly with CD44 (Rilla et al. 2014a; Mustonen et al. 2021; Babula et al. 2023). Significant amounts of HA were detected in EVs secreted by mesothelial cells (Rilla et al. 2014b), colorectal (Babula et al. 2023), and chondrosarcoma cancer cell lines (Rilla et al. 2014b). Lesley et al. postulated that the density of HA molecules may be related to the density of CD44 on the cells of origin (Lesley et al. 2000) and probably TEVs. However, it has to be mentioned that CD44 has no known impact on EVs formation and their coating with HA and that this process is rather HAS dependent (Kultti et al. 2014).

4.4 Interactions of TEVs-Carried HA with Cells of Monocytic Origin

Myeloid cells, including blood monocytes, tissue macrophages, and dendritic cells play a crucial role in host response to the developing tumor. Although dendritic cells (DCs) make up a relevant population of myeloid cells present in the tissue, this review focuses mainly on monocytes, macrophages, and myeloid-derived suppressor cells (MDSC), which are the principal subject of our group's research. In this section, we would like to present the impact of HA (in soluble or EVs-bounded form) on these populations of cells, mainly in the context of their anti- or pro-tumorigenic activity.

4.4.1 Monocytes

Monocytes develop in the bone marrow from a common myeloid progenitor, through the stage of monoblast and promonocyte. Following their egress from the bone marrow, monocytes circulate in peripheral blood for c.a. 72 hours, after which

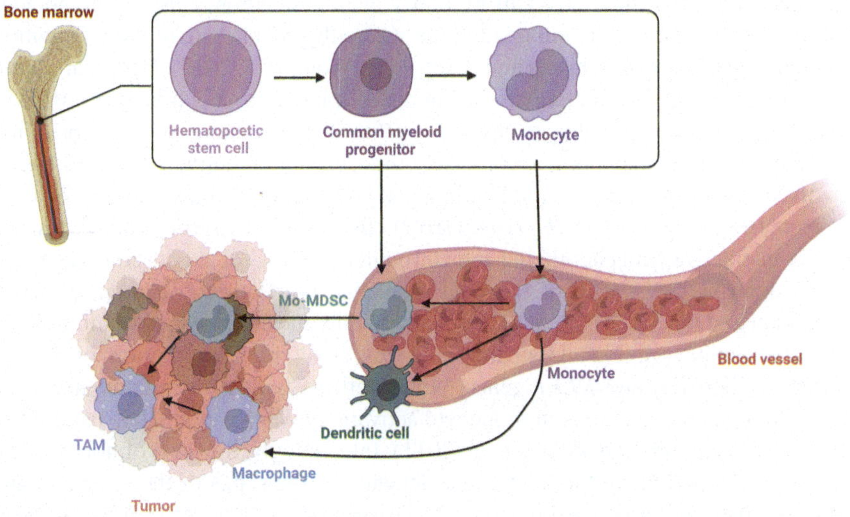

Fig. 4.3 Schematic representation of the maturation of monocytes and cells of monocytic origin. Monocytic cells are generated from hematopoietic stem cells in the bone marrow via a common myeloid progenitor. Monocytes are directly derived from common myeloid progenitors, whereas Mo-MDSC may come from monocytes or directly from common myeloid progenitors. Blood monocytes may also differentiate into macrophages or dendritic cells (after leaving the blood vessels). As shown in the figure, the tumor mass may be infiltrated by different cells of monocytic origin—macrophages that polarized into Tumor-Associated Macrophages (TAM, mainly M2) and Mo-MDSC (monocytic myeloid-derived suppressor cells)

most of them die by apoptosis. The overview of monocytes maturation process is presented in Fig. 4.3. In humans, monocytes are the largest leukocytes in circulation with a typical kidney-shaped nucleus. They represent 4–11% of blood leukocytes (Kiss et al. 2020) and play an important role in innate and adaptive immune response. Although monocytes share several common features, e.g., all are positive for CD14, they compose a heterogenous population, which differs in the expression of many surface molecules and functions, e.g., phagocytosis and killing of bacteria, destroying tumor cells and capacity to present antigens to T lymphocytes (Zembala and Asherson 1989). Prior to the flow cytometry era, monocytes were divided into subsets according to their size and density by centrifugation. Based on the differences in the composition of surface markers, many classifications of monocytes were implemented through the years, e.g., Fc receptor positive (FcR^+) and FcR^- subsets using rosetting, density gradient centrifugation, and adherence (Zembala et al. 1984) with the long-accepted division into $CD64^+$ ($FcgRI^+$) and $CD64^-$ ($FcgRI^-$) subsets (Grage-Griebenow et al. 1993, 1996, 2001). However, the currently accepted classification of monocytes proposed by a panel of leading experts in monocyte biology in 2010 is based on the expression of two surface molecules CD14 (coreceptor for LPS) and CD16 (FcgRIII) (Ziegler-Heitbrock et al. 2010). The major subset with a high expression of CD14 and the lack of CD16 ($CD14^{++}CD16^-$), referred to as classical monocytes, represents about 75–85% of the total number of blood monocytes and is highly active in phagocytosis and bacterial clearance (Cormican and Griffin 2020). During injury or inflammation, classical monocytes are rapidly recruited to the inflamed tissue where they contribute to the recognition and removal of microorganisms and dying cells (Wong et al. 2012). Classical monocytes, after leaving blood circulation, can differentiate into DCs (Qu et al. 2014) or long-living tissue macrophages (Rivollier et al. 2012). Gene expression profiling analysis indicates that human classical monocytes preferentially express genes involved in angiogenesis, wound healing, and coagulation (Wong et al. 2012).

Nonclassical monocytes ($CD14^+CD16^{++}$), also called patrolling monocytes, are highly interactive with endothelial cells through the CX3CR1 expression. They represent approximately 10% of the total monocyte population (Wong et al. 2012; Cormican and Griffin 2020) and are considered the most mature population of circulating monocytes. Nonclassical monocytes show high expression of *CD16*, *HMOX1*, *CX3CR1*, and *KLF1* genes, which reflects both their inflammatory and anti-inflammatory functions in monocyte activation (Wenzel et al. 2015) and interactions with endothelium (Wong et al. 2011). Interestingly, the level of nonclassical monocytes in peripheral blood strongly depends on the status of the organism and therefore they might be used as a potential diagnostic parameter (Selimoglu-Buet et al. 2015).

Intermediate monocytes, also called inflammatory, are characterized by $CD14^{++}CD16^+$ expression and are the least abundant population in the blood (approximately 5%) (Cormican and Griffin 2020). It was suggested that this population can be separated from nonclassical monocytes by the expression of 6-sulfo LacNAc (SLAN) antigen (Hofer et al. 2015). These monocytes are characterized by a high expression of HLA-class II antigens and are associated with antigen processing and presentation, secretion of inflammatory cytokines (TNF-α, IL-1β, IL-6) upon TLR stimulation, inflammation, monocyte activation, and differentiation (Zawada et al. 2011). At the transcriptional level, this cell subset was shown to express high levels of class II MHC genes, e.g., *CD74* and *HLA-DR*, genes that regulate chemotaxis and angiogenesis (*AIF1* and *TIE2*) as well as phagocytosis and tissue repair (*TGF-β* and *CD93*) (Wong et al. 2011; Beyer et al. 2012). Their level increases in systemic infections (Fingerle et al. 1993) and in cancer, e.g., breast, non-small cell lung carcinoma (Chen et al. 2023b), ovarian (Prat et al. 2020), and gastric cancers (Eljaszewicz et al. 2013).

Under pathological conditions, such as inflammation and cancer, the rapid recruitment of myeloid cells to the sites of injury stimulates an "emergency" hematopoiesis that might generate monocytes through different pathways or precursors (Wu et al. 2014; Yáñez et al. 2017).

Monocytes exposed to tumor cells or their derivates become activated in a narrow time window between 4 and 16 h after the exposition (Mytar et al. 2001; Kuang et al. 2007; Gärtner et al. 2018). Afterward, monocytes exposed to tumor cells become "exhausted" (Kuang et al. 2007) or "deactivated" (Mytar et al. 2003) and subsequently immunosuppressive, which is mainly manifested by the production of IL-10 (Mytar et al. 2003). Kuang et al. pointed to tumor-derived soluble factors (e.g., HA) as a reason for these changes in monocyte activity (Kuang et al. 2007). Current knowledge may expand this idea to the HA transfer in the membrane-bound form (e.g., TEVs) (Rilla et al. 2014a; Lenart et al. 2017; Babula et al. 2023). Monocytes' response to TEVs mimics/reflects their response to cancer cells. This may be due to the fact that TEVs resemble the chemical composition of cancer cells they originate from (Depciuch et al. 2020).

Monocytes may encounter TEVs in blood circulation and in the infiltrated tissues (Manning and Danielson 2018; Luong et al. 2021; Chen et al. 2023a). Upon contact, monocytes, as phagocytic cells, are eager to engulf TEVs (Popēna et al. 2018; Babula et al. 2023). By definition, the overall size of TEVs is in a range of monocyte phagocytic capacity, and phagocytosis is the major way of TEVs clearance but other forms of endocytosis and/or internalization can also be involved in this process (Popēna et al. 2018; Kwok et al. 2021). Recently, it has been suggested that TEVs engulfment may be facilitated by the presence of HA on their surface (Babula et al. 2023). It seems that EVs engulfment is CD44 dependent, as its blocking with monoclonal antibodies (mAbs) on monocytes significantly slowed down the whole process. Differences in the engulfment of HA-TEVs were also observed between monocyte subsets, which differ in the level of CD44 expression (Babula et al. 2023).

The wide spectrum of monocytes' response to TEVs of different origins was described by many authors, which in some cases resembles the response to HA or directly indicates the role of HA (see below). The role of HA carried by TEVs and their interactions with monocytes is only marginally described in the literature with particularly little data on the monocyte subpopulations. The current state of knowledge on this topic is summarized below and graphically presented in Fig. 4.4.

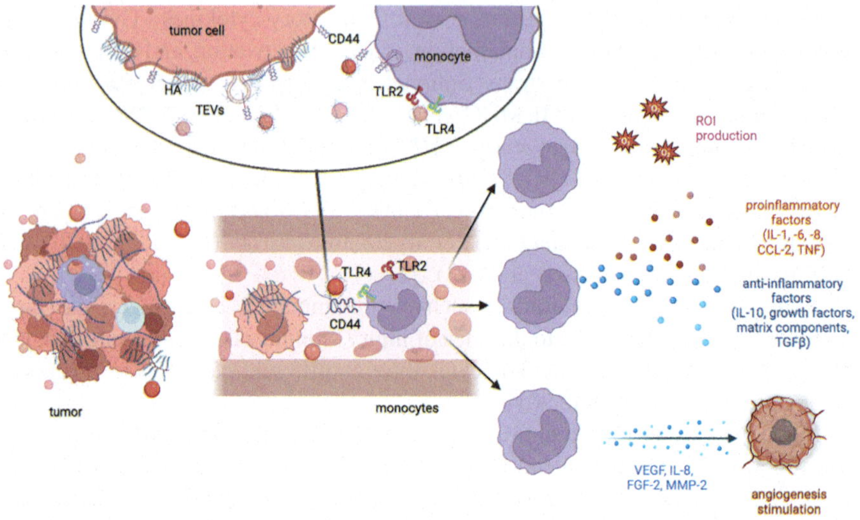

Fig. 4.4 Monocytes after contact with a tumor cell or TEVs-carried HA increase production of ROI and a number of pro- and anti-inflammatory factors as well as stimulate angiogenesis. Monocytes may contact directly with tumor cells or TEVs in the blood vessels and the tumor bed via their CD44 receptors, which can have multiple outcomes. Contact with HA or HA carried by TEVs with monocytes, may induce reactive oxygen intermediates (ROI) production. Following HA stimulation, monocytes may produce many soluble factors such as cytokines and chemokines with pleiotropic effects (pro-inflammatory, anti-inflammatory, and pro-angiogenic). FGF-2 (Fibroblast Growth Factor-2), MMP-2 (metalloproteinase-2), VEGF (Vascular Endothelial Growth Factor), CCL2 (chemokine (C-C motif) ligand 2 also referred to as Monocyte-Chemotactic Protein-1, MCP-1), IL—Interleukin, TGF-β—Transforming Growth Factor β, TNF-α—Tumor Necrosis Factor α

4.4.1.1 Impact of TEVs-Carried HA on Monocyte Phenotype

The impact of HA and HA carried by EVs on monocyte phenotype and function is still an open question. Spinelli and colleagues did not observe differences in the expression of CD14, HLA, CD80, and CD206 on monocytes after 24 h incubation with HA (HMW or LMW) (Spinelli et al. 2019). Similar observations were presented for tumor cell lysates (LoVo and MDA-MB-231 cell lines), however, expression of CD206 was observed to be slightly elevated (without significance). This data is in line with our results showing that monocytes after contact with TEVs (HCT116 cell line-derived) have increased the expression of CD206, but not that of HLA-DR (unpublished data). An increase in CD14 expression was observed on THP-1 cells after contact with TEVs derived from HA-producing SW480 and SW620 cell lines (Kim et al. 2004; Popēna et al. 2018). TEVs derived from glioblastoma cell lines significantly reduced HLA-DR and increased CD14 expression levels in monocytes (de Vrij et al. 2015). Only TEVs derived from a pancreatic cancer cell line (PANC-1 and HPC-4) were described to increase HLA-DR on classical monocytes, and at least in the case of PANC-1, the effect was HA-dependent (Baj-Krzyworzeka et al. 2007; Purushothaman et al. 2023). The expression of HLA-DR on monocytes was significantly lower after contact with HA-depleted TEVs (4-methyl umbelliferone treated) (Purushothaman et al. 2023). According to Purushothaman and colleagues, the presence of HA in TEVs did not impact the expression of PD-1, CD86, and CD64 on monocytes, independently of the monocytes' subset (Purushothaman et al. 2023).

Short incubation of monocytes with HA significantly decreased CD44 expression, suggesting that HA binds to CD44 present on monocytes (Mytar et al. 2001). In contrast, the incubation of monocytes with TEVs from HCT116 cell line (rich in HA) did not change the CD44 expression. One of the possible explanations of this phenomenon may be the spontaneous increase of membrane CD44 and CD44 variants expression on monocytes during cell culture (Levesque and Haynes 1996; Weiss et al. 1998; Braumüller et al. 2000; Babula et al. 2023) and different kinetics of HA and TEVs processing by monocytes. The ability of CD44 to bind HA is upregulated during in vitro culture (Levesque and Haynes 1996). Moreover, the binding of HA to surface CD44, and probably to TEVs, may be modulated by soluble CD44 molecules present in the cellular environment (body fluids or extracellular space) (Lesley and Hyman 1998). An increased CD44 cleavage (source of soluble CD44) has been described in many tumors, e.g. breast, ovarian, colon, and gliomas (Okamoto et al. 2002). Interaction of monocytes with HA-carried by TEVs may also be impacted by the transfer of CD44 via TEVs. For example, monocytes were shown to gain expression of CD44 through the transfer of CD44v7/8 from pancreatic adenocarcinoma cell line-derived EVs (Baj-Krzyworzeka et al. 2006). Although the biological activity of CD44 after the transfer from TEVs was not formally proven, a similar transfer involving chemokine receptors CCR6 (Baj-Krzyworzeka et al. 2006) and CCR5 (Mack et al. 2000) to monocytes was documented to be functionally effective.

4.4.1.2 Impact of TEVs-Carried HA on Cytokines, Chemokines, and ROI Production

As mentioned above, HA, depending on its molecular weight (but not its origin), may induce secretion of pro-inflammatory cytokines (IL-1, -6, -8, MCP-1, and TNF-α) or anti-inflammatory factors (IL-10, growth factors or matrix components, TGF-β) by monocytes (Mytar et al. 2003; Day and de la Motte 2005; Valenti et al. 2006; Yamawaki et al. 2009; Lenart et al. 2017; Spinelli et al. 2019). LMW-HA stimulates the expression of pro-inflammatory factors, activates intracellular inflammasomes, triggers sterile inflammation, activates cPLA2, and induces the release of arachidonic acid and eicosanoid production (Sokolowska et al. 2014). Sokołowska et al. presented that LWM-HA increased MAPK (ERK1/2, p38) phosphorylation in monocytes, among other types of cells, with the activation peak maximum occurring at 30 min (Sokolowska et al. 2014). Purified fragments of HA, such as Damage-Associated Molecular Patterns (DAMPs), induce pro-inflammatory signals mostly through TLR4, in contrast to larger HA molecules which exhibit cable structures and mostly interact through CD44 (Day and de la Motte 2005). Several studies revealed that ligation of CD44 molecules on monocytes leads to inflammatory cytokine release (Webb et al. 1990; Katoh et al. 1999; Bourcier et al. 2010).

Yamawaki et al. documented that secretion of IL-6 by mononuclear cells after a small-size HA exposure is CD44 and/or TLR4 dependent, while the CCL2 secretion in the same model was dependent only on CD44 expression. In both cases, the signaling pathway involved ERK and p38 MAPK kinases (Yamawaki et al. 2009). Only small-size HA fragments were found to activate NFκβ in murine macrophages, which resulted in restoring the production of TNF-α and IL-1β (Noble et al. 1996). Secretion of IL-6 and IL-8 by monocytes after their exposure to HA of platelet origin has been reported by Motte et al. (de la Motte et al. 2003). Moreover, small-size HA molecules were able to stimulate the release of arachidonic acid and induce different signaling pathways (including ERK1/2, p38, JNK) in primary human monocytes and murine RAW 264.7 cell lines (Sokolowska et al. 2014).

TEVs derived from different types of tumors, such as colon, pancreas, head and neck, and lung cancers, were shown to stimulate primary monocytes to produce proinflammatory (TNF-α, IL-1β, IL-6, IL-8, IL-12, CCL2, CCL3, CCL4, CCL5) and immunosuppressive (IL-10) cytokines and chemokines (Baj-Krzyworzeka et al. 2007; Baj-Krzyworzeka et al. 2011; Gärtner et al. 2018; Purushothaman et al. 2023). For example, TEVs derived from pancreatic cancer cell lines (PANC-1) were found to be enriched in HA and significantly induced secretion of IL-1β and CCL4 in monocytes (Purushothaman et al. 2023). Depletion of HA in the tested pancreatic TEVs resulted in a significant reduction of IL-1β and an increase in CCL4 secretion (Purushothaman et al. 2023). Also, suppression of autologous T cell proliferation was reduced following exposure of monocytes to HA-depleted TEVs (Purushothaman et al. 2023), which taken together may indicate the immunosuppressive role of HA found on TEVs. Production of TNF-α induced by TEVs of pancreatic origin (HPC-4 cell line) was significantly reduced by blocking CD44 molecules on monocytes (Baj-Krzyworzeka et al. 2007). Also, blocking of CD44 reduced the phosphorylation of STAT3, reported previously as an important factor in the transcription of TNF-α, IL-10, and chemokines (Baj-Krzyworzeka et al. 2019). These data are in line with Gärtner's observations that immediate/early response in monocytes, after contact with TEVs, activates the NFκβ pathway (Gärtner et al. 2018). Furthermore, TEVs induced secretion of chemokines CCL2, CCL3, CCL4, CCL5, and IL-8, which previously were described to be induced by HA (McKee et al. 1996; Jiang et al. 2011). Interestingly, secretion of CCL4 and CCL5, but not CCL2, CCL3, and IL-8, induced by TEVs was decreased after the blocking of CD44 molecules (Baj-Krzyworzeka et al. 2019).

The production of IL-10 after stimulation with tumor cells or TEVs was investigated in detail by Lenart and colleagues (Lenart et al. 2017). In this work, HA, TEVs, and HPC-4 cell lines induced the production of IL-10 by monocytes. Much higher production of IL-10 (secretion and mRNA expression) was observed in classical monocytes (Lenart et al. 2017), similarly for HPC-4, TEVs and HA. This observation supports previous reports suggesting that the major producers of TNF-α and IL-12 are $CD16^+$ monocytes whereas the major source of IL-10 after tumor cells or TEVs exposure are $CD16^-$ monocytes (Szaflarska et al. 2004;

Baj-Krzyworzeka et al. 2010). Moreover, the authors showed that treatment of TEVs with Hyal significantly decreased IL-10 production and IL-10 promoter activity in monocytes and that a similar effect was acquired when inhibitors of the PI3K/Akt/mTOR pathway were used. Interestingly, IL-10 expression initiated by the PI3K/Akt/mTOR pathway was reported to be mediated, among others, by STAT3 and IRF-1 transcription factors. In this paper, Lenart et al. elegantly provided evidence that IL-10 production in the classical monocyte subset was mediated via this signaling pathway (Lenart et al. 2017). In this case, however, HA more efficiently induced IL-10 secretion, while the IL-10 gene promoter was stimulated more effectively by TEVs. This may suggest that, apart from HA, TEVs may carry other component(s) capable of inducing the IL-10 gene.

The study by Mytar et al. presented that HA stimulated ROI (reactive oxygen intermediates) production in monocytes as measured by chemiluminescence (Mytar et al. 2001). ROI production was also elevated when monocytes were incubated with supernatant from cancer (but not normal) cell line culture. Moreover, the effect was abolished when culture supernatant was pretreated with Hyal or when monocytes were pretreated with anti-CD44 mAb. These results were one of the first in literature to report the important role of HA in interactions between tumor cells and the immune system. They showed that monocytes preincubated with HA responded to $CD44^{+}$ tumor cell lines by significantly lower production of ROI (Mytar et al. 2001), which may suggest one of the "deactivation" or "exhaustion" mechanisms occurring in cancer (Kuang et al. 2007). Pretreatment of monocytes with HA followed by PMA stimulation did not change monocyte response, which directly pointed to the role of the HA–CD44 axis in tumor cell–monocyte interaction.

Production of ROI by monocytes after stimulation with TEVs derived from pancreatic cell lines (PANC-1 and HPC-4), lung (A549) and colon cancer (DeTa) has been previously described (Baj-Krzyworzeka et al. 2007; Javeed et al. 2017). A direct implication of HA during these interactions has not been confirmed, however, the ROI production after TEVs stimulation has been mainly attributed to the classical subset of monocytes, reflecting their response to the entire tumor cells (Baj-Krzyworzeka et al. 2010). Taken together, CD44 is another surface molecule involved in the triggering of monocyte cytotoxic response to tumor cells.

It should be noted that the presented information should be viewed with caution as tumor cells and TEVs express many ligands (mucins, mannan, phosphatidylserine, etc.) recognized by scavenger receptors or mannose receptors present on monocytes which may, in parallel, trigger cytokines, and ROI production (Mytar et al. 2004).

4.4.1.3 Other Interactions Mediated by HA and/or TEVs-Carried HA

HA has been previously described to assist in tumor cell passage through endothelium (Tammi et al. 2019). It has been also reported that HA facilitates the recruitment of myeloid cells into the tumor bed (Tammi et al. 2019). Furthermore, the mobility of monocytes was shown to be induced by HA via the PKC signaling pathway (Tabata et al. 2007) and inhibited by anti-CD44 mAbs. This phenomenon was described mainly in the inflammatory environment; however, it cannot be excluded from occurring in the tumor bed. LMW-HA induces a variety of inflammatory factors such as TNF-α, IL-1β, CCL3, and CCL4 (Tabata et al. 2007), which may modulate cellular mobility.

In parallel, EVs were reported by many authors to be involved in the induction or enhancement of cell motility, however, little is known about the role of HA in this aspect (Akbar et al. 2017; Sung et al. 2021).

TEVs secreted by colorectal cancer cells were shown to carry mRNAs for HAS2 and HAS3 (Hong et al. 2009) and promoted cell protrusion formation, implying their possible involvement in cell motility. The synthesis of HA on the long protrusions accompanied by the shedding of HA-coated EVs could facilitate HA transport away from the parental cell (Aaltonen et al. 2022). The increasing number of EVs in plasma of colon cancer patients (Silva et al. 2012) and the increasing concentration of HA in patients' tissues (Ropponen et al. 1998) may be a direct consequence of this process. EVs release may result in horizontal transfer and deposition of components for HA synthesis even at sites quite far from the cells of its origin (Rilla et al. 2013).

EVs/TEVs were described to induce chemotaxis of different types of cells, however, there is no direct connection to HA as EVs/TEVs may carry many chemoattractants such as cAMP, chemokines (IL-8) (Baj-Krzyworzeka et al. 2011), fibronectin (Sung et al. 2015), and bioactive lipids (Skotland et al. 2020). EVs may also enhance a migratory response to soluble chemokines (CCL19 and CCL21) by inducing protrusion formation and directional motility toward chemoattractive targets (Sung et al. 2021). Interestingly, the role of EVs/TEVs in cell migration is not limited to paracrine signaling but can also involve an autocrine loop to enhance cell movement (Szatanek et al. 2020).

Apart from the above, EVs/TEVs by carrying components such as MMPs, metalloproteinase inducers or inhibitors, heparanases, Hyals, and aggrecanases may also modulate cell mobility (Nawaz et al. 2018; McAtee et al. 2019).

The available data concerning the pro-angiogenic potential of HA needs reexamination and careful analysis. Spinelli et al. presented that HA derived from breast cancer cells stimulated monocytes to produce angiogenic factors such as VEGF, IL-8, FGF-2, and MMP-2. The profile of the produced chemokines was tumor-type dependent (e.g., breast vs colon) (Spinelli et al. 2019). In this case, LMW-HA did not affect the angiogenic phenotype of monocytes, however, HMW-HA was able to induce an angiogenic potential in these cells, which was associated with the down-regulation of TSG-6 expression in a breast cancer model (Spinelli et al. 2019). Pro-angiogenic activity was described also in the case of HA oligomers (3–25 disaccharide) (Jiang et al. 2011). The ability of HA oligomers to induce proliferation of endothelial cells is being contributed to CD44 and RHAMM binding. In this process, CD44 plays a double role—as a receptor for HA and as a docking receptor for MMP-9, which is important for the activation of TGF-β.

Additionally, TEVs derived from melanoma or colorectal carcinoma cells were shown to stimulate TGF-β production by monocytes (Huber et al. 2018), supporting the claim of the possible involvement in angiogenesis. In accordance with these findings, our group has also reported that TEVs alone have induced angiogenesis by stimulating HUVEC cell proliferation and increased content of hemoglobin in matrigel in NOD-SCID mice. The strongest pro-angiogenic activity was observed in TEVs derived from the A549 lung carcinoma cell line, which correlated with IL-8 carried by TEVs. Contact of monocytes with TEVs resulted in the release of pro-angiogenic chemokines (CXCL8, CCL2, and CCL3) and higher hemoglobin concentration in the matrigel in NOD-SCID mice, however, the role of HA in this process is unclear (Baj-Krzyworzeka et al. 2020).

4.4.2 Macrophages

Macrophages are thought to be a double-edged sword in cancer. Due to their potential to kill tumor cells, mediating antibody-dependent cellular cytotoxicity (ADCC) and phagocytosis, and activating adaptive immune cells, they may act in an antitumor manner (Shapouri-Moghaddam et al. 2018). On the other hand, macrophages contribute to tumor progression by promoting cancer cell proliferation, suppression of pro-inflammatory immune response, and stimulating angiogenesis. The latter functions provided a basis for the characterization of tumor-associated macrophages (TAMs). TAMs are detected in a variety of solid tumors (Mantovani et al. 2002; Ngambenjawong et al. 2017) and they originate from monocytes recruited from circulation, as well as from monocytic precursors, residing in the local environment (Pan et al. 2020).

Macrophages can differentiate from resting, naive M0 macrophages either to M1- or M2-type, depending on the microenvironment signals. The M2-type macrophages can be further divided into M2a-M2d subgroups, depending on the polarization induction stimuli (Lavin et al. 2015). M1 macrophages resemble pro-inflammatory potential. In response to bacterial infection or IFN-γ stimulation, M1 macrophages mediate phagocytosis and killing of pathogens or target cells, as well as the production of nitric oxide synthase (iNOS), ROI and pro-inflammatory cytokines, including IL-12 (Zhang et al. 2019; Paul et al. 2020; Duan et al. 2021). M2 macrophage functions include tissue remodeling by promoting angiogenesis and debris removal (Ngambenjawong et al. 2017), and these activities are associated with upregulated expression of anti-inflammatory cytokines, such as IL-10, as well as tissue growth factors, including VEGF and matrix metalloproteinases (MMPs) (Tu et al. 2023). M1 macrophages possess antitumor activity, which is manifested by killing of malignant cells by direct cytotoxicity through the release of NO and ROI, or via ADCC, which requires tumor cell recognition by antibodies (Taniyama and Holden 1980; Ralph and Nakoinz 1984). Other nomenclature divides macrophages into classically activated macrophages, regulatory macrophages, and wound-healing macrophages (Mosser and Zhang 2008). Importantly, both classifications are largely generalized as macrophages phenotype shows high plasticity. Being able to sense a variety of environmental stimuli, macrophages' phenotype and function might share characteristics of more than one specific population (Mosser and Zhang 2008; Mosser et al. 2021).

TAMs resemble M2 macrophages, as they produce a number of cytokines stimulating tumor cell proliferation and survival, such as IL-6, IL-10, epithelial growth factor (EGF), platelet-derived growth factor (PDGF), and TGFβ (Galizia et al. 2002; Yadav et al. 2011; Yin et al. 2016). The other mechanism promoting tumor growth and metastases is associated with the destruction of matrix membrane by enzymes secreted by M2 macrophages, e.g., serine proteases or matrix metalloproteinases (Najafi et al. 2019; Qin et al. 2022). TAMs act pro-angiogenic, and thus promoting abnormal blood vessel formation within the tumor, by secretion of a number of factors, i.e., VEGF, EGF, angiopoietins, plasminogen activator, COX-2, osteopontin, and osteonectin (Fu et al. 2020; Larionova et al. 2021). TAMs were also shown to inhibit the killing of tumor cells by NK cell-mediated ADCC and T cell-mediated cytotoxicity (Su et al. 2018), stimulate immune checkpoint receptors on T cells (Sumitomo et al. 2019), and recruit regulatory T cells (Tregs) (Kouketsu et al. 2019), which further suppressed antitumor immune response. It needs to be pointed out that TAMs show tissue-specific transcriptome, that is different from those of blood monocytes and tissue-resident macrophages (Cassetta et al. 2019).

LMW-HA, sulfated HA derivatives and CD44 are important regulators in TME, which was shown to impact macrophage polarization and pro-tumorigenic TAMs activity. In phorbol 12-myristate 13-acetate (PMA)-induced THP-1 cell line model, CD44 was shown to be involved in monocyte differentiation. PMA stimulated CD44 expression and clustering via ERK1/2 signaling pathway, while blocking of CD44 inhibited CD14 expression (Zhang et al. 2014). Recently, in a breast cancer model, cancer cells that highly expressed CD44 formed clusters accompanied by $CD206^+$ TAMs. It was also detected that CD44 expression in cancer cell clusters is stimulated by MDM2/p53 pathway activation by CCL8 secreted from TAMs (Gao et al. 2022). Moreover, CD44 expression level in bladder cancer was associated with increased infiltration of $CD68^+$ macrophages, mostly of M2 phenotype. CD44 was also shown to upregulate PD-L1 expression at the mRNA level, suggesting a possible linkage between HA and immune checkpoint phenomenon (Liu et al. 2023).

HA-induced M2 macrophage differentiation was shown in a model using artificial extracellular matrix (aECM) composed of collagen I and HA or sulfated HA derivatives. Monocytes treated with aECM released IL-10 and upregulated CD163 expression, suggesting M2 macrophage polarization (Kajahn et al. 2012). The number of M2 macrophages highly correlated with HA levels in tumor samples from breast cancer patients. Breast cancer-derived HA stimulated TAMs development via HA-CD44-ERK1/2-STAT3 pathway (Zhang et al. 2016). Other studies showed that elevated numbers of TAMs correlated with an increased CD44 expression on stromal cells, high HA levels, and HA synthases in the tumor samples (Tiainen et al. 2020). Expression of HAS2 in tumor cells has been associated with their ability to induce macrophage dysfunctions (Kuang et al. 2007), while *Has2* gene-deficient fibroblasts have severely impaired the recruitment of macrophages toward tumor cells (Kuang et al. 2007; Kobayashi et al. 2010). HA deficiency limited tumor angiogenesis (Kobayashi et al. 2010). Recent data showed that TAMs exhibit high levels of *Hyal1* and *Hyal2*, suggesting macrophages' role in degradation of HA in the TME. In vitro studies showed that macrophages treated with mammary tumor cell conditioned media increased Hyal activity (Witschen et al. 2023). Blocking of HA synthesis or HA binding to CD44 on macrophages increased M1 macrophage proportion in glioblastoma model. This phenomenon was associated with upregulation of signal regulatory protein α (SIRPα) expression in macrophages via STAT1 phosphorylation (Yan et al. 2022). TAMs can also express lymphatic vessel endothelial hyaluronan receptor-1 (LYVE-1). In a murine model of spontaneous breast cancer, LYVE-1$^+$ TAMs formed coordinated multicellular "nests" characterized by immunosuppression and resistance to chemotherapy. Blocking of the LYVE-1$^+$ TAMs resulted in cytotoxic T-cell recruitment and induction of antitumor immune response (Anstee et al. 2023). The receptor for HA-mediated motility (RHAMM) on macrophages has been linked to chemotaxis in response to TGF-β (Foley et al. 2012) yet its role in TAMs is still unknown.

HA is also transported by TEVs, which are widely considered to modulate macrophage phenotype and functions in the TME, affecting their polarization and cytokine secretion profiles (Fig. 4.5). For example, colorectal TEVs stimulated M2

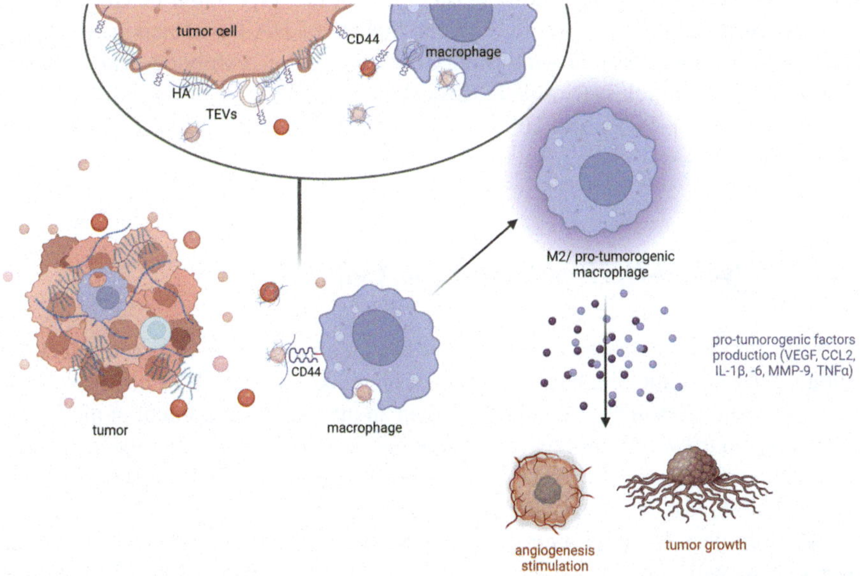

Fig. 4.5 Macrophage polarization initiated after contact with TEVs-carried HA. Macrophages infiltrate the tumor bed and upon contact with tumor cells may internalize TEVs (via phagocytosis and other types of endocytosis). The presence of HA in TEVs may facilitate polarization of macrophages into the pro-tumorigenic phenotype (M2) characterized by the secretion of factors inducing cell proliferation and angiogenesis (MMP-9—metalloproteinase -9; IL-10—Interleukin-10; TNF-α—Tumor Necrosis Factor α; VEGF—Vascular Endothelial Growth Factor; CCL2—chemokine (C-C motif) ligand 2 also referred to as Monocyte-Chemotactic Protein-1, MCP-1), which may result in further tumor progression

macrophage polarization by activating the PI3K/AKT signaling pathway and down-regulating PTEN expression (Zhao et al. 2020). Similarly, TEVs induced M2 macrophage polarization in canine melanoma (Kim et al. 2022), as well as in human hepatocellular carcinoma (Zhou et al. 2022). In the latter, TEVs transmitted long noncoding RNA (lncRNA) prostate androgen-regulated transcript 1 (PART1), which targeted miR-372-3p in macrophages, resulting in the overexpression of TLR4 (Zhou et al. 2022). TEVs isolated from ascites-derived human pancreatic ductal adenocarcinoma cell line polarized macrophages toward M2-like phenotype ($CD14^{hi}CD163^{hi}CD206^{hi}$) and enhanced the secretion of pro-tumorigenic cytokines, including VEGF, CCL-2, IL-6, IL-1β, MMP-9, and TNF-α (Linton et al. 2018). Interestingly, TAMs were shown to deliver and transfer the TEVs content. Macrophages that had absorbed TEVs showed increased invasiveness and dissemination capacity. Upon contact with stromal cells, macrophages released membrane blebs containing TEVs, which resulted in cancer-derived proteins and RNA, such as TGF-β, activated Src, Wnt3, and HIF1α, transmission into the surrounding stromal cells (Umakoshi et al. 2019).

In the mouse model of skin antiaging studies, stem cell-derived EVs bearing HA-based hydrogels induced macrophage polarization toward CD301b-positive population, which was shown to play a pivotal role in skin regeneration (You et al. 2022). In the study by Rios de la Rosa et al., higher expression of CD44 on M1 macrophages resulted in a more efficient capture of HA-based nanoparticles, yet M2 macrophages, although with lower CD44 expression, exhibited better internalization of such particles (Rios de la Rosa et al. 2017).

The impact of TEVs and HA carried by TEVs in macrophage polarization is summarized in Fig. 4.5.

4.4.3 Myeloid-Derived Suppressor Cell (MDSCs)

In solid tumors, hematopoiesis is generally impaired. This is due to the fact that tumor-derived factors, including cytokines, chemokines, and TEVs not only induce the "emergency" hematopoiesis, resulting in an increased recruitment of monocytes from bone marrow to tumor location (Wu et al. 2014), but also favor the acquisition of immunosuppressive features of myeloid cells (Canè et al. 2019; Ugel et al. 2021; Long et al. 2022).

Myeloid-derived suppressor cells (MDSCs) compose a heterogeneous population of cells with suppressive activity. Their role has been widely described in infections, chronic inflammation, and cancers (Veglia et al. 2021). Being most extensively studied in oncological disorders, these cells, in conjunction with other mechanisms, propagate cancer development, progression (Umansky et al. 2019) and tumor recurrence (Siemińska et al. 2022). The mode of MDSCs action covers several direct and indirect effects on other immune cells, targeting mainly T effector lymphocytes. These include, among others, the expression of "checkpoint" molecules, e.g., PD-L1 and VISTA on their surface, activity of arginase-1 (ARG-1), iNOS, and indoleamine 2, 3-dioxygenase (IDO) production and secretion of immunosuppressive cytokines, e.g., IL-10 and TGF-β (Tyrinova et al. 2022).

Despite almost two decades of research on MDSCs (Gabrilovich et al. 2007), the aspects of their development, differentiation, and activation are still not fully elucidated. Up to now, three MDSCs subsets, differing in their origin, have been identified, namely: monocytic (Mo-MDSCs), granulocytic (PMN-MDSCs) and, the so called, "early-stage" MDSCs, which are believed to have the ability to differentiate into the two other types. Understanding the interactions between MDSCs and HA is relevant, especially in the context of cancer, however, there are only a few studies covering this topic. These studies show that MDSCs express receptors for HA and Hyal2. For example, in the study by Dominguez-Gutierrez, Hyal2 was confirmed in Mo-MDSCs detected in peripheral blood of patients with bladder cancer together with the co-expression of CD44 (Dominguez-Gutierrez et al. 2021). However, the cell phenotype used in this work to define MDSCs ($CD11b^+CD33^+CCR2^+M\text{-}CSF^+GM\text{-}CSFR^+$ and $HLA\text{-}DR^+$) seems to identify a subset of monocytes rather than MDSCs, as the latter are characterized by lack or low expression of HLA-DR molecules. In another study concerning kidney cancer, the presence of $CD33^+CD11b^+$ MDSCs with the co-expression of PD-L1 and Hyal2 was documented in peripheral blood alongside the detection of $PD\text{-}L1^+HLA\text{-}DR^+$ macrophage-like or dendritic cell-like antigen-presenting cells in the tumor stroma (Kusmartsev et al. 2022). This observation may suggest the possible transition of tumor-recruited $Hyal2^+$ MDSCs from peripheral blood to TME, where they differentiate into $PD\text{-}L1^+$ TAMs (Donelan et al. 2022). This scenario is consistent with the theory that MDSCs are just in a transition state to other tumor-associated myeloid cells, e.g., TAMs (Hegde et al. 2021).

TLR2, another HA receptor, may play a dual role in MDSCs functioning, depending on the MDSCs subset. On one side, its expression on PMN-MDSCs may be associated with their lower suppressive activity, while Mo-MDSCs expressing TLR2 may become highly immunosuppressive (Fu et al. 2021). Interestingly, this observation was dependent on the type of breast cancer (human 4T1 and mouse 4T07, or EMT6) cells used in the study. The activation of TLR2 on Mo-MDSCs was shown to increase their viability and potential to differentiate into macrophages (Shime et al. 2017).

Furthermore, the elevated presence of HA in tumor tissues, along with the recruitment of cells expressing HA receptors and enzymes like Hyal2 within MDSCs, contributes to the subsequent degradation of tumor-associated HA. This enzymatic breakdown results in the formation of smaller HA fragments with a lower molecular weight, thereby fostering conditions conducive to cancer-related inflammation and angiogenesis (Dominguez-Gutierrez et al. 2022).

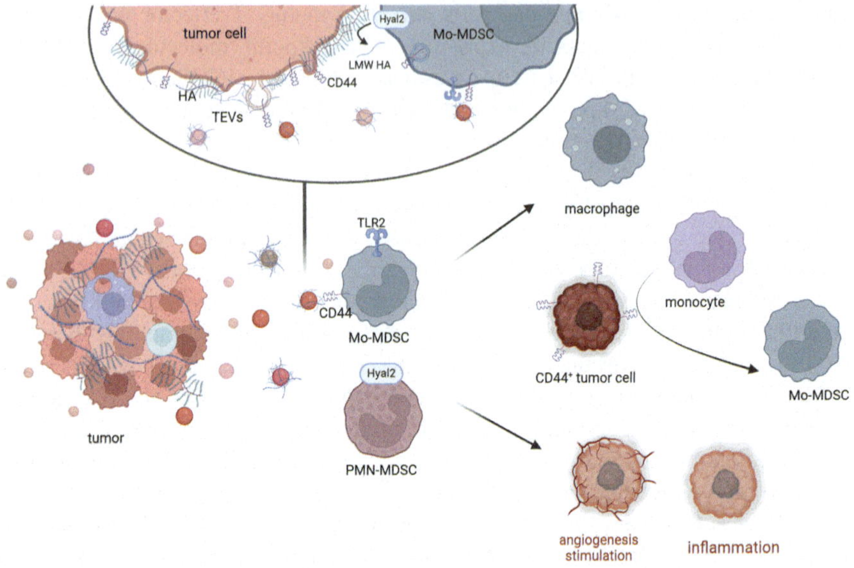

Fig. 4.6 Possible interactions of MDSC with HA and TEVs-carried HA. MDSCs (myeloid-derived suppressor cells) of monocytic and polymorphonuclear origins (Mo-MDSCs and PMN-MDSCc) express receptors for HA (CD44 and TLR2) and Hyal2 (hyaluronidase 2), which may play a role in interactions with cancer cells. First, tumor cells via CD44 may induce the differentiation of monocytes to Mo-MDSC. In parallel, activation of TLR2 facilitates Mo-MDSC differentiation into macrophages. Due to the expression of Hyal2, MDSCs contribute to the degradation of tumor-associated HA into low molecular weight HA (LMW-HA) fragments, which supports cancer-related inflammation and angiogenesis. The direct impact of HA and TEVs in these processes is unknown

Data on the expression of CD44 on MDSCs is limited. More is known about the role of CD44 in MDSCs functioning. In this context, it was already shown that its expression on activated hepatic stellate cells in contact with monocytes causes trans-differentiation of the latter to MDSCs of monocytic origin (Höchst et al. 2013). This role of CD44 was also confirmed by other studies showing that $CD44^{+}$ cancer stem cells isolated from head and neck squamous cell carcinoma were more efficient in the induction of regulatory cells, including MDSCs, than were $CD44^{-}$ cells (Chikamatsu et al. 2011). Currently, there are no reports considering the involvement of TEVs-carried HA or HA receptors in MDSCs crosstalk, however, their possible interactions are presented in Fig. 4.6.

4.5 Conclusion and Future Perspectives

Given the pro-tumorigenic role of HA, TEVs, and TEVs-carried HA, HA-targeted therapies represent a potentially promising approach to cancer treatment. Therapeutic strategies can be various, such as inhibiting HA synthesis, depletion of stromal HA, or blocking of HA receptors and HA-mediated signaling (Jordan et al. 2015; Sato et al. 2016; Wessels et al. 2017; Ye et al. 2020). Another method takes advantage of the strong association between HA and CD44 receptor to directly deliver synthetic nanoparticles coated with HA, which are loaded with chemotherapeutics to cancer cells (Yang et al. 2019; Mahira et al. 2019; Li et al. 2019). Interestingly, there are also attempts to utilize HA-coated, milk-derived exosomes loaded with potential drugs to treat liver fibrosis (Gong et al. 2023). However, it is important to realize that HA manipulation and/or blockage therapy may also act as a double-edged sword, affecting both the pro-inflammatory and antitumorigenic activities of cells of monocyte origin. Therefore, a more detailed understanding of the possible impact associated with the interference in this HA-receptor binding process is needed, to fully address its influence on myeloid cells.

The association between HA molecular weight and cellular functions plays a relevant role, especially in the context of cancer (Bourguignon et al. 2014). Since tumor progression and invasion were correlated with LMW-HA (Chen et al. 2018), specific molecular characterization of HA content and size in the TME may offer insights into the mechanisms driving cancer aggressiveness. Moreover, the mechanisms of verifying HA molecular weight by myeloid cells remain unclear. Recent studies suggest that binding of HA to CD44, regardless of its size, is unlikely to be the sole source of the differential cell response (Kim and Owen 2020). Thus, understanding these mechanisms and predicting the size of HA that binds to CD44 can contribute to the development of targeted, anticancer therapies. The current, widespread usage of HA-based therapeutics, i.e., in orthopedy or cosmetology, only confirms the need for caution since different HA sizes/molecular weights may exert unwanted responses from the immune cells.

Moreover, HA receptors, e.g., CD44, might become a promising marker in disease monitoring or potential targets in antitumor immunotherapy approaches (Xu et al. 2020). A number of studies show that CD44 may be a predictor of cancer patient outcomes, including gastric cancer, colorectal cancer, and endometrial cancer (Szczepanik et al. 2019; Sadeghi et al. 2019; Klarić et al. 2019). CD44 inhibitors were recently introduced into clinical trials of ovarian epithelium cancer therapy, with promising initial results (Girda et al. 2022).

Finally, in this work we have provided an insight into the interactions between monocyte and TEVs-carried HA and their influence on monocyte functions. Macrophage and MDSC response to TEVs and TEVs-carried HA is still largely unknown. An overview of their interactions is presented in Fig. 4.7. Thus, further studies are needed to provide a better understanding of the interactions between immune cells of monocytic origin and TEVs.

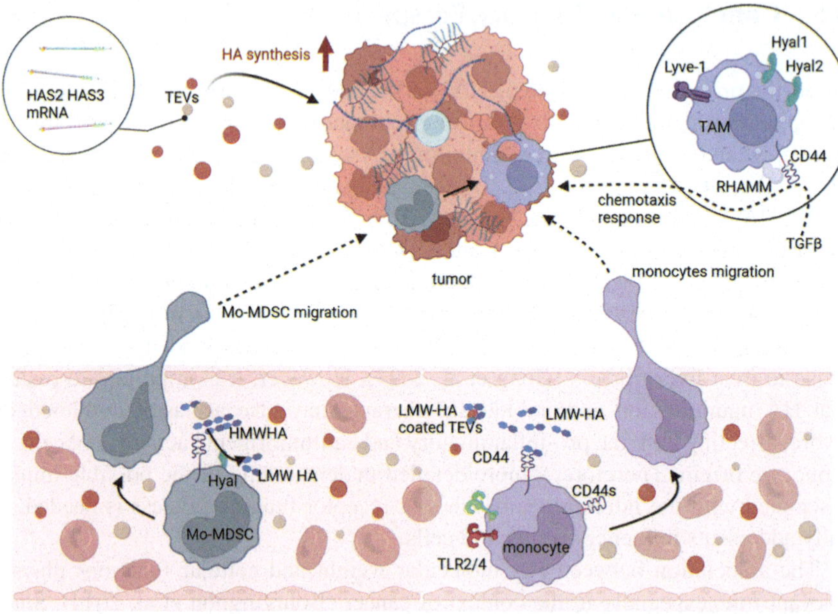

Fig. 4.7 Schematic representation of interactions between cells of monocytic origin and tumor/tumor derivatives. Monocytes (Mo) and monocytic myeloid-derived suppressor cells (Mo-MDSC) represent two monocytic populations of cells circulating in the blood vessels. These cells are exposed to tumor derivates, including hyaluronan (HA) in its soluble form (low molecular weight HA, LMW-HA, and high molecular weight HA, HMW-HA) directly or carried by Tumor-derived Extracellular Vesicles (TEVs). Interactions of HA with these cells may occur via CD44 (mainly in standard CD44s form) or other (TLR2, TLR4, RHAMM, and Lyve-1) receptors. Contact with HA may induce migration toward the tumor site and modulate the functional activity of monocytic cells. Upon arrival at the tumor site, Mo and Mo-MDSC differentiate into Tumor-Associated Macrophages (TAM). TAM are mostly considered to be pro-tumorigenic, however, their function strongly depends on the tumor type. TEVs are universal transporters of macromolecules including HA and mRNA for HA synthases (HAS2 and HAS3), which may support HA synthesis by tumor cells. Cell expression of hyaluronidases is indicated by Hyal1 and Hyal2

Acknowledgment This work was supported by the National Science Centre, Poland, grant number 2019/33/B/NZ5/00647.

Compliance with Ethical Standards This manuscript adheres to the highest ethical standards in all stages of its development and execution. The authorship was determined in accordance with the proper guidelines. All authors made substantial intellectual contributions to the manuscript, and the order of authorship reflects the degree of their contributions. The manuscript complies with the principles of responsible and ethical publishing. All sources of funding and support are acknowledged, and proper citation practices have been followed. The authors declare that the work was conducted in the absence of any commercial or financial relationships that could be construed as a potential conflict of interest. All presented figures were created with BioRender.com.

References

Aaltonen N, Kyykallio H, Tollis S, Capra J, Hartikainen JM, Matilainen J, Oikari S, Rilla K (2022) MCF10CA breast cancer cells utilize hyaluronan-coated EV-rich trails for coordinated migration. Front Oncol 12:869417. https://doi.org/10.3389/fonc.2022.869417

Abatangelo G, Vindigni V, Avruscio G, Pandis L, Brun P (2020) Hyaluronic acid: redefining its role. Cells 9(7):1743. https://doi.org/10.3390/cells9071743

Adib-Conquy M, Cavaillon J-M (2002) Gamma interferon and granulocyte/monocyte colony-stimulating factor prevent endotoxin tolerance in human monocytes by promoting interleukin-1 receptor-associated kinase expression and its association to MyD88 and not by modulating TLR4 expression. J Biol Chem 277(31):27927–27934. https://doi.org/10.1074/jbc.M200705200

Akbar N, Digby JE, Cahill TJ, Tavare AN, Corbin AL, Saluja S, Dawkins S, Edgar L, Rawlings N, Ziberna K, McNeill E, Johnson E, Aljabali AA, Dragovic RA, Rohling M, Belgard TG, Udalova IA, Greaves DR, Channon KM, Riley PR, Anthony DC, Choudhury RP (2017) Endothelium-derived extracellular vesicles promote splenic monocyte mobilization in myocardial infarction. JCI Insight 2(17). https://doi.org/10.1172/jci.insight.93344

Al-Nedawi K, Meehan B, Micallef J, Lhotak V, May L, Guha A, Rak J (2008) Intercellular transfer of the oncogenic receptor EGFRvIII by microvesicles derived from tumour cells. Nat Cell Biol 10(5):619–624. https://doi.org/10.1038/ncb1725

Anstee JE, Feehan KT, Opzoomer JW, Dean I, Muller HP, Bahri M, Cheung TS, Liakath-Ali K, Liu Z, Choy D, Caron J, Sosnowska D, Beatson R, Muliaditan T, An Z, Gillett CE, Lan G, Zou X, Watt FM, Ng T, Burchell JM, Kordasti S, Withers DR, Lawrence T, Arnold JN (2023) LYVE-1+ macrophages form a collaborative CCR5-dependent perivascular niche that influences chemotherapy responses in murine breast cancer. Dev Cell 58(17):1548–1561.e10. https://doi.org/10.1016/j.devcel.2023.06.006

Babula A, Gałuszka-Bulaga A, Węglarczyk K, Siedlar M, Baj-Krzyworzeka M (2023) CD44-hyaluronan axis plays a role in the interactions between colon cancer-derived extracellular vesicles and human monocytes. Oncol Lett 26(3):413. https://doi.org/10.3892/ol.2023.13999

Baj-Krzyworzeka M, Szatanek R, Weglarczyk K, Baran J, Urbanowicz B, Brański P, Ratajczak MZ, Zembala M (2006) Tumour-derived microvesicles carry several surface determinants and mRNA of tumour cells and transfer some of these determinants to monocytes. Cancer Immunol Immunother 55(7):808–818. https://doi.org/10.1007/s00262-005-0075-9

Baj-Krzyworzeka M, Szatanek R, Weglarczyk K, Baran J, Zembala M (2007) Tumour-derived microvesicles modulate biological activity of human monocytes. Immunol Lett 113(2):76–82. https://doi.org/10.1016/j.imlet.2007.07.014

Baj-Krzyworzeka M, Baran J, Weglarczyk K, Szatanek R, Szaflarska A, Siedlar M, Zembala M (2010) Tumour-derived microvesicles (TMV) mimic the effect of tumour cells on monocyte subpopulations. Anticancer Res 30(9):3515–3519

Baj-Krzyworzeka M, Weglarczyk K, Mytar B, Szatanek R, Baran J, Zembala M (2011) Tumour-derived microvesicles contain interleukin-8 and modulate production of chemokines by human monocytes. Anticancer Res 31(4):1329–1335

Baj-Krzyworzeka M, Weglarczyk K, Szatanek R, Mytar B, Baran J, Siedlar M (2019) The role of CD44H molecule in the interactions between human monocytes and pancreatic adenocarcinoma-derived microvesicles. Folia Histochem Cytobiol 57(1):28–34. https://doi.org/10.5603/FHC.a2019.0005

Baj-Krzyworzeka M, Mytar B, Weglarczyk K, Szatanek R, Kijowski J, Siedlar M (2020) Protumorogenic potential of pancreatic adenocarcinoma-derived extracellular vesicles. Folia Biol (Praha) 66(3):104–110

Beyer M, Mallmann MR, Xue J, Staratschek-Jox A, Vorholt D, Krebs W, Sommer D, Sander J, Mertens C, Nino-Castro A, Schmidt SV, Schultze JL (2012) High-resolution transcriptome of human macrophages. PLoS One 7(9):e45466. https://doi.org/10.1371/journal.pone.0045466

Bourcier S, Sansonetti A, Durand L, Chomienne C, Robert-Lézénés J, Smadja-Joffe F (2010) CD44-ligation induces, through ERK1/2 pathway, synthesis of cytokines TNF-α and IL-6 required for differentiation of THP-1 monoblastic leukemia cells. Leukemia 24(7):1372–1375. https://doi.org/10.1038/leu.2010.100

Bourguignon LYW, Shiina M, Li J-J (2014) Hyaluronan-CD44 interaction promotes oncogenic signaling, microRNA functions, chemoresistance, and radiation resistance in cancer stem cells leading to tumor progression. Adv Cancer Res 123:255–275. https://doi.org/10.1016/B978-0-12-800092-2.00010-1

Braumüller H, Gansauge S, Ramadani M, Gansauge F (2000) CD44v6 cell surface expression is a common feature of macrophages and macrophage-like cells - implication for a natural macrophage extravasation mechanism mimicked by tumor cells. FEBS Lett 476(3):240–247. https://doi.org/10.1016/s0014-5793(00)01737-3

Briand J, Garnier D, Nadaradjane A, Clément-Colmou K, Potiron V, Supiot S, Bougras-Cartron G, Frenel J-S, Heymann D, Vallette FM, Cartron P-F (2020) Radiotherapy-induced overexpression of exosomal miRNA-378a-3p in cancer cells limits natural killer cells cytotoxicity. Epigenomics 12(5):397–408. https://doi.org/10.2217/epi-2019-0193

Brown RL, Reinke LM, Damerow MS, Perez D, Chodosh LA, Yang J, Cheng C (2011) CD44 splice isoform switching in human and mouse epithelium is essential for epithelial-mesenchymal transition and breast cancer progression. J Clin Invest 121(3):1064–1074. https://doi.org/10.1172/JCI44540

Buzás EI, Tóth EÁ, Sódar BW, Szabó-Taylor KÉ (2018) Molecular interactions at the surface of extracellular vesicles. Semin Immunopathol 40(5):453–464. https://doi.org/10.1007/s00281-018-0682-0

Canè S, Ugel S, Trovato R, Marigo I, De Sanctis F, Sartoris S, Bronte V (2019) The endless saga of monocyte diversity. Front Immunol 10

Cassetta L, Fragkogianni S, Sims AH, Swierczak A, Forrester LM, Zhang H, Soong DYH, Cotechini T, Anur P, Lin EY, Fidanza A, Lopez-Yrigoyen M, Millar MR, Urman A, Ai Z, Spellman PT, Hwang ES, Dixon JM, Wiechmann L, Coussens LM, Smith HO, Pollard JW (2019) Human tumor-associated macrophage and monocyte transcriptional landscapes reveal cancer-specific reprogramming, biomarkers, and therapeutic targets. Cancer Cell 35(4):588–602.e10. https://doi.org/10.1016/j.ccell.2019.02.009

Chen J-WE, Pedron S, Shyu P, Hu Y, Sarkaria JN, Harley BAC (2018) Influence of hyaluronic acid transitions in tumor microenvironment on glioblastoma malignancy and invasive behavior. Front Mater 5:39. https://doi.org/10.3389/fmats.2018.00039

Chen K-L, Li D, Lu T-X, Chang S-W (2020) Structural characterization of the CD44 stem region for standard and cancer-associated isoforms. Int J Mol Sci 21(1):336. https://doi.org/10.3390/ijms21010336

Chen QY, Gao B, Tong D, Huang C (2023a) Crosstalk between extracellular vesicles and tumor-associated macrophage in the tumor microenvironment. Cancer Lett 552:215979. https://doi.org/10.1016/j.canlet.2022.215979

Chen X, Li Y, Xia H, Chen YH (2023b) Monocytes in tumorigenesis and tumor immunotherapy. Cells 12(13):1673. https://doi.org/10.3390/cells12131673

Chikamatsu K, Takahashi G, Sakakura K, Ferrone S, Masuyama K (2011) Immunoregulatory properties of CD44+ cancer stem-like cells in squamous cell carcinoma of the head and neck. Head Neck 33(2):208–215. https://doi.org/10.1002/hed.21420

Choi BH, Quan YH, Rho J, Hong S, Park Y, Choi Y, Park J-H, Yong HS, Han KN, Choi YH, Kim HK (2020) Levels of extracellular vesicles in pulmonary and peripheral blood correlate with stages of lung cancer patients. World J Surg 44(10):3522–3529. https://doi.org/10.1007/s00268-020-05630-y

Clancy J, Sedgwick A, Rosse C, Muralidharan-Chari V, Raposo G, Method M, Chavrier P, D'Souza-Schorey C (2015) Regulated delivery of molecular cargo to invasive tumor-derived microvesicles. Nat Commun 6. https://doi.org/10.1038/ncomms7919

Cormican S, Griffin MD (2020) Human monocyte subset distinctions and function: insights from gene expression analysis. Front Immunol 11

Culty M, Nguyen HA, Underhill CB (1992) The hyaluronan receptor (CD44) participates in the uptake and degradation of hyaluronan. J Cell Biol 116(4):1055–1062. https://doi.org/10.1083/jcb.116.4.1055

Day AJ, de la Motte CA (2005) Hyaluronan cross-linking: a protective mechanism in inflammation? Trends Immunol 26(12):637–643. https://doi.org/10.1016/j.it.2005.09.009

de la Motte CA, Hascall VC, Drazba J, Bandyopadhyay SK, Strong SA (2003) Mononuclear leukocytes bind to specific hyaluronan structures on colon mucosal smooth muscle cells treated with polyinosinic acid:polycytidylic acid: inter-α-trypsin inhibitor is crucial to structure and function. Am J Pathol 163(1):121–133. https://doi.org/10.1016/S0002-9440(10)63636-X

de Vrij J, Maas SLN, Kwappenberg KMC, Schnoor R, Kleijn A, Dekker L, Luider TM, de Witte LD, Litjens M, van Strien ME, Hol EM, Kroonen J, Robe PA, Lamfers ML, Schilham MW, Broekman MLD (2015) Glioblastoma-derived extracellular vesicles modify the phenotype of monocytic cells. Int J Cancer 137(7):1630–1642. https://doi.org/10.1002/ijc.29521

Depciuch J, Klębowski B, Stec M, Szatanek R, Węglarczyk K, Baj-Krzyworzeka M, Parlińska-Wojtan M, Baran J (2020) Similarities in the general chemical composition of colon cancer cells and their microvesicles investigated by spectroscopic methods-potential clinical relevance. Int J Mol Sci 21(5):1826. https://doi.org/10.3390/ijms21051826

Dicker KT, Gurski LA, Pradhan-Bhatt S, Witt RL, Farach-Carson MC, Jia X (2014) Hyaluronan: a simple polysaccharide with diverse biological functions. Acta Biomater 10(4):1558–1570. https://doi.org/10.1016/j.actbio.2013.12.019

Domanegg K, Sleeman JP, Schmaus A (2022) CEMIP, a promising biomarker that promotes the progression and metastasis of colorectal and other types of cancer. Cancers (Basel) 14(20):5093. https://doi.org/10.3390/cancers14205093

Dominguez-Gutierrez PR, Kwenda EP, Donelan W, O'Malley P, Crispen PL, Kusmartsev S (2021) Hyal2 expression in tumor-associated myeloid cells mediates cancer-related inflammation in bladder cancer. Cancer Res 81(3):648–657. https://doi.org/10.1158/0008-5472.CAN-20-1144

Dominguez-Gutierrez PR, Kwenda EP, Donelan W, Miranda M, Doty A, O'Malley P, Crispen PL, Kusmartsev S (2022) Detection of PD-L1-expressing myeloid cell clusters in the hyaluronan-enriched stroma in tumor tissue and tumor-draining lymph nodes. J Immunol 208(12):2829–2836. https://doi.org/10.4049/jimmunol.2100026

Donelan W, Dominguez-Gutierrez PR, Kusmartsev S (2022) Deregulated hyaluronan metabolism in the tumor microenvironment drives cancer inflammation and tumor-associated immune suppression. Front Immunol 13:971278. https://doi.org/10.3389/fimmu.2022.971278. PMID: 36238286; PMCID: PMC9550864

Duan B, Shao L, Liu R, Msuthwana P, Hu J, Wang C (2021) Lactobacillus rhamnosus GG defense against Salmonella enterica serovar Typhimurium infection through modulation of M1 macrophage polarization. Microb Pathog 156:104939. https://doi.org/10.1016/j.micpath.2021.104939

Eljaszewicz A, Jankowski M, Gackowska L, Helmin-Basa A, Wiese M, Kubiszewska I, Kaszewski W, Michalkiewicz J, Zegarski W (2013) Clinical immunology. Gastric cancer increase the percentage of intermediate (CD14++CD16+) and nonclassical (CD14+CD16+) monocytes. Cent Eur J Immunol 37(4):355–361. https://doi.org/10.5114/ceji.2012.32725

Fingerle G, Pforte A, Passlick B, Blumenstein M, Ströbel M, Ziegler-Heitbrock HW (1993) The novel subset of CD14+/CD16+ blood monocytes is expanded in sepsis patients. Blood 82(10):3170–3176

Foley JP, Lam D, Jiang H, Liao J, Cheong N, McDevitt TM, Zaman A, Wright JR, Savani RC (2012). Toll-like receptor 2 (TLR2), transforming growth factor-β, hyaluronan (HA), and receptor for HA-mediated motility (RHAMM) are required for surfactant protein A-stimulated macrophage chemotaxis) J Biol Chem 287(44):37406–37419. https://doi.org/10.1074/jbc.M112.360982

Fu L-Q, Du W-L, Cai M-H, Yao J-Y, Zhao Y-Y, Mou X-Z (2020) The roles of tumor-associated macrophages in tumor angiogenesis and metastasis. Cell Immunol 353:104119. https://doi.org/10.1016/j.cellimm.2020.104119

Fu C, Fu Z, Jiang C, Xia C, Zhang Y, Gu X, Zheng K, Zhou D, Tang S, Lyu S, Ma S (2021) CD205+ polymorphonuclear myeloid-derived suppressor cells suppress antitumor immunity by overexpressing GLUT3. Cancer Sci 112(3):1011–1025. https://doi.org/10.1111/cas.14783

Fujita N, Yaegashi N, Ide Y, Sato S, Nakamura M, Ishiwata I, Yajima A (1994) Expression of CD44 in normal human versus tumor endometrial tissues: possible implication of reduced expression of CD44 in lymph-vascular space involvement of cancer cells. Cancer Res 54(14):3922–3928

Gabrilovich DI, Bronte V, Chen S-H, Colombo MP, Ochoa A, Ostrand-Rosenberg S, Schreiber H (2007) The terminology issue for myeloid-derived suppressor cells. Cancer Res 67(1):425–426. https://doi.org/10.1158/0008-5472.CAN-06-3037

Galizia G, Orditura M, Romano C, Lieto E, Castellano P, Pelosio L, Imperatore V, Catalano G, Pignatelli C, De Vita F (2002) Prognostic significance of circulating IL-10 and IL-6 serum levels in colon cancer patients undergoing surgery. Clin Immunol 102(2):169–178. https://doi.org/10.1006/clim.2001.5163

Gamperl H, Plattfaut C, Freund A, Quecke T, Theophil F, Gieseler F (2016) Extracellular vesicles from malignant effusions induce tumor cell migration: inhibitory effect of LMWH tinzaparin. Cell Biol Int 40(10):1050–1061. https://doi.org/10.1002/cbin.10645

Gao F, Zhang G, Liu Y, He Y, Sheng Y, Sun X, Du Y, Yang C (2022) Activation of CD44 signaling in leader cells induced by tumor-associated macrophages drives collective detachment in luminal breast carcinomas. Cell Death Dis 13(6):1–14. https://doi.org/10.1038/s41419-022-04986-4

Gärtner K, Battke C, Dünzkofer J, Hüls C, von Neubeck B, Kellner M-K, Fiestas E, Fackler S, Lang S, Zeidler R (2018) Tumor-derived extracellular vesicles activate primary monocytes. Cancer Med 7(5):2013–2020. https://doi.org/10.1002/cam4.1465

Girard N, Maingonnat C, Bertrand P, Tilly H, Vannier J-P, Delpech B (2002) Human monocytes synthesize hyaluronidase. Br J Haematol 119(1):199–203. https://doi.org/10.1046/j.1365-2141.2002.03733.x

Girda E, Hou J, Nelson D, Finlayson M, de Meritens AB, Chekmareiva M, Leiser A, Song M, Stephenson R, Chan N, Tergas AI, Vattakalam R, Wright JD, Yu H, Martincuks A, Kohut A, Palmer J, Rodriguez-Rodriguez L (2022) Phase I trial of daily subcutaneous SPL-108 injections in combination with paclitaxel in patients with platinum resistant CD44+ advanced ovarian epithelial cancer. Int J Gynecol Cancer 32(8):1032–1038. https://doi.org/10.1136/ijgc-2021-003316

Gong L, Zhou H, Zhang S, Wang C, Fu K, Ma C, Zhang Y, Peng C, Li Y (2023) CD44-targeting drug delivery system of exosomes loading forsythiaside A combats liver fibrosis via regulating NLRP3-mediated pyroptosis. Adv Healthc Mater 12(11):e2202228. https://doi.org/10.1002/adhm.202202228

Goodfellow PN, Banting G, Wiles MV, Tunnacliffe A, Parkar M, Solomon E, Dalchau R, Fabre JW (1982) The gene, MIC4, which controls expression of the antigen defined by monoclonal antibody F10.44.2, is on human chromosome 11. Eur J Immunol 12(8):659–663. https://doi.org/10.1002/eji.1830120807

Grage-Griebenow E, Lorenzen D, Fetting R, Flad H-D, Ernst M (1993) Phenotypical and functional characterization of Fcγ receptor I (CD64)-negative monocytes, a minor human monocyte subpopulation with high accessory and antiviral activity. Eur J Immunol 23(12):3126–3135. https://doi.org/10.1002/eji.1830231213

Grage-Griebenow E, Flad HD, Ernst M (1996) Fc gamma receptor I (CD64)-negative human monocytes are potent accessory cells in viral antigen-induced T cell activation and exhibit high IFN-alpha-producing capacity. J Leukoc Biol 60(3):389–396. https://doi.org/10.1002/jlb.60.3.389

Grage-Griebenow E, Flad H-D, Ernst M (2001) Heterogeneity of human peripheral blood monocyte subsets. J Leukoc Biol 69(1):11–20. https://doi.org/10.1189/jlb.69.1.11

Ham S, Lima LG, Chai EPZ, Muller A, Lobb RJ, Krumeich S, Wen SW, Wiegmans AP, Möller A (2018) Breast cancer-derived exosomes alter macrophage polarization via gp130/STAT3 signaling. Front Immunol 9:871. https://doi.org/10.3389/fimmu.2018.00871

Härkönen K, Oikari S, Kyykallio H, Capra J, Hakkola S, Ketola K, Thanigai Arasu U, Daaboul G, Malloy A, Oliveira C, Jokelainen O, Sironen R, Hartikainen JM, Rilla K (2019) CD44s assembles hyaluronan coat on filopodia and extracellular vesicles and induces tumorigenicity of MKN74 gastric carcinoma cells. Cells 8(3):276. https://doi.org/10.3390/cells8030276

Hassn Mesrati M, Syafruddin SE, Mohtar MA, Syahir A (2021) CD44: a multifunctional mediator of cancer progression. Biomolecules 11(12):1850. https://doi.org/10.3390/biom11121850

Hegde S, Leader AM, Merad M (2021) MDSC: markers, development, states, and unaddressed complexity. Immunity 54(5):875–884. https://doi.org/10.1016/j.immuni.2021.04.004

Heldin P, Lin C-Y, Kolliopoulos C, Chen Y-H, Skandalis SS (2019) Regulation of hyaluronan biosynthesis and clinical impact of excessive hyaluronan production. Matrix Biol 78–79:100–117. https://doi.org/10.1016/j.matbio.2018.01.017

Hernández JRH, García JMG, Muñíz MAM, Monclus MTA, Morell AR (1995) Clinical utility of hyaluronic acid values in serum and bronchoalveolar lavage fluid as tumor marker for bronchogenic carcinoma. Int J Biol Markers 10(3):149–155. https://doi.org/10.1177/172460089501000304

Höchst B, Schildberg FA, Sauerborn P, Gäbel YA, Gevensleben H, Goltz D, Heukamp LC, Türler A, Ballmaier M, Gieseke F, Müller I, Kalff J, Kurts C, Knolle PA, Diehl L (2013) Activated human hepatic stellate cells induce myeloid derived suppressor cells from peripheral blood monocytes in a CD44-dependent fashion. J Hepatol 59(3):528–535. https://doi.org/10.1016/j.jhep.2013.04.033

Hofer TP, Zawada AM, Frankenberger M, Skokann K, Satzl AA, Gesierich W, Schuberth M, Levin J, Danek A, Rotter B, Heine GH, Ziegler-Heitbrock L (2015) slan-defined subsets of CD16-positive monocytes: impact of granulomatous inflammation and M-CSF receptor mutation. Blood 126(24):2601–2610. https://doi.org/10.1182/blood-2015-06-651331

Hong BS, Cho J-H, Kim H, Choi E-J, Rho S, Kim J, Kim JH, Choi D-S, Kim Y-K, Hwang D, Gho YS (2009) Colorectal cancer cell-derived microvesicles are enriched in cell cycle-related mRNAs that promote proliferation of endothelial cells. BMC Genomics 10:556. https://doi.org/10.1186/1471-2164-10-556

Huang J, Wang X, Wen J, Zhao X, Wu C, Wang L, Cao X, Dong H, Xu X, Huang F, Zhu W, Wang M (2023) Gastric cancer cell-originated small extracellular vesicle induces metabolic reprogramming of BM-MSCs through ERK-PPARγ-CPT1A signaling to potentiate lymphatic metastasis. Cancer Cell Int 23(1):87. https://doi.org/10.1186/s12935-023-02935-5

Huber V, Vallacchi V, Fleming V, Hu X, Cova A, Dugo M, Shahaj E, Sulsenti R, Vergani E, Filipazzi P, De Laurentiis A, Lalli L, Di Guardo L, Patuzzo R, Vergani B, Casiraghi E, Cossa M, Gualeni A, Bollati V, Arienti F, De Braud F, Mariani L, Villa A, Altevogt P, Umansky V, Rodolfo M, Rivoltini L (2018) Tumor-derived microRNAs induce myeloid suppressor cells and predict immunotherapy resistance in melanoma. J Clin Invest 128(12):5505–5516. https://doi.org/10.1172/JCI98060

Hughes EN, Mengod G, August JT (1981) Murine cell surface glycoproteins. Characterization of a major component of 80,000 daltons as a polymorphic differentiation antigen of mesenchymal cells. J Biol Chem 256(13):7023–7027. https://doi.org/10.1016/S0021-9258(19)69093-9

Itano N, Kimata K (2002) Mammalian hyaluronan synthases. IUBMB Life 54(4):195–199. https://doi.org/10.1080/15216540214929

Javeed N, Gustafson MP, Dutta SK, Lin Y, Bamlet WR, Oberg AL, Petersen GM, Chari ST, Dietz AB, Mukhopadhyay D (2017) Immunosuppressive CD14+HLA-DRlo/neg monocytes are elevated in pancreatic cancer and "primed" by tumor-derived exosomes. Oncoimmunology 6(1):e1252013. https://doi.org/10.1080/2162402X.2016.1252013

Jiang D, Liang J, Noble PW (2011) Hyaluronan as an immune regulator in human diseases. Physiol Rev 91(1):221–264. https://doi.org/10.1152/physrev.00052.2009

Jordan AR, Racine RR, Hennig MJP, Lokeshwar VB (2015) The role of CD44 in disease pathophysiology and targeted treatment. Front Immunol 6:182. https://doi.org/10.3389/fimmu.2015.00182

Kaisho T, Akira S (2000) Critical roles of toll-like receptors in host defense. Crit Rev Immunol 20(5). https://doi.org/10.1615/CritRevImmunol.v20.i5.30

Kajahn J, Franz S, Rueckert E, Forstreuter I, Hintze V, Moeller S, Simon JC (2012) Artificial extracellular matrices composed of collagen I and high sulfated hyaluronan modulate monocyte to macrophage differentiation under conditions of sterile inflammation. Biomatter 2(4):226–236. https://doi.org/10.4161/biom.22855

Katoh S, Miyagi T, Taniguchi H, Matsubara Y, Kadota J, Tominaga A, Kincade PW, Matsukura S, Kohno S (1999) Cutting edge: an inducible sialidase regulates the hyaluronic acid binding ability of CD44-bearing human monocytes. J Immunol 162(9):5058–5061. https://doi.org/10.4049/jimmunol.162.9.5058

Kim SJ, Owen SC (2020) Hyaluronic acid binding to CD44S is indiscriminate of molecular weight. Biochim Biophys Acta Biomembr 1862(9):183348. https://doi.org/10.1016/j.bbamem.2020.183348

Kim H-R, Wheeler MA, Wilson CM, Iida J, Eng D, Simpson MA, McCarthy JB, Bullard KM (2004) Hyaluronan facilitates invasion of colon carcinoma cells in vitro via interaction with CD44. Cancer Res 64(13):4569–4576. https://doi.org/10.1158/0008-5472.CAN-04-0202

Kim N-H, An J-H, Lee J-H, Park S-M, Kim K-B, Kim T-H, Oh Y-I, Seo K-W, Youn H-Y (2022) Effects of cyclooxygenase-2 in canine melanoma-derived extracellular vesicles on tumor microenvironment in vitro. Anticancer Res 42(11):5397–5405. https://doi.org/10.21873/anticanres.16044

Kiss M, Caro AA, Raes G, Laoui D (2020) Systemic reprogramming of monocytes in cancer. Front Oncol 10

Klarić M, Haller H, Brnčić Fischer A, Babarović E, Prijić A, Eminović S (2019) The role of CD44 and RHAMM in endometrial (endometrioid type) cancer: an immunohistochemical study. Appl Immunohistochem Mol Morphol 27(8):606–612. https://doi.org/10.1097/PAI.0000000000000673

Klingbeil P, Isacke CM (2011) The "alternative" EMT switch. Breast Cancer Res 13(4):313. https://doi.org/10.1186/bcr2915

Kobayashi N, Miyoshi S, Mikami T, Koyama H, Kitazawa M, Takeoka M, Sano K, Amano J, Isogai Z, Niida S, Oguri K, Okayama M, McDonald JA, Kimata K, Taniguchi S, Itano N (2010) Hyaluronan deficiency in tumor stroma impairs macrophage trafficking and tumor neovascularization. Cancer Res 70(18):7073–7083. https://doi.org/10.1158/0008-5472.CAN-09-4687

Koistinen V, Kärnä R, Koistinen A, Arjonen A, Tammi M, Rilla K (2015) Cell protrusions induced by hyaluronan synthase 3 (HAS3) resemble mesothelial microvilli and share cytoskeletal features of filopodia. Exp Cell Res 337(2):179–191. https://doi.org/10.1016/j.yexcr.2015.06.016

Kouketsu A, Sato I, Oikawa M, Shimizu Y, Saito H, Tashiro K, Yamashita Y, Takahashi T, Kumamoto H (2019) Regulatory T cells and M2-polarized tumour-associated macrophages are associated with the oncogenesis and progression of oral squamous cell carcinoma. Int J Oral Maxillofac Surg 48(10):1279–1288. https://doi.org/10.1016/j.ijom.2019.04.004

Kowal J, Arras G, Colombo M, Jouve M, Morath JP, Primdal-Bengtson B, Dingli F, Loew D, Tkach M, Théry C (2016) Proteomic comparison defines novel markers to characterize heterogeneous populations of extracellular vesicle subtypes. Proc Natl Acad Sci U S A 113(8):E968–E977. https://doi.org/10.1073/pnas.1521230113

Kuang D-M, Wu Y, Chen N, Cheng J, Zhuang S-M, Zheng L (2007) Tumor-derived hyaluronan induces formation of immunosuppressive macrophages through transient early activation of monocytes. Blood 110(2):587–595. https://doi.org/10.1182/blood-2007-01-068031

Kultti A, Zhao C, Singha NC, Zimmerman S, Osgood RJ, Symons R, Jiang P, Li X, Thompson CB, Infante JR, Jacobetz MA, Tuveson DA, Frost GI, Shepard HM, Huang Z (2014) Accumulation of extracellular hyaluronan by hyaluronan synthase 3 promotes tumor growth and modulates the pancreatic cancer microenvironment. Biomed Res Int 2014:817613. https://doi.org/10.1155/2014/817613

Kusmartsev S, Kwenda E, Dominguez-Gutierrez PR, Crispen PL, O'Malley P (2022) High levels of PD-L1+ and Hyal2+ myeloid-derived suppressor cells in renal cell carcinoma. J Kidney Cancer VHL 9(2):1–6. https://doi.org/10.15586/jkcvhl.v9i2.208

Kwok ZH, Wang C, Jin Y (2021) Extracellular vesicle transportation and uptake by recipient cells: a critical process to regulate human diseases. Processes 9(2):273. https://doi.org/10.3390/pr9020273

Larionova I, Kazakova E, Gerashchenko T, Kzhyshkowska J (2021) new angiogenic regulators produced by TAMs: perspective for targeting tumor angiogenesis. Cancers (Basel) 13(13):3253. https://doi.org/10.3390/cancers13133253

Lavin Y, Mortha A, Rahman A, Merad M (2015) Regulation of macrophage development and function in peripheral tissues. Nat Rev Immunol 15(12):731–744. https://doi.org/10.1038/nri3920

Lee-Sayer SSM, Dong Y, Arif AA, Olsson M, Brown KL, Johnson P (2015) The where, when, how, and why of hyaluronan binding by immune cells. Front Immunol 6

Lenart M, Rutkowska-Zapala M, Baj-Krzyworzeka M, Szatanek R, Węglarczyk K, Smallie T, Ziegler-Heitbrock L, Zembala M, Siedlar M (2017) Hyaluronan carried by tumor-derived microvesicles induces IL-10 production in classical (CD14++CD16−) monocytes via PI3K/Akt/mTOR-dependent signalling pathway. Immunobiology 222(1):1–10. https://doi.org/10.1016/j.imbio.2015.06.019

Lesley J, Hyman R (1998) CD44 structure and function. Front Biosci 3:d616–d630. https://doi.org/10.2741/a306

Lesley J, Hascall VC, Tammi M, Hyman R (2000) Hyaluronan binding by cell surface CD44. J Biol Chem 275(35):26967–26975. https://doi.org/10.1074/jbc.M002527200

Levesque MC, Haynes BF (1996) In vitro culture of human peripheral blood monocytes induces hyaluronan binding and up-regulates monocyte variant CD44 isoform expression. J Immunol 156(4):1557–1565

Li Y, Duy Le TM, Nam Bui Q, Yang HY, Lee DS (2019) Tumor acidity and CD44 dual targeting hyaluronic acid-coated gold nanorods for combined chemo- and photothermal cancer therapy. Carbohydr Polym 226:115281. https://doi.org/10.1016/j.carbpol.2019.115281

Linton SS, Abraham T, Liao J, Clawson GA, Butler PJ, Fox T, Kester M, Matters GL (2018) Tumor-promoting effects of pancreatic cancer cell exosomes on THP-1-derived macrophages. PLoS One 13(11):e0206759. https://doi.org/10.1371/journal.pone.0206759

Liu S, Liu Z, Shang A, Xun J, Lv Z, Zhou S, Liu C, Zhang Q, Yang Y (2023) CD44 is a potential immunotherapeutic target and affects macrophage infiltration leading to poor prognosis. Sci Rep 13(1):9657. https://doi.org/10.1038/s41598-023-33915-4

Loh TJ, Moon H, Cho S, Jang H, Liu YC, Tai H, Jung D-W, Williams DR, Kim H-R, Shin M-G, Liao DJ, Zhou J, Shi W, Zheng X, Shen H (2015) CD44 alternative splicing and hnRNP A1 expression are associated with the metastasis of breast cancer. Oncol Rep 34(3):1231–1238. https://doi.org/10.3892/or.2015.4110

Long H, Jia Q, Wang L, Fang W, Wang Z, Jiang T, Zhou F, Jin Z, Huang J, Zhou L, Hu C, Wang X, Zhang J, Ba Y, Gong Y, Zeng X, Zeng D, Su X, Alexander PB, Wang L, Wang L, Wan YY, Wang X-F, Zhang L, Li Q-J, Zhu B (2022) Tumor-induced erythroid precursor-differentiated myeloid cells mediate immunosuppression and curtail anti-PD-1/PD-L1 treatment efficacy. Cancer Cell 40(6):674–693.e7. https://doi.org/10.1016/j.ccell.2022.04.018

Luong N, Lenz JA, Modiano JF, Olson JK (2021) Extracellular vesicles secreted by tumor cells promote the generation of suppressive monocytes. Immunohorizons 5(8):647–658. https://doi.org/10.4049/immunohorizons.2000017

Mack M, Kleinschmidt A, Brühl H, Klier C, Nelson PJ, Cihak J, Plachý J, Stangassinger M, Erfle V, Schlöndorff D (2000) Transfer of the chemokine receptor CCR5 between cells by membrane-derived microparticles: a mechanism for cellular human immunodeficiency virus 1 infection. Nat Med 6(7):769–775. https://doi.org/10.1038/77498

Mahira S, Kommineni N, Husain GM, Khan W (2019) Cabazitaxel and silibinin co-encapsulated cationic liposomes for CD44 targeted delivery: a new insight into nanomedicine based combinational chemotherapy for prostate cancer. Biomed Pharmacother 110:803–817. https://doi.org/10.1016/j.biopha.2018.11.145

Manning S, Danielson KM (2018) The immunomodulatory role of tumor-derived extracellular vesicles in colorectal cancer. Immunol Cell Biol 96(7):733–741. https://doi.org/10.1111/imcb.12038

Mantovani A, Sozzani S, Locati M, Allavena P, Sica A (2002) Macrophage polarization: tumor-associated macrophages as a paradigm for polarized M2 mononuclear phagocytes. Trends Immunol 23(11):549–555. https://doi.org/10.1016/s1471-4906(02)02302-5

Marostica G, Gelibter S, Gironi M, Nigro A, Furlan R (2021) Extracellular vesicles in neuroinflammation. Front Cell and Dev Biol 8

McAtee CO, Booth C, Elowsky C, Zhao L, Payne J, Fangman T, Caplan S, Henry MD, Simpson MA (2019) Prostate tumor cell exosomes containing hyaluronidase Hyal1 stimulate prostate stromal cell motility by engagement of FAK-mediated integrin signaling. Matrix Biol 78–79:165–179. https://doi.org/10.1016/j.matbio.2018.05.002

McKee CM, Penno MB, Cowman M, Burdick MD, Strieter RM, Bao C, Noble PW (1996) Hyaluronan (HA) fragments induce chemokine gene expression in alveolar macrophages. The role of HA size and CD44. J Clin Invest 98(10):2403–2413. https://doi.org/10.1172/JCI119054

Mishra MN, Chandavarkar V, Sharma R, Bhargava D (2019) Structure, function and role of CD44 in neoplasia. J Oral Maxillofac Pathol 23(2):267–272. https://doi.org/10.4103/jomfp.JOMFP_246_18

Misra S, Hascall VC, Markwald RR, Ghatak S (2015) Interactions between hyaluronan and its receptors (CD44, RHAMM) regulate the activities of inflammation and cancer. Front Immunol 6:201. https://doi.org/10.3389/fimmu.2015.00201

Momen-Heravi F, Bala S (2018) Extracellular vesicles in oral squamous carcinoma carry oncogenic miRNA profile and reprogram monocytes via NF-κB pathway. Oncotarget 9(78):34838–34854. https://doi.org/10.18632/oncotarget.26208

Mosser DM, Zhang X (2008) Activation of murine macrophages. Curr Protoc Immunol 14:14.2.1–14.2.8. https://doi.org/10.1002/0471142735.im1402s83

Mosser DM, Hamidzadeh K, Goncalves R (2021) Macrophages and the maintenance of homeostasis. Cell Mol Immunol 18(3):579–587. https://doi.org/10.1038/s41423-020-00541-3

Mustonen A-M, Capra J, Rilla K, Lehenkari P, Oikari S, Kääriäinen T, Joukainen A, Kröger H, Paakkonen T, Matilainen J, Nieminen P (2021) Characterization of hyaluronan-coated extracellular vesicles in synovial fluid of patients with osteoarthritis and rheumatoid arthritis. BMC Musculoskelet Disord 22(1):247. https://doi.org/10.1186/s12891-021-04115-w

Mytar B, Siedlar M, Woloszyn M, Colizzi V, Zembala M (2001) Cross-talk between human monocytes and cancer cells during reactive oxygen intermediates generation: the essential role of hyaluronan. Int J Cancer 94(5):727–732. https://doi.org/10.1002/ijc.1530

Mytar B, Wołoszyn M, Szatanek R, Baj-Krzyworzeka M, Siedlar M, Ruggiero I, Więckiewicz J, Zembala M (2003) Tumor cell-induced deactivation of human monocytes. J Leukoc Biol 74(6):1094–1101. https://doi.org/10.1189/jlb.0403140

Mytar B, Woloszyn M, Macura-Biegun A, Hajto B, Ruggiero I, Piekarska B, Zembala M (2004) Involvement of pattern recognition receptors in the induction of cytokines and reactive oxygen intermediates production by human monocytes/macrophages stimulated with tumour cells. Anticancer Res 24(4):2287–2293

Najafi M, Farhood B, Mortezaee K (2019) Extracellular matrix (ECM) stiffness and degradation as cancer drivers. J Cell Biochem 120(3):2782–2790. https://doi.org/10.1002/jcb.27681

Nakai W, Yoshida T, Diez D, Miyatake Y, Nishibu T, Imawaka N, Naruse K, Sadamura Y, Hanayama R (2016) A novel affinity-based method for the isolation of highly purified extracellular vesicles. Sci Rep 6(1):33935. https://doi.org/10.1038/srep33935

Nakamura K, Sawada K, Kinose Y, Yoshimura A, Toda A, Nakatsuka E, Hashimoto K, Mabuchi S, Morishige K-I, Kurachi H, Lengyel E, Kimura T (2017) Exosomes promote ovarian cancer cell invasion through transfer of CD44 to peritoneal mesothelial cells. Mol Cancer Res 15(1):78–92. https://doi.org/10.1158/1541-7786.MCR-16-0191

Nawaz M, Shah N, Zanetti BR, Maugeri M, Silvestre RN, Fatima F, Neder L, Valadi H (2018) Extracellular vesicles and matrix remodeling enzymes: the emerging roles in extracellular

matrix remodeling, progression of diseases and tissue repair. Cells 7(10):167. https://doi.org/10.3390/cells7100167
Németh A, Orgovan N, Sódar BW, Osteikoetxea X, Pálóczi K, Szabó-Taylor KÉ, Vukman KV, Kittel Á, Turiák L, Wiener Z, Tóth S, Drahos L, Vékey K, Horvath R, Buzás EI (2017) Antibiotic-induced release of small extracellular vesicles (exosomes) with surface-associated DNA. Sci Rep 7(1):8202. https://doi.org/10.1038/s41598-017-08392-1
Ngambenjawong C, Gustafson HH, Pun SH (2017) Progress in tumor-associated macrophage (TAM)-targeted therapeutics. Adv Drug Deliv Rev 114:206–221. https://doi.org/10.1016/j.addr.2017.04.010
Noble PW, McKee CM, Cowman M, Shin HS (1996) Hyaluronan fragments activate an NF-kappa B/I-kappa B alpha autoregulatory loop in murine macrophages. J Exp Med 183(5):2373–2378. https://doi.org/10.1084/jem.183.5.2373
Okamoto I, Tsuiki H, Kenyon LC, Godwin AK, Emlet DR, Holgado-Madruga M, Lanham IS, Joynes CJ, Vo KT, Guha A, Matsumoto M, Ushio Y, Saya H, Wong AJ (2002) Proteolytic cleavage of the CD44 adhesion molecule in multiple human tumors. Am J Pathol 160(2):441–447. https://doi.org/10.1016/S0002-9440(10)64863-8
Pan Y, Yu Y, Wang X, Zhang T (2020) Tumor-associated macrophages in tumor immunity. Front Immunol 11
Paul D, Roy A, Nandy A, Datta B, Borar P, Pal SK, Senapati D, Rakshit T (2020) Identification of biomarker hyaluronan on colon cancer extracellular vesicles using correlative AFM and spectroscopy. J Phys Chem Lett 11(14):5569–5576. https://doi.org/10.1021/acs.jpclett.0c01018
Paul D, Paul A, Mukherjee D, Saroj S, Ghosal M, Pal S, Senapati D, Chakrabarti J, Pal SK, Rakshit T (2022) A mechanoelastic glimpse on hyaluronan-coated extracellular vesicles. J Phys Chem Lett 13(36):8564–8572. https://doi.org/10.1021/acs.jpclett.2c01629
Peinado H, Alečković M, Lavotshkin S, Matei I, Costa-Silva B, Moreno-Bueno G, Hergueta-Redondo M, Williams C, García-Santos G, Ghajar CM, Nitadori-Hoshino A, Hoffman C, Badal K, Garcia BA, Callahan MK, Yuan J, Martins VR, Skog J, Kaplan RN, Brady MS, Wolchok JD, Chapman PB, Kang Y, Bromberg J, Lyden D (2012) Melanoma exosomes educate bone marrow progenitor cells toward a pro-metastatic phenotype through MET. Nat Med 18(6):883–891. https://doi.org/10.1038/nm.2753
Ponting J, Howell A, Pye D, Kumar S (1992) Prognostic relevance of serum hyaluronan levels in patients with breast cancer. Int J Cancer 52(6):873–876. https://doi.org/10.1002/ijc.2910520607
Popēna I, Ābols A, Saulīte L, Pleiko K, Zandberga E, Jēkabsons K, Endzeliņš E, Llorente A, Linē A, Riekstiņa U (2018) Effect of colorectal cancer-derived extracellular vesicles on the immunophenotype and cytokine secretion profile of monocytes and macrophages. Cell Commun Signal 16(1):17. https://doi.org/10.1186/s12964-018-0229-y
Prat M, Naour AL, Coulson K, Lemée F, Leray H, Jacquemin G, Rahabi MC, Lemaitre L, Authier H, Ferron G, Barret J-M, Martinez A, Ayyoub M, Delord J-P, Gladieff L, Tabah-Fisch I, Prost J-F, Couderc B, Coste A (2020) Circulating $CD14^{high}$ $CD16^{low}$ intermediate blood monocytes as a biomarker of ascites immune status and ovarian cancer progression. J Immunother Cancer 8(1):e000472. https://doi.org/10.1136/jitc-2019-000472
Purushothaman A, Oliva-Ramírez J, Treekitkarnmongkol W, Sankaran D, Hurd MW, Putluri N, Maitra A, Haymaker C, Sen S (2023) Differential effects of pancreatic cancer-derived extracellular vesicles driving a suppressive environment. Int J Mol Sci 24(19):14652. https://doi.org/10.3390/ijms241914652
Qin S, Wang Z, Huang C, Huang P, Li D (2022) Serine protease PRSS23 drives gastric cancer by enhancing tumor associated macrophage infiltration via FGF2. Front Immunol 13:955841. https://doi.org/10.3389/fimmu.2022.955841
Qu C, Brinck-Jensen N-S, Zang M, Chen K (2014) Monocyte-derived dendritic cells: targets as potent antigen-presenting cells for the design of vaccines against infectious diseases. Int J Infect Dis 19:1–5. https://doi.org/10.1016/j.ijid.2013.09.023
Rall CJ, Rustgi AK (1995) CD44 isoform expression in primary and metastatic pancreatic adenocarcinoma. Cancer Res 55(9):1831–1835

Ralph P, Nakoinz I (1984) Cooperation of IgG monoclonal antibodies in macrophage antibody-dependent cellular cytotoxicity (ADCC) to tumor targets. J Leukoc Biol 35(1):131–139. https://doi.org/10.1002/jlb.35.1.131
Rilla K (2021) Diverse plasma membrane protrusions act as platforms for extracellular vesicle shedding. J Extracell Vesicles 10(11):e12148. https://doi.org/10.1002/jev2.12148
Rilla K, Pasonen-Seppänen S, Deen AJ, Koistinen VVT, Wojciechowski S, Oikari S, Kärnä R, Bart G, Törrönen K, Tammi RH, Tammi MI (2013) Hyaluronan production enhances shedding of plasma membrane-derived microvesicles. Exp Cell Res 319(13):2006–2018. https://doi.org/10.1016/j.yexcr.2013.05.021
Rilla K, Siiskonen H, Tammi M, Tammi R (2014a) Hyaluronan-coated extracellular vesicles—a novel link between hyaluronan and cancer. Adv Cancer Res 123:121–148. https://doi.org/10.1016/B978-0-12-800092-2.00005-8
Rilla K, Siiskonen H, Tammi M, Tammi R (2014b) Chapter Five - Hyaluronan-coated extracellular vesicles—a novel link between hyaluronan and cancer. In: Simpson MA, Heldin P (eds) Advances in cancer research. Academic Press, pp 121–148
Rios de la Rosa JM, Tirella A, Gennari A, Stratford IJ, Tirelli N (2017) The CD44-mediated uptake of hyaluronic acid-based carriers in macrophages. Adv Healthc Mater 6(4). https://doi.org/10.1002/adhm.201601012
Rivollier A, He J, Kole A, Valatas V, Kelsall BL (2012) Inflammation switches the differentiation program of Ly6Chi monocytes from antiinflammatory macrophages to inflammatory dendritic cells in the colon. J Exp Med 209(1):139–155. https://doi.org/10.1084/jem.20101387
Ropert C (2019) How toll-like receptors reveal monocyte plasticity: the cutting edge of antiinflammatory therapy. Cell Mol Life Sci 76(4):745–755. https://doi.org/10.1007/s00018-018-2959-9
Ropponen K, Tammi M, Parkkinen J, Eskelinen M, Tammi R, Lipponen P, Ågren U, Alhava E, Kosma V (1998) Tumor cell-associated hyaluronan as an unfavorable prognostic factor in colorectal cancer. Cancer Res
Ropponen K, Kellokoski J, Pirinen R, Moisio K, Eskelinen M, Alhava E, Kosma V (2001) Expression of transcription factor AP-2 in colorectal adenomas and adenocarcinomas; comparison of immunohistochemistry and in situ hybridisation. J Clin Pathol 54(7):533–538. https://doi.org/10.1136/jcp.54.7.533
Sadeghi A, Roudi R, Mirzaei A, Zare Mirzaei A, Madjd Z, Abolhasani M (2019) CD44 epithelial isoform inversely associates with invasive characteristics of colorectal cancer. Biomark Med 13(6):419–426. https://doi.org/10.2217/bmm-2018-0337
Saito S, Hiemori K, Kiyoi K, Tateno H (2018) Glycome analysis of extracellular vesicles derived from human induced pluripotent stem cells using lectin microarray. Sci Rep 8:3997. https://doi.org/10.1038/s41598-018-22450-2
Sato N, Cheng X-B, Kohi S, Koga A, Hirata K (2016) Targeting hyaluronan for the treatment of pancreatic ductal adenocarcinoma. Acta Pharm Sin B 6(2):101–105. https://doi.org/10.1016/j.apsb.2016.01.002
Schanté CE, Zuber G, Herlin C, Vandamme TF (2011) Chemical modifications of hyaluronic acid for the synthesis of derivatives for a broad range of biomedical applications. Carbohydr Polym 85(3):469–489. https://doi.org/10.1016/j.carbpol.2011.03.019
Schyns J, Bai Q, Ruscitti C, Radermecker C, De Schepper S, Chakarov S, Farnir F, Pirottin D, Ginhoux F, Boeckxstaens G, Bureau F, Marichal T (2019) Non-classical tissue monocytes and two functionally distinct populations of interstitial macrophages populate the mouse lung. Nat Commun 10(1):3964. https://doi.org/10.1038/s41467-019-11843-0
Selimoglu-Buet D, Wagner-Ballon O, Saada V, Bardet V, Itzykson R, Bencheikh L, Morabito M, Met E, Debord C, Benayoun E, Nloga A-M, Fenaux P, Braun T, Willekens C, Quesnel B, Adès L, Fontenay M, Rameau P, Droin N, Koscielny S, Solary E, Francophone Myelodysplasia Group (2015) Characteristic repartition of monocyte subsets as a diagnostic signature of chronic myelomonocytic leukemia. Blood 125(23):3618–3626. https://doi.org/10.1182/blood-2015-01-620781

Shapouri-Moghaddam A, Mohammadian S, Vazini H, Taghadosi M, Esmaeili S-A, Mardani F, Seifi B, Mohammadi A, Afshari JT, Sahebkar A (2018) Macrophage plasticity, polarization, and function in health and disease. J Cell Physiol 233(9):6425–6440. https://doi.org/10.1002/jcp.26429

Sherman L, Sleeman J, Herrlich P, Ponta H (1994) Hyaluronate receptors: key players in growth, differentiation, migration and tumor progression. Curr Opin Cell Biol 6(5):726–733. https://doi.org/10.1016/0955-0674(94)90100-7

Shime H, Maruyama A, Yoshida S, Takeda Y, Matsumoto M, Seya T (2017) Toll-like receptor 2 ligand and interferon-γ suppress anti-tumor T cell responses by enhancing the immunosuppressive activity of monocytic myeloid-derived suppressor cells. Oncoimmunology 7(1):e1373231. https://doi.org/10.1080/2162402X.2017.1373231

Siemińska I, Węglarczyk K, Walczak M, Czerwińska A, Pach R, Rubinkiewicz M, Szczepanik A, Siedlar M, Baran J (2022) Mo-MDSCs are pivotal players in colorectal cancer and may be associated with tumor recurrence after surgery. Transl Oncol 17:101346. https://doi.org/10.1016/j.tranon.2022.101346

Silva J, Garcia V, Rodriguez M, Compte M, Cisneros E, Veguillas P, Garcia JM, Dominguez G, Campos-Martin Y, Cuevas J, Peña C, Herrera M, Diaz R, Mohammed N, Bonilla F (2012) Analysis of exosome release and its prognostic value in human colorectal cancer. Genes Chromosomes Cancer 51(4):409–418. https://doi.org/10.1002/gcc.21926

Skelton TP, Zeng C, Nocks A, Stamenkovic I (1998) Glycosylation provides both stimulatory and inhibitory effects on cell surface and soluble CD44 binding to hyaluronan. J Cell Biol 140(2):431–446

Skotland T, Sagini K, Sandvig K, Llorente A (2020) An emerging focus on lipids in extracellular vesicles. Adv Drug Deliv Rev 159:308–321. https://doi.org/10.1016/j.addr.2020.03.002

Sódar BW, Kittel Á, Pálóczi K, Vukman KV, Osteikoetxea X, Szabó-Taylor K, Németh A, Sperlágh B, Baranyai T, Giricz Z, Wiener Z, Turiák L, Drahos L, Pállinger É, Vékey K, Ferdinandy P, Falus A, Buzás EI (2016) Low-density lipoprotein mimics blood plasma-derived exosomes and microvesicles during isolation and detection. Sci Rep 6:24316. https://doi.org/10.1038/srep24316

Sokolowska M, Chen L-Y, Eberlein M, Martinez-Anton A, Liu Y, Alsaaty S, Qi H-Y, Logun C, Horton M, Shelhamer JH (2014) Low molecular weight hyaluronan activates cytosolic phospholipase A2α and eicosanoid production in monocytes and macrophages. J Biol Chem 289(7):4470–4488. https://doi.org/10.1074/jbc.M113.515106

Spinelli FM, Vitale DL, Icardi A, Caon I, Brandone A, Giannoni P, Saturno V, Passi A, García M, Sevic I, Alaniz L (2019) Hyaluronan preconditioning of monocytes/macrophages affects their angiogenic behavior and regulation of TSG-6 expression in a tumor type-specific manner. FEBS J 286(17):3433–3449. https://doi.org/10.1111/febs.14871

Stamenkovic I, Amiot M, Pesando JM, Seed B (1989) A lymphocyte molecule implicated in lymph node homing is a member of the cartilage link protein family. Cell 56(6):1057–1062. https://doi.org/10.1016/0092-8674(89)90638-7

Stamenkovic I, Aruffo A, Amiot M, Seed B (1991) The hematopoietic and epithelial forms of CD44 are distinct polypeptides with different adhesion potentials for hyaluronate-bearing cells. EMBO J 10(2):343–348. https://doi.org/10.1002/j.1460-2075.1991.tb07955.x

Stefanius K, Servage K, de Souza SM, Gray HF, Toombs JE, Chimalapati S, Kim MS, Malladi VS, Brekken R, Orth K (2019) Human pancreatic cancer cell exosomes, but not human normal cell exosomes, act as an initiator in cell transformation. eLife 8:e40226. https://doi.org/10.7554/eLife.40226

Stern R, Asari AA, Sugahara KN (2006) Hyaluronan fragments: an information-rich system. Eur J Cell Biol 85(8):699–715. https://doi.org/10.1016/j.ejcb.2006.05.009

Su S, Zhao J, Xing Y, Zhang X, Liu J, Ouyang Q, Chen J, Su F, Liu Q, Song E (2018) Immune checkpoint inhibition overcomes ADCP-induced immunosuppression by macrophages. Cell 175(2):442–457.e23. https://doi.org/10.1016/j.cell.2018.09.007

Sumitomo R, Hirai T, Fujita M, Murakami H, Otake Y, Huang C-L (2019) PD-L1 expression on tumor-infiltrating immune cells is highly associated with M2 TAM and aggressive malignant potential in patients with resected non-small cell lung cancer. Lung Cancer 136:136–144. https://doi.org/10.1016/j.lungcan.2019.08.023

Sung BH, Ketova T, Hoshino D, Zijlstra A, Weaver AM (2015) Directional cell movement through tissues is controlled by exosome secretion. Nat Commun 6(1):7164. https://doi.org/10.1038/ncomms8164

Sung BH, Parent CA, Weaver AM (2021) Extracellular vesicles: critical players during cell migration. Dev Cell 56(13):1861–1874. https://doi.org/10.1016/j.devcel.2021.03.020

Szaflarska A, Baj-Krzyworzeka M, Siedlar M, Węglarczyk K, Ruggiero I, Hajto B, Zembala M (2004) Antitumor response of CD14+/CD16+ monocyte subpopulation. Exp Hematol 32(8):748–755. https://doi.org/10.1016/j.exphem.2004.05.027

Szatanek R, Weglarczyk K, Stec M, Baran J, Parlinska-Wojtan M, Siedlar M, Baj-Krzyworzeka M (2020) Autologous tumor-derived microvesicles influence gene expression profiles and enhance protumorigenic chemotactic potential, signal transduction and cellular respiration in gastric cancer cells. Int J Oncol 56(1):359–367. https://doi.org/10.3892/ijo.2019.4923

Szczepanik A, Sierzega M, Drabik G, Pituch-Noworolska A, Kołodziejczyk P, Zembala M (2019) CD44+ cytokeratin-positive tumor cells in blood and bone marrow are associated with poor prognosis of patients with gastric cancer. Gastric Cancer 22(2):264–272. https://doi.org/10.1007/s10120-018-0858-2

Tabata T, Mine S, Okada Y, Tanaka Y (2007) Low molecular weight hyaluronan increases the uptaking of oxidized LDL into monocytes. Endocr J 54(5):685–693. https://doi.org/10.1507/endocrj.K05-120

Tammi R, Rilla K, Pienimaki JP, MacCallum DK, Hogg M, Luukkonen M, Hascall VC, Tammi M (2001) Hyaluronan enters keratinocytes by a novel endocytic route for catabolism. J Biol Chem 276(37):35111–35122. https://doi.org/10.1074/jbc.M103481200

Tammi MI, Oikari S, Pasonen-Seppänen S, Rilla K, Auvinen P, Tammi RH (2019) Activated hyaluronan metabolism in the tumor matrix - causes and consequences. Matrix Biol 78–79:147–164. https://doi.org/10.1016/j.matbio.2018.04.012

Tanabe KK, Ellis LM, Saya H (1993) Expression of CD44R1 adhesion molecule in colon carcinomas and metastases. Lancet 341(8847):725–726. https://doi.org/10.1016/0140-6736(93)90490-8

Taniyama T, Holden HT (1980) Cytolytic activity against tumor cells by macrophage cell lines and augmentation by macrophage stimulants. Int J Cancer 26(1):61–69. https://doi.org/10.1002/ijc.2910260110

Theodoraki M-N, Matsumoto A, Beccard I, Hoffmann TK, Whiteside TL (2020) CD44v3 protein-carrying tumor-derived exosomes in HNSCC patients' plasma as potential noninvasive biomarkers of disease activity. Oncoimmunology 9(1):1747732. https://doi.org/10.1080/2162402X.2020.1747732

Tiainen S, Masarwah A, Oikari S, Rilla K, Hämäläinen K, Sudah M, Sutela A, Vanninen R, Ikonen J, Tammi R, Tammi M, Auvinen P (2020) Tumor microenvironment and breast cancer survival: combined effects of breast fat, M2 macrophages and hyaluronan create a dismal prognosis. Breast Cancer Res Treat 179(3):565–575. https://doi.org/10.1007/s10549-019-05491-7

Tóth EÁ, Turiák L, Visnovitz T, Cserép C, Mázló A, Sódar BW, Försönits AI, Petővári G, Sebestyén A, Komlósi Z, Drahos L, Kittel Á, Nagy G, Bácsi A, Dénes Á, Gho YS, Szabó-Taylor KÉ, Buzás EI (2021) Formation of a protein corona on the surface of extracellular vesicles in blood plasma. J Extracell Vesicles 10(11):e12140. https://doi.org/10.1002/jev2.12140

Tu B, Li J, Sun Z, Zhang T, Liu H, Yuan F, Fan C (2023) Macrophage-derived TGF-β and VEGF promote the progression of trauma-induced heterotopic ossification. Inflammation 46(1):202–216. https://doi.org/10.1007/s10753-022-01723-z

Tyrinova TV, Batorov EV, Aristova TA, Ushakova GY, Sizikova SA, Denisova VV, Ostanin AA, Chernykh ER (2022) Expression of inhibitory molecules (arginase-1, IDO, and PD-L1) by myeloid-derived suppressor cells in multiple myeloma patients in remission. Bull Exp Biol Med 174(1):71–75. https://doi.org/10.1007/s10517-022-05651-8

Ugel S, Canè S, De Sanctis F, Bronte V (2021) Monocytes in the tumor microenvironment. Annu Rev Pathol 16:93–122. https://doi.org/10.1146/annurev-pathmechdis-012418-013058

Umakoshi M, Takahashi S, Itoh G, Kuriyama S, Sasaki Y, Yanagihara K, Yashiro M, Maeda D, Goto A, Tanaka M (2019) Macrophage-mediated transfer of cancer-derived components to stromal cells contributes to establishment of a pro-tumor microenvironment. Oncogene 38(12):2162–2176. https://doi.org/10.1038/s41388-018-0564-x

Umansky V, Adema GJ, Baran J, Brandau S, Van Ginderachter JA, Hu X, Jablonska J, Mojsilovic S, Papadaki HA, Pico de Coaña Y, Santegoets KCM, Santibanez JF, Serre K, Si Y, Sieminska I, Velegraki M, Fridlender ZG (2019) Interactions among myeloid regulatory cells in cancer. Cancer Immunol Immunother 68(4):645–660. https://doi.org/10.1007/s00262-018-2200-6

Valenti R, Huber V, Filipazzi P, Pilla L, Sovena G, Villa A, Corbelli A, Fais S, Parmiani G, Rivoltini L (2006) Human tumor-released microvesicles promote the differentiation of myeloid cells with transforming growth factor-β–mediated suppressive activity on T lymphocytes. Cancer Res 66(18):9290–9298. https://doi.org/10.1158/0008-5472.CAN-06-1819

Veglia F, Sanseviero E, Gabrilovich DI (2021) Myeloid-derived suppressor cells in the era of increasing myeloid cell diversity. Nat Rev Immunol 21(8):485–498. https://doi.org/10.1038/s41577-020-00490-y

Vinik Y, Ortega FG, Mills GB, Lu Y, Jurkowicz M, Halperin S, Aharoni M, Gutman M, Lev S (2020) Proteomic analysis of circulating extracellular vesicles identifies potential markers of breast cancer progression, recurrence, and response. Sci Adv 6(40):eaba5714. https://doi.org/10.1126/sciadv.aba5714

Webb DS, Shimizu Y, Van Seventer GA, Shaw S, Gerrard TL (1990) LFA-3, CD44, and CD45: physiologic triggers of human monocyte TNF and IL-1 release. Science 249(4974):1295–1297. https://doi.org/10.1126/science.1697984

Weigel PH, Hascall VC, Tammi M (1997) Hyaluronan synthases. J Biol Chem 272(22): 13997–14000. https://doi.org/10.1074/jbc.272.22.13997

Weiss JM, Renkl AC, Ahrens T, Mai BH, Denfeld RW, Schöpf E, Simon JC, Moll J, Ponta H, Herrlich P (1998) Activation-dependent modulation of hyaluronate-receptor expression and of hyaluronate-avidity by human monocytes. J Investig Dermatol 111(2):227–232. https://doi.org/10.1046/j.1523-1747.1998.00286.x

Wenzel P, Rossmann H, Müller C, Kossmann S, Oelze M, Schulz A, Arnold N, Simsek C, Lagrange J, Klemz R, Schönfelder T, Brandt M, Karbach SH, Knorr M, Finger S, Neukirch C, Häuser F, Beutel ME, Kröller-Schön S, Schulz E, Schnabel RB, Lackner K, Wild PS, Zeller T, Daiber A, Blankenberg S, Münzel T (2015) Heme oxygenase-1 suppresses a pro-inflammatory phenotype in monocytes and determines endothelial function and arterial hypertension in mice and humans. Eur Heart J 36(48):3437–3446. https://doi.org/10.1093/eurheartj/ehv544

Wessels D, Lusche DF, Voss E, Kuhl S, Buchele EC, Klemme MR, Russell KB, Ambrose J, Soll BA, Bossler A, Milhem M, Goldman C, Soll DR (2017) Melanoma cells undergo aggressive coalescence in a 3D Matrigel model that is repressed by anti-CD44. PLoS One 12(3):e0173400. https://doi.org/10.1371/journal.pone.0173400

Willemen Y, Van den Bergh JMJ, Bonte SM, Anguille S, Heirman C, Stein BMH, Goossens H, Kerre T, Thielemans K, Peeters M, Van Tendeloo VFI, Smits ELJ, Berneman ZN (2016) The tumor-associated antigen RHAMM (HMMR/CD168) is expressed by monocyte-derived dendritic cells and presented to T cells. Oncotarget 7(45):73960–73970. https://doi.org/10.18632/oncotarget.12170

Witschen PM, Elfstrum AK, Nelson AC, Schwertfeger KL (2023) Characterization of hyaluronan localization in the developing mammary gland and mammary tumors. J Mammary Gland Biol Neoplasia 28(1):1. https://doi.org/10.1007/s10911-023-09528-y

Wong KL, Tai JJ-Y, Wong W-C, Han H, Sem X, Yeap W-H, Kourilsky P, Wong S-C (2011) Gene expression profiling reveals the defining features of the classical, intermediate, and nonclassical human monocyte subsets. Blood 118(5):e16–e31. https://doi.org/10.1182/blood-2010-12-326355

Wong KL, Yeap WH, Tai JJY, Ong SM, Dang TM, Wong SC (2012) The three human monocyte subsets: implications for health and disease. Immunol Res 53(1–3):41–57. https://doi.org/10.1007/s12026-012-8297-3

Wu W-C, Sun H-W, Chen H-T, Liang J, Yu X-J, Wu C, Wang Z, Zheng L (2014) Circulating hematopoietic stem and progenitor cells are myeloid-biased in cancer patients. Proc Natl Acad Sci U S A 111(11):4221–4226. https://doi.org/10.1073/pnas.1320753111

Xing R, Chang S, Li J, Li H, Han Z (2008) Serum hyaluronan levels in oral cancer patients. Chin Med J (Engl) 121(4):327–330

Xu H, Niu M, Yuan X, Wu K, Liu A (2020) CD44 as a tumor biomarker and therapeutic target. Exp Hematol Oncol 9(1):36. https://doi.org/10.1186/s40164-020-00192-0

Yadav A, Kumar B, Datta J, Teknos TN, Kumar P (2011) IL-6 promotes head and neck tumor metastasis by inducing epithelial-mesenchymal transition via the JAK-STAT3-SNAIL signaling pathway. Mol Cancer Res 9(12):1658–1667. https://doi.org/10.1158/1541-7786.MCR-11-0271

Yamaguchi Y, Yamamoto H, Tobisawa Y, Irie F (2019) TMEM2: a missing link in hyaluronan catabolism identified? Matrix Biol 78–79:139–146. https://doi.org/10.1016/j.matbio.2018.03.020

Yamawaki H, Hirohata S, Miyoshi T, Takahashi K, Ogawa H, Shinohata R, Demircan K, Kusachi S, Yamamoto K, Ninomiya Y (2009) Hyaluronan receptors involved in cytokine induction in monocytes. Glycobiology 19(1):83–92. https://doi.org/10.1093/glycob/cwn109

Yan T, Wang K, Li J, Hu H, Yang H, Cai M, Liu R, Li H, Wang N, Shi Y, Hua W, Liu H (2022) Suppression of the hyaluronic acid pathway induces M1 macrophages polarization via STAT1 in glioblastoma. Cell Death Discov 8(1):1–13. https://doi.org/10.1038/s41420-022-00973-y

Yáñez A, Coetzee SG, Olsson A, Muench DE, Berman BP, Hazelett DJ, Salomonis N, Grimes HL, Goodridge HS (2017) Granulocyte-monocyte progenitors and monocyte-dendritic cell progenitors independently produce functionally distinct monocytes. Immunity 47(5):890–902.e4. https://doi.org/10.1016/j.immuni.2017.10.021

Yang R-B, Mark MR, Gray A, Huang A, Xie MH, Zhang M, Goddard A, Wood WI, Gurney AL, Godowski PJ (1998) Toll-like receptor-2 mediates lipopolysaccharide-induced cellular signalling. Nature 395(6699):284–288. https://doi.org/10.1038/26239

Yang C, Cao M, Liu H, He Y, Xu J, Du Y, Liu Y, Wang W, Cui L, Hu J, Gao F (2012) The high and low molecular weight forms of hyaluronan have distinct effects on CD44 clustering. J Biol Chem 287(51):43094–43107. https://doi.org/10.1074/jbc.M112.349209

Yang F, Zheng Z, Xue X, Zheng L, Qin J, Li H, Zhou Y, Fang G (2019) Targeted eradication of gastric cancer stem cells by CD44 targeting USP22 small interfering RNA-loaded nanoliposomes. Future Oncol 15(3):281–295. https://doi.org/10.2217/fon-2018-0295

Ye S, Liu Y, Fuller AM, Katti R, Ciotti GE, Chor S, Alam MZ, Devalaraja S, Lorent K, Weber K, Haldar M, Pack MA, Eisinger-Mathason TSK (2020) TGFβ and Hippo pathways cooperate to enhance sarcomagenesis and metastasis through the hyaluronan-mediated motility receptor (HMMR). Mol Cancer Res 18(4):560–573. https://doi.org/10.1158/1541-7786.MCR-19-0877

Yerneni SS, Solomon T, Smith J, Campbell PG (2022) Radioiodination of extravesicular surface constituents to study the biocorona, cell trafficking and storage stability of extracellular vesicles. Biochim Biophys Acta Gen Subj 1866(2):130069. https://doi.org/10.1016/j.bbagen.2021.130069

Yin M, Li X, Tan S, Zhou HJ, Ji W, Bellone S, Xu X, Zhang H, Santin AD, Lou G, Min W (2016) Tumor-associated macrophages drive spheroid formation during early transcoelomic metastasis of ovarian cancer. J Clin Invest 126(11):4157–4173. https://doi.org/10.1172/JCI87252

You DG, An JY, Um W, Jung JM, Oh BH, Nguyen VQ, Jeon J, Lee J, Jo D-G, Cho YW, Park JH (2022) Stem cell-derived extracellular vesicle-bearing dermal filler ameliorates the dermis microenvironment by supporting CD301b-expressing macrophages. ACS Nano 16(1):251–260. https://doi.org/10.1021/acsnano.1c06096

Yu Q, Stamenkovic I (1999) Localization of matrix metalloproteinase 9 to the cell surface provides a mechanism for CD44-mediated tumor invasion. Genes Dev 13(1):35–48

Zawada AM, Rogacev KS, Rotter B, Winter P, Marell R-R, Fliser D, Heine GH (2011) SuperSAGE evidence for CD14++CD16+ monocytes as a third monocyte subset. Blood 118(12):e50–e61. https://doi.org/10.1182/blood-2011-01-326827
Zembala M, Asherson GL (eds) (1989) Human monocytes, 1st edn. Academic Press, London
Zembala M, Uracz W, Ruggiero I, Mytar B, Pryjma J (1984) Isolation and functional characteristics of FcR+ and FcR- human monocyte subsets. J Immunol 133(3):1293–1299
Zembala M, Siedlar M, Ruggiero I, Wieckiewicz J, Mytar B, Mattei M, Colizzi V (1994) The MHC class-II and CD44 molecules are involved in the induction of tumour necrosis factor (TNF) gene expression by human monocytes stimulated with tumour cells. Int J Cancer 56(2):269–274. https://doi.org/10.1002/ijc.2910560221
Zhang G, Zhang H, Liu Y, He Y, Wang W, Du Y, Yang C, Gao F (2014) CD44 clustering is involved in monocyte differentiation. Acta Biochim Biophys Sin (Shanghai) 46(7):540–547. https://doi.org/10.1093/abbs/gmu042
Zhang G, Guo L, Yang C, Liu Y, He Y, Du Y, Wang W, Gao F (2016) A novel role of breast cancer-derived hyaluronan on inducement of M2-like tumor-associated macrophages formation. Oncoimmunology 5(6):e1172154. https://doi.org/10.1080/2162402X.2016.1172154
Zhang H, Brown RL, Wei Y, Zhao P, Liu S, Liu X, Deng Y, Hu X, Zhang J, Gao XD, Kang Y, Mercurio AM, Goel HL, Cheng C (2019) CD44 splice isoform switching determines breast cancer stem cell state. Genes Dev 33(3–4):166–179. https://doi.org/10.1101/gad.319889.118
Zhao P, Damerow MS, Stern P, Liu AH, Sweet-Cordero A, Siziopikou K, Neilson JR, Sharp PA, Cheng C (2013) CD44 promotes Kras-dependent lung adenocarcinoma. Oncogene 32(43):5186–5190. https://doi.org/10.1038/onc.2012.542
Zhao S, Mi Y, Guan B, Zheng B, Wei P, Gu Y, Zhang Z, Cai S, Xu Y, Li X, He X, Zhong X, Li G, Chen Z, Li D (2020) Tumor-derived exosomal miR-934 induces macrophage M2 polarization to promote liver metastasis of colorectal cancer. J Hematol Oncol 13(1):156. https://doi.org/10.1186/s13045-020-00991-2
Zhou J, Che J, Xu L, Yang W, Zhou W, Zhou C (2022) Tumor-derived extracellular vesicles containing long noncoding RNA PART1 exert oncogenic effect in hepatocellular carcinoma by polarizing macrophages into M2. Dig Liver Dis 54(4):543–553. https://doi.org/10.1016/j.dld.2021.07.005
Ziegler-Heitbrock L, Ancuta P, Crowe S, Dalod M, Grau V, Hart DN, Leenen PJM, Liu Y-J, MacPherson G, Randolph GJ, Scherberich J, Schmitz J, Shortman K, Sozzani S, Strobl H, Zembala M, Austyn JM, Lutz MB (2010) Nomenclature of monocytes and dendritic cells in blood. Blood 116(16):e74–e80. https://doi.org/10.1182/blood-2010-02-258558

Chapter 5
Extracellular Vesicles in Inflammation

Saara Hämälistö and Lujain AlGhadir

Abstract The various immune cells patrol in the body in the search for foreign substances. They follow chemotactic cues, sent by other cells, to enter adjacent or distant tissues and help resolve foreign pathogen attack. The cues are a mixture of secreted well-known bioreactive cytokines and chemokines. The recent years has demonstrated that extracellular vesicles (EVs) are also actively involved in delivering these chemokines and various other inflammatory messages between cells and tissues. This has opened up several lines of research for developing EV-mediated immunotherapeutics, diagnostics, and drug delivery in immune-mediated diseases. The excitement in the EV field has been high for a decade, but only the recent technical developments and increasingly standardizing methods to isolate and analyze EVs have accelerated the research. The complexity of this research field comes from EV heterogeneity and spatiotemporal context dependency of EV secretion; the EV-mediated cargo can e.g. support pro-inflammatory or anti-inflammatory response in the recipient cells or tissues. Additionally, several reports indicate the increase of EVs for instance in patients with autoimmune diseases, findings that are being explored for potential biomarker discovery. This chapter summarizes the current view of EVs in promoting basic processes of innate and adaptive immunity: we also discuss the inflammatory conditions such as autoimmune diseases and antitumor immune responses where EVs are reported to take part.

5.1 Introduction

Innate immune cells (e.g., mast cells, neutrophils, macrophages, eosinophils, basophils, and natural killer NK cells) are characterized by their immediate responses against invading pathogens or harmful agents (Marshall et al. 2018; Chaplin 2010;

S. Hämälistö (✉) · L. AlGhadir
Institute of Biomedicine and FICAN WEST Cancer Center, University of Turku, Turku, Finland
e-mail: saara.hamalisto@utu.fi

K. Rilla (ed.), *Extracellular Vesicles as Matrix Messengers*, Biology of Extracellular Matrix 15, https://doi.org/10.1007/978-3-031-68811-9_5

Häder et al. 2023). Some subsets of the innate immune cells recognize pathogen-associated molecular patterns (PAMPs) on harmful invaders, such as bacteria, by their pattern recognition receptors (PRRs); binding of pathogen to these receptors initiates the secretion of inflammatory cytokines and chemokines to recruit other immune cells (Marshall et al. 2018). This first line of defense is further amplified by the cells of the adaptive immune system, T and B lymphocytes (referred to as T and B cells from here on), which are able to modulate and improve their inflammatory responses or binding affinity toward antigen/pathogen epitopes resulting in target cell killing and antibody production (Yuseff et al. 2013; Hämälistö et al. 2024; Blum et al. 2013). Thus, to elicit a full response to eliminate the pathogen threat, the immune cells must communicate and coordinate their actions with rapid and enduring responses.

For long, the intercellular communication between immune cells was thought to be mediated mainly by transient cell–cell contacts and the following secretion of soluble mediators to the extracellular space and toward recipient cells. However, nowadays it is widely accepted that EVs contribute to this communication and immune cell-derived EVs carry cytokines, receptors, enzymes, and nucleic acid cargo both on their surfaces and inside the vesicle (Groot Kormelink et al. 2018; Buzas 2023; Bui et al. 2018). The EV cargo may for example mediate the down-regulation of the inflammatory response or apoptosis in the target cells; however, current research data indicate that the target cell response is enormously affected by the experimental conditions such as EV amount, incubation time, and EV cell of origin (Buzas 2023; Rayamajhi et al. 2023). It is also not yet known why some molecules e.g. chemokines or proteins are transported to target cells using vehicle (EV)-assisted transport and some in soluble form. It may be that, in contrast to soluble molecules, some proteins are resistant to inhibition when shielded by EVs or they can induce more potent responses in target cells: e.g., EV-associated Hsp70 was able to induce 260 times higher TNF-α production than free-form Hsp70 (Vega et al. 2008). Additionally, the transport routes via endothelium or lymphatic vessels are not uniform throughout various tissue areas especially in inflammatory conditions where the surface is altered: this may represent yet another layer to the comprehensive picture of EV release, uptake, travel, and effect in the body (Iannotta et al. 2024). Understanding the underlying cellular choices, routes, and responses that specific EV populations elicit in target cells would also help in the design of therapeutic vehicles for specific inflammatory conditions. Here, we will cover research from what is currently known of immune cell-derived EV functions during inflammatory responses and the current clinical studies utilizing immune cell EVs. As the literature is vast with various disease categories, we will not be able to cover all aspects of immune cell EV functions, but rather give an overview of the main points. As per the renewed MISEV2023 guidelines (Welsh et al. 2024), we will refer to the released vesicles as "EVs" throughout the text, irrespective of the nomenclature used in respective original articles.

5.1.1 Discovery of EVs in the Immune Regulation

Originally considered as dust or garbage, the EVs have been slowly recognized as important biological entities, although the detailed functions are still to be discovered. EVs exhibit immunomodulatory features such as antigen presentation, immunological stimulation, immune repression, and immune tolerance (Buzas 2023; Shefler et al. 2021). What initially got the attention of immunologists back in 1996 was the finding that membrane-enclosed vesicles released by B cells carried major histocompatibility complex (MHC)-II-peptide complexes and possibly transported them to other immune cells (Raposo et al. 1996). At the same time, membrane vesiculation or microparticles produced by platelets, monocytes, or neutrophils were investigated, largely for their function in modulating vascular homeostasis (Sims et al. 1989; Satta et al. 1994). In the late 1990s and early twenty-first century, EV-mediated intercellular communication was postulated by several researchers examining the effects of plasma membrane (PM)-derived microparticles: they released anti-inflammatory mediators affecting adhesion, survival, and endothelial cell activation (Sims et al. 1989, Mesri and Altieri 1998, Baj-Krzyworzeka et al. 2002, Gasser and Schifferli 2004). These promising functions of EVs as immune response mediators, as well as the suggestion to utilize EVs as immunotherapeutic agents, have pushed a new branch into immunological research with several findings to be expected in the coming years both in vitro and in vivo.

5.1.2 Intercellular Communication by Immune Cells: Traditional Ways

If a foreign pathogen gets through the physical skin barrier and enters the body, then the human immune system is responsible for reacting to the incoming antigenic foreign particles. The immune cells eliminate the threat in coordination with other biological systems in the body, utilizing contact-independent or contact-dependent mechanisms (Miskolci et al. 2021) in differentially structured tissue (i.e., altered tissue stiffness) environments (Du et al. 2023). Transient cell–cell contacts mediated e.g. via integrin adhesion receptors during leukocyte rolling on the endothelial wall are needed to allow the adhesion and the subsequent cytokine release and intravasation of the cells into the underlying tissue (Mitroulis et al. 2015; Miskolci et al. 2021). Contact-dependent interactions between immune cells allow the boosted chemical release: for example, activated human NK cells and immature dendritic cells (DC) grown in 1:5 ratio were shown to result in exponential increases in DC cytokine production and were completely dependent on cell-to-cell contact (Piccioli et al. 2002). Macrophages were shown to undergo epithelial reprogramming during granuloma formation, which includes the upregulation of E-cadherin expression that subsequently mediates the intercellular adhesion between macrophages (Cronan et al. 2016). Also, a specialized structure called immune synapse (discussed in more detail in Chap. 8) represents a site of intercellular communication with a key role in physical contact between cells and exchange of soluble mediators during antigen presentation (Yuseff et al. 2013; Céspedes et al. 2022; Kuokkanen et al. 2015).

5.1.3 Intercellular Communication by Immune Cell EVs

All cells of the immune system secrete EVs in resting state and frequently increase the release of EVs upon activation via receptors or by cell–cell contacts (Groot Kormelink et al. 2016, 2018). During their receptor FcεRI activation, the innate immune mast cells both increase the release of EVs and alter the EV lipid composition, density, and size (Groot Kormelink et al. 2016). The cargo of mast cell EVs is dependent on the activation status of the cells: upon IgE-mediated stimulus, the release of EVs is significantly increased and the EVs contain for example tetraspanin CD63 and beta-hexosaminidase (Groot Kormelink et al. 2016). Furthermore, the OX40-ligation mediated release of CD63-positive mast cell EVs is demonstrated to induce differentiation of naïve CD4 T cells to Th2 helper subtype (Groot Kormelink et al. 2016; Li et al. 2016).

Even if several molecules are common between different immune cell EV populations, the molecular composition is largely dependent on the status and origin of the donor cell, thus affecting the impact in the recipient cell (van Niel et al. 2018). A plethora of molecules that bind to their cognate receptors and provoke immune regulation are identified on the surface of immune cell EVs (Buzas 2023). For example, the death signal FasL or immune checkpoint molecules, e.g., cytotoxic late antigen-4 (CTLA-4) and programmed cell death (PD-1) molecules, are found on the surface of for example T regulatory (Treg) cells or T helper cells (Buzas 2023; Cai et al. 2012; Smyth et al. 2013; Qiu et al. 2021). In Treg cells, the ectoenzyme CD73 promotes adenosine production from ATP which then suppresses immune reactions (Smyth et al. 2013; Clayton et al. 2011). The key molecules that convey messages to other cells, cytokines and chemokines, work via binding to their cognate receptors on the cell surface: however, these soluble factors can also travel on the surface of or inside EVs (Fitzgerald et al. 2018). With several proteins decorating their surfaces, the EVs themselves may also bind with specificity to the target cell proteins: for example, the B cell-derived EVs bind to Siglec-1/CD169 receptor on macrophages via B-cell EV surface-expressed α2,3-linked sialic acids (Saunderson et al. 2014).

The EV entry into recipient cells may affect the inflammatory response in multiple ways (Buzas 2023; Fig. 5.1). For example, in sepsis, a condition where the body reacts massively to an infection, the plasma EVs carry a monomeric form of the soluble C-reactive protein CRP that triggers an inflammatory response via the release of CXCL8 from monocytes (Fendl et al. 2021). This in turn triggers the recruitment of neutrophils to the site, thus exaggerating the inflammatory response. This and a multitude of other examples nicely underline the interactions between

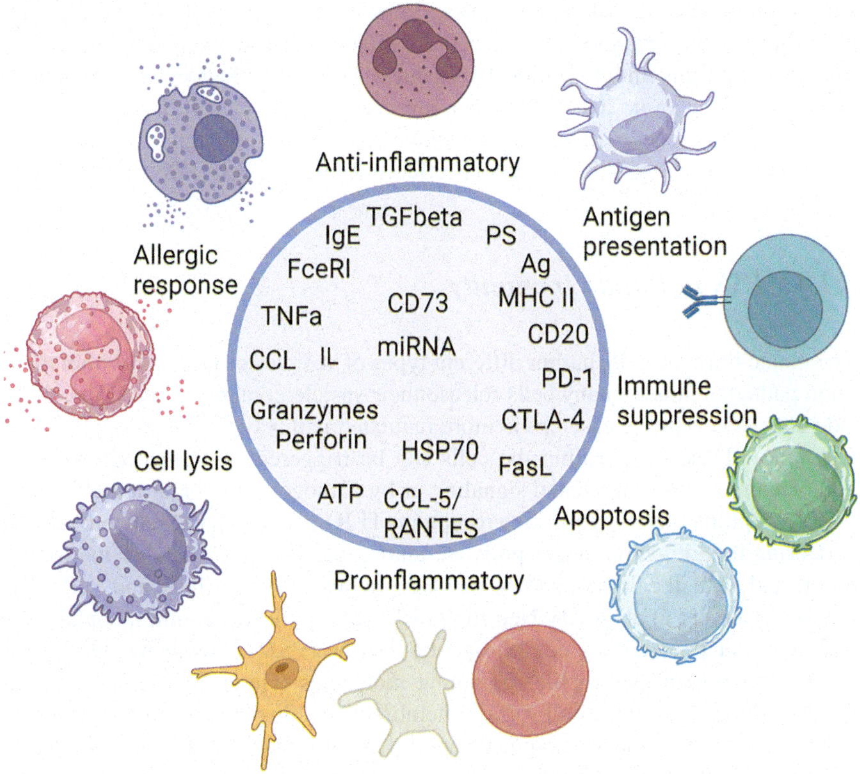

Fig. 5.1 The various immune cells and examples of cargo molecules and possible effects on target cells. The innate immune cells, neutrophils, release TGF-beta and thereby downregulate macrophage inflammation. The dendritic cells and B cells release EVs that express antigen (Ag) and MHC-I or MHC-II molecules for antigen presentation. Malignant B-cell EVs upregulate CD20 receptor on the EV surface. Immune checkpoint molecules CTLA-4 and PD-1 are found on the T helper cell EVs and they mediate immune suppression in their target cells. The death signal FasL is shed on the T regulatory cell EVs and can induce apoptosis in target cells. In regulatory T cells, the ectoenzyme CD73 promotes adenosine production from ATP which suppresses immune reactions. Platelets secrete EVs containing pro-inflammatory cytokines such as CCL5 (also known as RANTES). Red blood cells are decorated with phosphatidylserine (PS) and contain e.g. Hsp70 and reportedly over 78 different miRNAs that modulate inflammatory response in target cells. Microglia (the specialized brain macrophages) EVs contain ATP that signals danger to other microglia cells in the brain microenvironment and promote inflammation. Human NK cells contain lytic enzymes such as perforin and granzymes as well as FasL that all mediate efficient target cell killing. Antigen-presenting cells (APC) and CD8+ T cells decorate their EVs with PS and can modulate adaptive immune cells (dendritic cells) or boost the immune response in the recipient cells (T cells). Eosinophils and mast cells promote allergic responses and contain for example interleukins (IL)-4, -5, -6, chemokine ligands CCL-2 and CCL-17, and tumor necrosis factor TNF-α inside their EVs

various immune cells with secreted cues. Previous research has shown that DCs and macrophages, among other immune cells, are the primary recipients of immune cell EVs; however, this finding is most likely due to their high phagocytic cellular characteristics (Feng et al. 2010). Table 5.1 summarizes examples of different types of immune cell EVs, cargo molecules, and their tissue targets, though it is good to note that the literature regarding the various effects of immune cell EVs is far broader.

5.1.4 EVs in Innate Immunity

The innate immune cells harbor different types of vesicles or granules within them: upon pathogen attack, many cells release their vesicle contents to the extracellular side either via rapid release or via more regulated pathway (Lacy and Stow 2011). Cytokine release from within the cells can be triggered by immunoglobulin- or complement receptor-mediated signaling or by pathogens via a multitude of membrane receptors, including toll-like receptors (TLR) (Iwasaki and Medzhitov 2010).

During inflammation, neutrophils are among the first to respond to the foreign threat and take the inflammatory message to wider dimensions and cell types (Herrero-Cervera et al. 2022). Neutrophil EVs, also called ectosomes, contain several common EV markers e.g. phosphatidylserine (PS), Annexin A1, tetraspanin CD63, and several neutrophil-originating molecules (Hong 2018). The EVs are released directly from the cell surface membrane and display intriguing spatiotemporal features in EV deposition: neutrophil-derived trails (NDTRs) are released by migrating neutrophils, whereas neutrophil-derived microvesicles (NDMVs) are generated by neutrophils that already migrated to the site of inflammation (Hong 2018). According to some studies, the resting state neutrophil EVs promote anti-inflammatory reaction by downregulating ROS production in the recipient cells whereas the activated neutrophil EVs contribute to increased inflammation, increased ROS production, and upregulation of endothelial cell activation markers E-selectin and VCAM-1 (Kolonics et al. 2021). Neutrophil EVs expressing myeloperoxidase have been demonstrated to inhibit wound healing in the intestinal epithelium (Slater et al. 2017). Some studies indicate that neutrophil EVs have no pro-inflammatory activity on human macrophages as assessed by the release of interleukin 8 (IL-8) and tumor necrosis factor alpha (TNF-α), but instead they increase the release of transforming growth factor beta1 (TGFβ1), which suggests that EVs downregulate activation in macrophages (Gasser and Schifferli 2004).

Monocyte EVs can enter endothelial cells and modulate TLR-4 receptor and NF-κB regulation leading to increased CCL (chemokine ligand)-2, IL-6 mRNA, and intercellular adhesion molecule 1 (ICAM-1) expression on the endothelial cells (Tang et al. 2016). Stimulation of monocyte EVs with endotoxin on the other hand can alter the glycans and glucose transporter-1 (GLUT-1) expression in the endothelial cells (Yang et al. 2022). NK cells, when stimulated for example via the cell surface receptor CD16 (Jong et al. 2017), release EVs filled with cytotoxic enzymes granzyme A&B, granulysin, and perforin: the EV enzymes can then attack against

Table 5.1 Examples of immune cell EV-targeted functions

EV origin	Target cell	Reported cargo/mode of activation	Effect	Phenotype in target	References
NK cells	T cells	miR-10b-5p, miR-92a-3p, and miR-155-5p	Downregulation of GATA3 mRNA	Th1 polarization and IFN-γ and IL-2 production	Dosil et al. (2022)
	Monocytes	MHC-II, CD86	Increased IFN-γ and IL-12 secretion	Dendritic cell polarization	Dosil et al. (2022)
	Hematological and other cancer cells	Perforin, granzymes, FasL	Initiate direct killing pathway	Cell death	Jong et al. (2017)
Neutrophils	Endothelial cells	EVs from activated neutrophils	ROS production	Upregulation of E-selectin and VCAM-1	Kolonics et al. (2021)
	Intestinal epithelial cells	Myeloperoxidase	Impeded actin dynamics and cell spreading	Inhibition of wound healing	Slater et al. (2017)
Monocytes	Endothelial cells	IFNα or LPS stimulation	miRNA cargo alteration, activation of NF-kappaB	Increased adhesion and cytokine release	Tang et al. (2016)
	Endothelial cell	Endotoxin-stimulated EV	Increased glucose transporter-1, alterations in glycans	Pro-inflammatory reaction	Yang et al. (2022)
Dendritic cells	T cells or other dendritic cells	MHC-II HLA-DQ	Several, e.g., cytokine release	CD4+ T-cell proliferation	Théry et al. (2009)
	n/a	TGF-beta1 overexpression	Inhibition of Th17, upregulation of Treg	EAE decreased disease burden	Yu et al. (2013b)
Mast cells	CD4+ T cells	OX40L, CD63	OX40 ligation to receptor	Differentiation of naive CD4+ T cells to TH2 cells	Li et al. (2016)
	n/a	CD63, β-hexosaminidase	Depending on mast cell stimulus	n/a	Groot Kormelink et al. (2016)
Platelets	Monocytes	miR-223	Downregulate IL6 and NLRP3	Reduction in inflammation by CBP	Poon et al. (2017)
	Leucocyte/endothelial cells	CCL-5 / RANTES	Pro-inflammatory to endothelium	Leucocyte adhesion to endothelium	Mause et al. (2005), Hwang et al. (2021)

(continued)

Table 5.1 (continued)

EV origin	Target cell	Reported cargo/mode of activation	Effect	Phenotype in target	References
Red blood cells	Monocytes	Small EVs	Pro-inflammatory signaling, TNF-alpha	T cell proliferation	Danesh et al. (2014)
	Blood transfusion	EV scavenging of NO	Nitric oxide homeostasis	Hemolysis	Thangaraju et al. (2020)
Macrophages	Naive macrophages	LPS stimulated macrophages	NF-kappaB activation	Reduced inflammatory paw swelling	McDonald et al. (2014)
	Dendritic cells	Modulation with hyaluronic acid (HA), 3-(diethylamino) propylamine (DEAP), monophosphoryl lipid A (MPLA), and MUC1	Triggered EV uptake in DCs, TAA release in endocytic compartment	Improved antigen presentation and T-cell activation	Lee et al. (2018)
B cells	B cells	Functional B-cell receptor and CD24	Increase in receptor activation	Antigen-mediated apoptosis	Phan et al. (2021)
	Cervical cancer cells	Monomeric IgM	Sustained antigen binding	Accumulation in target cells	Gutknecht et al. (2023)
T cells	Dendritic cells	DNA	cGAS-STING cytosolic DNA-sensing pathway	Protection against pathogen	Torralba et al. (2018)
	B cells	miRNAs miR-20a-5p, miR-25-3p, and miR-155-3p	Gene silencing	PTEN and BIM downregulation	Fernández-Messina et al. (2020)
Treg cells	Dendritic cell	miRNAs miR-150-5p, miR-142-3p	Cytokine production	IL-10 increased, IL-6 decreased	Tung et al. (2018)
Microglia	Microglia	ATP	Danger signal	Increased inflammation	Hwang et al. (2021) (Review)
SLE plasma	n/a	Mitochondria, IgG	n/a	Higher disease state	Mobarrez et al. (2019)
Sepsis plasma	Monocyte	CRP-positive EVs	IL-8 increased secretion	Pro-inflammatory response	Fendl et al. (2021)

the pathogenic particles or enter cancer cells where the uptake may promote caspase activation and cell death (Jong et al. 2017; Vojdani et al. 2024). The NK cell EVs can also modulate the polarization of T cells and monocytes via delivering a mixture of miRNAs to T cells or MHC-II and activatory molecule CD86 to monocytes (Dosil et al. 2022).

Macrophages, the phagocytic cells of the immune system, when infected with microorganisms, release EVs harboring antigens with pathogen-associated molecular patterns (PAMPs). These EVs can boost the production of pro-inflammatory cytokines or chemokine TNF-α by engaging toll-like receptors (Volgers et al. 2017). Also, when stimulated by lipopolysaccharides, the macrophages release EVs that can relieve inflammatory swelling and pain when introduced into naïve macrophages (McDonald et al. 2014). Additionally, with parasite infections, they can elicit an EV-dependent response that either dampens or activates immune cells in a stage-specific manner as was shown for helminth parasite EVs deactivating macrophages (Coakley et al. 2017). In the confined area within the brain, the microglia, also considered as the resident macrophage cells of the nervous system, are the first to respond to pathogenic signals upon neuroinflammatory reaction (reviewed in Hwang et al. 2021). Upon activation, the phagocytic microglial cells multiply in numbers and each release EVs filled with ATP, thus sending danger signals across the brain (Hwang et al. 2021). Thus, nearby microglial cells and astrocytes are stimulated and secrete inflammatory signals which further disseminate the inflammation.

One interesting aspect to the diversity of the innate immune cell population are the erythrocytes and platelets, which are anucleated cells that possess many immune cell-like features. Erythrocytes, i.e., red blood cell EVs, loaded with proteins, lipids, and different miRNAs communicate with e.g. the endothelial cells and regulate nitric oxide (NO) and O_2 homeostasis and vasodilation at the endothelium, redox balance, and immunomodulation (Ma et al. 2023a; Thangaraju et al. 2020). On the other hand, even though debate is still ongoing on whether platelets are real cells or not, it is clear that they secrete EVs and modulate both innate and adaptive immune cells (Garraud and Cognasse 2015). The platelet EVs (PLT-EVs) contain a multitude of biomarkers e.g. CCL5 (or RANTES), and this cargo may induce pro-inflammation in for example eosinophils (Mause et al. 2005) or downregulate inflammation in monocytes (Poon et al. 2017). The platelet EVs can go beyond barriers in their quest to modulate inflammatory responses: it was shown by French et al. that during inflammation, platelet-like particles enter the bone marrow and platelet EVs induce the biogenesis of hematopoietic cells (French et al. 2020).

5.1.5 EVs in Antigen-Presenting Cells

Dendritic cells, B cells, and macrophages are specialized antigen-presenting cells (APCs) whose primary purpose is to capture, process, and present antigens to other immune cells (Rastogi et al. 2022). Recent emerging data show that virtually all innate immune cells communicate with DC cells, the main APC type of the innate immunity, and with T and B cells, thereby influencing the adaptive immune responses. The APC cells release EVs that target T lymphocytes, natural killer (NK) cells, and NK T cells and alter their immune functions with distinct cargo molecules (Théry et al. 2009). EVs generated by APCs include membrane proteins that are involved in antigen presentation, e.g., the MHC-I and MHC-II complexes, as well as several costimulatory molecules and for instance the phosphatidylserine (PS) that can bind and modulate the adaptive immune cells (Kranich et al. 2020). Several studies have shown that when DCs and B cells interact with activated CD4+ T cells, they secrete more EVs and proteins including MHC-II and costimulatory molecules that are transported from DCs to CD4+ cells via EVs (Muntasell et al. 2007; Buschow et al. 2009; Nolte-'t Hoen et al. 2009). Furthermore, B or T cells secrete relatively little EVs in the steady state (i.e., without activation), but activation significantly increases multivesicular body (MVB)-derived EV secretion through interactions with T- or B-cell receptors (Muntasell et al. 2007; Mittelbrunn et al. 2011). Such EVs can activate cognate T-cell clones or lines, as well as CD4+ T cells that have already been activated (Buschow et al. 2009). Upon activation of their cell surface P2 receptors, mature DCs were shown to release EVs containing high levels of IL-1β, a pro-inflammatory cytokine, and enzymes such as caspase-1 and -3 and cathepsin D (Pizzirani et al. 2007). One study showed that EVs from mature DCs can activate naïve antigen-specific T cells, after the DCs have first collected the EVs with their lymphocyte function-associated antigen 1 (LFA-1) receptor (Segura et al. 2007).

EVs secreted by DCs, loaded with specific peptides or antigens, can activate CD8+ or CD4+ T cells in an antigen-specific manner, even in the absence of DCs. Similar CD8+ T-cell responses were observed in EVs generated from cells that received a viral peptide (Luketic et al. 2007). Furthermore, macrophage-derived EVs can transfer their surface antigens to DCs, increasing the response of CD4+ T cells (Xu et al. 2016). DCs participate in "cross-dressing," which is the transition of preformed functional peptide–MHC complexes from the surface of the peptide donor cells to the cell surface of an acceptor APC or another type of recipient cell (reviewed in Campana et al. 2015). EVs produced from APCs carrying antigenic peptide–MHC complexes have been shown to be delivered to T cells in different ways: (1) a subset of the transferred EVs remain adhered to the DC surface and are directly presented to T cells with DC co-stimulation or (2) DCs can internalize the EVs and present the antigen to acceptor cells or (3) autologous EVs are formed that carry the antigen and are then presented to T cells (reviewed in Campana et al. 2015). A model of cross-dressing is nicely demonstrated in acute allograft transplantation rejection, where cross-dressing occurs between recipient APCs and EVs carrying donor MHC-II or MHC-I (Bracamonte-Baran et al. 2017; Burlingham 2017). Overall, the research data thus far confirm the ability of immune cell-derived EVs to present antigens and produce immunomodulatory effects in target cells.

5.1.6 EV Functions in Adaptive Immunity: T Cells

The various T-cell subsets influence and govern cellular and humoral immunity. The T cells are classified into two types based on their phenotype: CD4+ helper T (Th) cells and CD8+ cytotoxic T cells (CTL). CD4+ T cells can be further characterized into several subsets: Th1, Th2, Th9, Th17, Th22, follicular helper T cells, and Tregs, based on their particular effector cytokine expression profiles produced under unique transcriptional control (Harrington et al. 2005). T cell-derived EVs are thought to function similarly to their cell of origin, thereby either efficiently activating or inhibiting the immune response, or engaging in autoimmune and infectious illnesses (Mittelbrunn et al. 2011). Studies have also indicated that EVs secreted by T cells may play a role in insufficient T-cell activities that characterize various human autoimmune disorders, allergies, and tumor immunity (Zhang et al. 2011; Gutiérrez-Vázquez et al. 2013). As a result, studying T cell-derived EVs could have a direct and practical impact on using engineered EVs to treat e.g. autoimmune disorders or malignancies.

T-cell membranes include a large number of functionally active molecules, e.g.. TCR, CD40L, RNA-binding proteins, and miRNAs with key roles in cell activation, differentiation, antigen presentation, and effector activities (Buzas 2023; Céspedes et al. 2022; Durlanik et al. 2016). T cell-derived EVs, like EVs formed by other cells, exhibit identical molecules to their parental cells on their surface, implying that EVs can share the functional capacities of their original cells. The presence of DNA on the extracellular side of the EVs of T cells has been shown to mediate interaction to the extracellular environment molecule fibronectin (Németh et al. 2017). During infection, pathogens not only cause cell damage at the infection site, but they can also produce EVs that carry immune-modulating antigens. This could potentially be a means for the infected cell to indicate danger to immune system cells. This has been demonstrated with HIV in T cells and human cytomegalovirus-infected endothelial cells (Walker et al. 2009; Martin-Jaular et al. 2021).

Treg cells are unique T cells distinguished by the expression of the transcription factor forkhead box P3 (Foxp3) and immunosuppressive actions that provide mechanisms for immunological tolerance and promote tissue repair (Baek et al. 2016). However, several preclinical studies suggest that Treg cells fail to upkeep the needed surveillance, thus leading to hampered cancer immunity and cancer progression (Shitara and Nishikawa 2018). Treg-derived EVs, on the other hand, have enormous promise for regulating the body's immunity and can also be used to prevent transplant rejection (Agarwal et al. 2014). For example, adoptive transfer of autologous Treg-derived exosomes from rats can improve kidney function and extend the survival of kidney allografts after transplantation by preventing immune cells from mounting an inappropriate response against the transplant (Yu et al. 2013a).

5.1.7 EV Functions in Adaptive Immunity: B Cells

B cells are unique in two ways: first, each B cell carries a one-of-a-kind B-cell receptor (BCR) on its surface: upon encountering an antigen (on a pathogen) the BCR gets activated and it triggers multiple intracellular signaling cascades (Kuokkanen et al. 2015; Hämälistö et al. 2024). This subsequently initiates the mutation-driven adaption of the BCR for it to more efficiently bind to the antigen. Additionally, it creates immunological memory that will speed up the response in the event the same pathogen would enter the body again (Nouri and Kleinstein 2020). The second major distinction to any other cell type is the capability of launching antibody production in plasma B cells upon antigen encounter (Sharonov et al. 2020).

B cells secrete EVs like other cells and proteomic analyses of B lymphoma cell EVs have revealed EV enrichment of proteins related to ubiquitination (DUBs) or cell adhesion (CD44 antigen, ICAM1, tyrosine-protein kinase JAK1) (Carvalho et al. 2021). B cell EVs are capable of stimulating CD4+ T cells with the antigenic peptide–MHC-II complex which then further invigorates the B-cell activation. Recent research showed that IgM was deposited from several B cell lines in a monomeric form that was incorporated into EVs and taken up by recipient cells whereas the pentameric form of IgM was secreted as soluble form into the extracellular space (Gutknecht et al. 2023). This indicates that the packing of intracellular effector molecules occurs via predetermined routing. Research on murine B cell models have shown that EVs can also deliver functional receptors to recipient B cells and trigger enhanced antigen-mediated apoptotic response via CD24 (Phan et al. 2021). Already decades ago it was postulated that B cells release vesicles with functional MHC-II molecules and other costimulatory signals on their surface (Raposo et al. 1996), making them ideal for use in exosome-based cell-free therapeutic vaccines. This also implies that these EVs can sustain a disease-free state by possibly targeting tumor antigenic peptides and boosting the immune system.

5.1.8 Specialized Structures in Adaptive Immunity: Immune Synapses

In order to mount an effective immune response, the B or T cells and APC need to physically interact in order to create a specific cell–cell contact site between them, called the immune synapse (IS; Yuseff et al. 2013; Céspedes et al. 2022; Kuokkanen et al. 2015). The synapse can also form between an immune cell and a target cell that needs to be eliminated, such as in the case of cytotoxic T cells or NK cells (Angus and Griffiths 2013; Zhang et al. 2011). This involves the reorganization of both the transmembrane and membrane-associated molecules into an elegant structure where actin cytoskeleton provides the physical context for IS formation and microtubules assist with the delivery of the vesicle populations toward the cell–cell contact site (Yuseff et al. 2013; Mittelbrunn et al. 2011; Bello-Gamboa et al. 2020; Gutiérrez-Vázquez et al. 2013). In the synapse between B cells and APC, the antigen exposure induces typically B-cell actin-driven spreading within 5 min and creates actin-rich signaling hot spots in the synapse (Kuokkanen et al. 2015). In these sites, the antigen interacts with the cognate receptor and is also internalized for further processing (Kuokkanen et al. 2015).

T cells, stimulated by T-cell receptor (TCR) activation, release EVs toward the immune synaptic cleft via fusion of the multivesicular body to the plasma membrane (Bello-Gamboa et al. 2020) and the T-cell EVs containing the TCR traffic within 10–30 min from synapse formation to APC cells (Céspedes et al. 2022). In a synapse involving T and B cell, the CD40L ligand is transferred from T cells to B cells via its corresponding CD40 receptor (Gardell and Parker 2017). It is intriguing that the EVs at the trans-synaptic cleft are more condensed with effector content (e.g., RNA-binding proteins and microRNAS) than EVs that are constitutively released from T cells (Céspedes et al. 2022). Indeed, it has been shown that the release of miRNA-loaded EVs from T cells to APC cells was shown to modify target cell gene expression (Mittelbrunn et al. 2011; Tung et al. 2018). Also, immune contacts between T and B cells allow for transfer of a very restricted set of T-cell EV-microRNAs (e.g., miR20-a-5p, miR-25-3p, and miR-155-3p) to the B cell (Fernández-Messina et al. 2020). These EV-microRNAs target genes related to B-cell function, including pro-apoptotic Bcl-2 interacting mediator of cell death (BIM) and the cell cycle regulator phosphatase and TENsin homolog deleted on chromosome 10 (PTEN), and promote survival, proliferation, and antibody class switching (Fernández-Messina et al. 2020).

To understand the complexity of the released particles, the repertoire of various vesicular pools in the synapse also consists of other similarly defined particles that are not to be mixed with EVs which are released to the synaptic cleft area. Namely the cytotoxic (killing) T cells release perforin-1 and granzyme B-positive carbon-dense particles to the synaptic cleft to mediate the killing of the adjacent cell (Bálint et al. 2020).

5.1.9 Traveling of EVs

EVs can influence cells in their immediate microenvironment or at distant locations via circulation. At the moment, scarce information exists about the distances traveled by the EVs in vivo. EVs were recently discovered to be capable of reaching a wide range of lymphoid organs through contact with phenotypically matching cells (Wiklander et al. 2015). In situ, EVs can cross the blood–brain (Busatto et al. 2021) and blood–testis barrier (Ma et al. 2023b) but e.g. the total distance and regulation of EV movements are currently unknown. The EV research field has taken advantage of for instance the zebrafish models and analysis of the live movements of the fluorescently labeled EVs to understand their traveling to possible target tissues (Hyenne et al. 2019; Verweij et al. 2021). The research has shown for instance that circulating tumor EVs are preferentially taken up by endothelial cells and macrophages (Hyenne et al. 2019). It is currently unknown how the EVs choose to travel short or long distances in vivo; also, it remains to be studied to what extent the epithelial lining and its heterogeneity affects the entry of EVs into different sites and tissues (Iannotta et al. 2024). The endothelial lining of the vessel walls is disturbed in various disease conditions with leakage or local permeability on the endothelial wall that may facilitate the passage of plasma proteins (Claesson-Welsh et al. 2021; Edrissi et al. 2016). One elegant study, performed in mice, looked into the biodistribution of labeled dendritic cell-derived EVs in mouse tissues. They found out that the targeting of EVs into tissues was affected by the dose of injected EVs and the route of administration (Wiklander et al. 2015). Additionally, the cargo proteins on the EVs affected the tissue targeting in a way that while most EVs were targeted to the spleen and liver, the addition of a glycoprotein to the EV surface deviated the route to acetylcholine-rich organs (Wiklander et al. 2015). Another study analyzed the biodistribution of mir-155-loaded EVs in Mir-155 KO mice and was able to demonstrate that most EVs with their mir-155 cargo was uptaken by liver hepatocytes and adipose tissue, lung, muscle, and kidney with the liver taking the most (Bala et al. 2015). These EVs carried cargo that modulates inflammation and immune cell functions at several levels, such as the abovementioned miR-155, an inflammation-associated miRNA whose abnormal expression has been linked to a variety of illnesses (Bala et al. 2015).

5.1.10 EVs in Immune-Mediated Diseases: Allergy

EVs released from damaged or infected cells have been shown to influence the disease state of numerous cells, primarily inflammatory disorders such as asthma, allergies, or infections (Sanwlani and Gangoda 2021). Due to their small size, the EVs can be airborne and inhaled, and in allergic responses, EVs are recognized both as carriers of the allergens and as modulators of the responses (Buzas 2023). Mast cells are the main innate immune cell type to elicit inflammatory responses during allergic reactions. They originate from hematopoietic cells and along vasculature they can reach any tissue, where they then fully mature (Shefler et al. 2021). The mast cells are activated e.g. via allergen-specific IgE or microbial protein binding. Upon activation, they secrete regulators like histamines, proteases, or proteoglycans from their granules and mast cell-derived EVs are decorated with immune complexes and IgE (Molfetta et al. 2020) which are substantially elevated for instance in the plasma of atopy patients. Mast cell EVs are uptaken by e.g. T cells and can modulate their differentiation into specific subsets (Li et al. 2016). However, the role of mast cell EVs in eliciting allergic reactions is still controversial (Molfetta et al. 2020).

In addition to mast cells, eosinophils take part in the allergic response by releasing chemokines and cytokines from the preexisting or de novo regulated granules (Davoine and Lacy 2014). Eosinophil EV secretion is augmented by IFN-gamma induction, and in asthmatic patient samples, they are characterized as being e.g. positive for Annexin V, CD9, and CD63 expression; upon stimulation of eosinophils by TNF-alpha they secrete relatively more EVs than by stimulating with CCL11 for example (Mazzeo et al. 2015; Akuthota et al. 2016). However, the research data on eosinophil EVs are still at its infancy.

5.1.11 EVs in Autoimmune Diseases

Based on research on cellular models, EVs can regulate immunological processes and cause inflammation (Buzas 2023). Clinical studies demonstrate a high concentration of EVs in the bronchoalveolar lavage fluid of autoimmune patients with sarcoidosis (Qazi et al. 2010): here, the EVs were shown to significantly increase the IFN-γ and IL-13 production in PBMCs and drive IL-8 production in epithelial cells, potentially leading to inflammation (Qazi et al. 2010). Also, patients with rheumatoid arthritis, another autoimmune disease, the EVs in the synovial fluid were shown to boost the immune inflammatory response and worsen the condition by e.g. upregulated effector matrix metalloproteinase (MMP-13) and A disintegrin and metalloproteinase with thrombospondin motifs 5 (ADAMTS-5) in the chondrocytes (Kato et al. 2014). In systemic erythematosus lupus (SLE), the patient plasma EVs were larger in size and carried mitochondria and had IgG decorating their surfaces (Mobarrez et al. 2019).

In diseased conditions, EV concentrations are reportedly raised in the systemic circulation or other bodily fluids, and for example, during hypoxia-induced inflammation, miRNA cargo in EVs modulates transcriptional regulation, and depending on the pathway, they may either promote or downregulate inflammation via affecting macrophage polarization (Kalluri 2016; Couch et al. 2017; Li et al. 2021). Pathogen-associated molecular patterns (PAMPs) on EV surfaces can enhance immune surveillance (Bhatnagar et al. 2007): EVs from infected macrophages can trigger TNF-α production by macrophages and neutrophil or macrophage EVs can promote IL-8 production in renal tubular cells and activate neutrophil migration into renal interstitium (Singhto and Thongboonkerd 2018). On the other hand, in murine models of experimental autoimmune encephalomyelitis (EAE), EVs loaded with TGF-β1 were able to inhibit disease progression via decreasing T helper Th17 cells in the mouse (Yu et al. 2013b).

5.1.12 EVs in Cancer Immunity

Dendritic cell EVs, in particular, have attracted attention for their ability to invigorate tumor-specific immune responses (Pitt et al. 2016). Also, attempts to modify EVs for more potent function have been made: for example, IFN-γ -maturated DC EVs enhance the tumor-specific immune response effect by NK cells or pH-responsive EVs bind specific receptors thereby lowering tumor growth (Besse et al. 2015; Lee et al. 2018). Macrophages on the other hand can suppress tumor immune evasion by taking up melanoma EVs to prevent their tumor-promoting interactions with B cells (Pucci et al. 2016). EVs can also migrate through the lymphatic system and reach lymph nodes, as demonstrated by subcutaneous injection of melanoma EVs (Hood et al. 2011). EV-mediated transfer of tumor antigens to distant macrophages and DCs leads to the secretion of high levels of pro-inflammatory cytokines. They then promote tumor invasion and metastasis by stimulating NF-κB activation in macrophages or toll-like receptor (TLR) 2 or TLR4 in DCs (Shen et al. 2017).

EV-associated tumor antigens are detected in a variety of body fluids. Melanoma patient-derived EVs contain the melanoma-associated antigen recognized by T cells 1 (MART-1) antigen, and when these EVs are pulsed with DCs, they produce and grow clones of tumor-specific cytotoxic T cells against the MART-1 antigen (Andre et al. 2002). Similarly, integrin alphaVbeta6 transfer from prostate cancer EVs to monocytes can modulate their adaptation toward M2 phenotype (Lu et al. 2018).

Intriguingly, the EVs shed by metastatic melanoma cells boost, via interferon (IFN)-γ, the EV expression of PD-1, thereby leading to loss of function in CD8+ T cells and tumor outgrowth (Chen et al. 2018). Of interest is also that in human B-cell lymphomas, the EVs shed by the malignant cells upregulate the expression of CD20 and facilitate the binding of anti-CD20 rituximab antibody to the EVs, thereby preventing the optimal treatment to eliminate the lymphoma cells (Aung et al. 2011). In prostate cancer, on the other hand, Src signaling molecules are enriched in EVs shed by the cancer cells and are promoted for biomarker studies (DeRita et al. 2017). The accumulating evidence on specific cargo expression or functional impact in tumorigenesis has led to an increase in clinically oriented EV research.

5.1.13 Clinical Implications of Immune Cell EVs

The enriched target expression of markers on the surface of EVs, such as CD20 and PD-L1, are hypothesized to contribute to increased therapeutic resistance in several cancers (reviewed in Marar et al. 2021). Recent clinical work was include e.g. removing these resistance-conferring EVs from the patient circulation (patients with head and neck carcinoma) using a specific device, called the hemopurifier before treatment with a-PD-1 pembrolizumab immunotherapy (https://clinicaltrials.gov/show/NCT04453046). Similar approach is tested for sepsis patients with elevated CRP or CRP-positive EVs, PentraSorb, which efficiently removes CRP-positive particles from the plasma (Fendl et al. 2021). Whether these attempts prove as efficient strategies to remove excess malignant EVs from the patients, it remains to be seen if it can be adapted for use in e.g. autoimmune diseases where the EV count is often significantly increased (Qazi et al. 2010; Kato et al. 2014; Mobarrez et al. 2019). The significance of understanding the intercellular communication

Table 5.2 Ongoing or recent clinical trials based on EV immunotherapy or diagnosis

ID	Location	Disease	EV source	Biomarkers	Endpoint
NCT03985696	France	B-non-Hodgkin lymphoma	Blood	CD20, PD-L1	EV role in immunotherapy escape
NCT05705583	China	Renal cell carcinoma	Blood and urine	PD-L1	Predictive biomarker
NCT05575622	China	Hepatocellular carcinoma	Blood	PD-L1, LAG-3	Predictive biomarker
NCT04664738	n/a	Skin graft	Platelet	Anti-inflammatory and angiogenic factors	Wound healing
NCT02565264	Japan	Various	Plasma	n/a	Wound healing

between different cells' EVs in the body is exemplified by the ongoing or recently performed clinical trials where the EVs are tested as vehicles in wound healing or as biomarkers in specific cancers such as renal or liver cancer or B-cell lymphoma (Table 5.2; Yang et al. 2021).

5.2 Conclusion and Future Perspectives

It is nowadays well accepted that immune cell EVs take part in mediating signals to other immune cells. Many aspects still remain to be addressed regarding the immune cell-specific (spatiotemporal) mechanism of EV release, delivery into target tissues, and the impact therein. For example, IgM from B cells is transported as pentameric vs monomeric isoform when soluble or inside EVs, respectively, but the reasons for these types of different modes are not known (Gutknecht et al. 2023). Also, cytokines may be located on the surface of EVs or transported hidden inside EVs of for instance T cells or monocytes (Vega et al. 2008; Fitzgerald et al. 2018). Is it possible that the molecules destined to distant locations or more challenging tissue environments are packed in a way that shields them from degradation?

Attention should also be paid to the subcellular origin of EVs and the impact they may have on the signaling outcome. Neutrophils for example have created peculiar ways to reroute the production of EVs: in addition to the NDTRs (neutrophil-derived trails) discussed earlier, the neutrophils can form EVs from nuclear envelope budding: leukotriene 4-positive EVs form at the ceramide-rich microdomains on the nuclear lamina (Arya et al. 2022). These types of novel biogenesis routes could provide some explanation for the possible presence of nuclear proteins in EV mass spectrometry analyses, and it is likely that nuclear proteins are found in other immune EV analyses as well.

Special emphasis should be given to the possible and likely delivery of immunogenic material within EVs in blood transfusion. For instance, the EVs released by erythrocytes boosted the pro-inflammatory response in human monocytes (Danesh et al. 2014) and in murine macrophages via TLR4 activation (Gao et al. 2022). It is of importance to understand the EV interactions in blood products as the red blood cell EVs and monocytes could trigger post-transfusion complications via increased pro-inflammatory cytokine secretion and through interactions between neutrophils and platelets (Fischer et al. 2017; Danesh et al. 2014; Straat et al. 2016).

EV-mediated intercellular communication may have a critical role in the development and progression of autoimmunity. Currently, the autoimmune diseases are treated with palliative options only i.e. treating the symptoms. Thus, the recent findings that dendritic or mesenchymal stem cell-derived EVs reduce inflammation (Lu et al. 2021) have created an exciting momentum to develop EV-based therapeutics for autoimmune diseases.

However, to be able to link the in vitro studies, no matter how consistently performed, to treatment modalities is still far. The field is limited by the lack of knowledge e.g. on the co-entry and cooperation between EVs from different immune cell types in target cells. During inflammation, when various types of immune cells come together and communicate, it is likely that EVs from multiple cell types are affecting one cell type and understanding the molar ratios of these is yet another level that is to be researched (Rayamajhi et al. 2023). Another aspect which is largely unknown is the crosstalk and possible fusion between EVs from different cells as was recently demonstrated with ovarian cancer cell line-derived EVs and cryoelectron microscopy (Morandi et al. 2022). Understanding these unexplored aspects of EV biology would eventually pave the way for the design of immune EV therapeutics and biomarkers, in addition to the fundamentally complemented view on intercellular communication.

References

Agarwal A, Fanelli G, Letizia M, Tung SL, Boardman D, Lechler R, Lombardi G, Smyth LA (2014) Regulatory T cell-derived exosomes: possible therapeutic and diagnostic tools in transplantation. Front Immunol 5:555. https://doi.org/10.3389/fimmu.2014.00555

Akuthota P, Carmo LA, Bonjour K, Murphy RO, Silva TP, Gamalier JP, Capron KL, Tigges J, Toxavidis V, Camacho V, Ghiran I, Ueki S, Weller PF, Melo RC (2016) Extracellular microvesicle production by human eosinophils activated by "inflammatory" stimuli. Front Cell Dev Biol 4:117. https://doi.org/10.3389/fcell.2016.00117

Andre F, Schartz NE, Movassagh M, Flament C, Pautier P, Morice P, Pomel C, Lhomme C, Escudier B, Le Chevalier T, Tursz T, Amigorena S, Raposo G, Angevin E, Zitvogel L (2002) Malignant effusions and immunogenic tumour-derived exosomes. Lancet 360(9329):295–305. https://doi.org/10.1016/S0140-6736(02)09552-1

Angus KL, Griffiths GM (2013) Cell polarisation and the immunological synapse. Curr Opin Cell Biol 25(1):85–91. https://doi.org/10.1016/j.ceb.2012.08.013

Arya SB, Chen S, Jordan-Javed F, Parent CA (2022) Ceramide-rich microdomains facilitate nuclear envelope budding for non-conventional exosome formation. Nat Cell Biol 24(7):1019–1028. https://doi.org/10.1038/s41556-022-00934-8

Aung T, Chapuy B, Vogel D, Wenzel D, Oppermann M, Lahmann M, Weinhage T, Menck K, Hupfeld T, Koch R, Trümper L, Wulf GG (2011) Exosomal evasion of humoral immunotherapy in aggressive B-cell lymphoma modulated by ATP-binding cassette transporter A3. Proc Natl Acad Sci USA 108(37):15336–15341. https://doi.org/10.1073/pnas.1102855108

Baek H, Ye M, Kang GH, Lee C, Lee G, Choi DB, Jung J, Kim H, Lee S, Kim JS, Lee HJ, Shim I, Lee JH, Bae H (2016) Neuroprotective effects of CD4+CD25+Foxp3+ regulatory T cells in a 3xTg-AD Alzheimer's disease model. Oncotarget 7(43):69347–69357. https://doi.org/10.18632/oncotarget.12469

Baj-Krzyworzeka M, Majka M, Pratico D, Ratajczak J, Vilaire G, Kijowski J, Reca R, Janowska-Wieczorek A, Ratajczak MZ (2002) Platelet-derived microparticles stimulate proliferation, survival, adhesion, and chemotaxis of hematopoietic cells. Exp Hematol 30(5):450–459. https://doi.org/10.1016/s0301-472x(02)00791-9

Bala S, Csak T, Momen-Heravi F, Lippai D, Kodys K, Catalano D, Satishchandran A, Ambros V, Szabo G (2015) Biodistribution and function of extracellular miRNA-155 in mice. Sci Rep 5:10721. https://doi.org/10.1038/srep10721

Bálint Š, Müller S, Fischer R, Kessler BM, Harkiolaki M, Valitutti S, Dustin ML (2020) Supramolecular attack particles are autonomous killing entities released from cytotoxic T cells. Science 368(6493):897–901. https://doi.org/10.1126/science.aay9207

Bello-Gamboa A, Velasco M, Moreno S, Herranz G, Ilie R, Huetos S, Dávila S, Sánchez A, Bernardino De La Serna J, Calvo V, Izquierdo M (2020) Actin reorganization at the centrosomal area and the immune synapse regulates polarized secretory traffic of multivesicular bodies in T lymphocytes. J Extracell Vesicles 9(1):1759926. https://doi.org/10.1080/20013078.2020.1759926

Besse B, Charrier M, Lapierre V, Dansin E, Lantz O, Planchard D, Le Chevalier T, Livartoski A, Barlesi F, Laplanche A, Ploix S, Vimond N, Peguillet I, Théry C, Lacroix L, Zoernig I, Dhodapkar K, Dhodapkar M, Viaud S, Soria JC et al (2015) Dendritic cell-derived exosomes as maintenance immunotherapy after first line chemotherapy in NSCLC. Onco Targets Ther 5(4):e1071008. https://doi.org/10.1080/2162402X.2015.1071008

Bhatnagar S, Shinagawa K, Castellino FJ, Schorey JS (2007) Exosomes released from macrophages infected with intracellular pathogens stimulate a proinflammatory response in vitro and in vivo. Blood 110(9):3234–3244. https://doi.org/10.1182/blood-2007-03-079152

Blum JS, Wearsch PA, Cresswell P (2013) Pathways of antigen processing. Annu Rev Immunol 31:443–473. https://doi.org/10.1146/annurev-immunol-032712-095910

Bracamonte-Baran W, Florentin J, Zhou Y, Jankowska-Gan E, Haynes WJ, Zhong W, Brennan TV, Dutta P, Claas FH, van Rood JJ, Burlingham WJ (2017) Modification of host dendritic cells by microchimerism-derived extracellular vesicles generates split tolerance. Proc Natl Acad Sci USA 114(5):1099–1104. https://doi.org/10.1073/pnas.1618364114

Bui TM, Mascarenhas LA, Sumagin R (2018) Extracellular vesicles regulate immune responses and cellular function in intestinal inflammation and repair. Tissue Barriers 6(2):e1431038. https://doi.org/10.1080/21688370.2018.1431038

Burlingham WJ (2017) "Cross-dressing" becomes fashionable among transplant recipients. Am J Transplant 17(1):5–6. https://doi.org/10.1111/ajt.14032

Busatto S, Morad G, Guo P, Moses MA (2021) The role of extracellular vesicles in the physiological and pathological regulation of the blood–brain barrier. FASEB Bioadv 3:665–675. https://doi.org/10.1096/fba.2021-00045

Buschow SI, Nolte-'t Hoen EN, van Niel G, Pols MS, ten Broeke T, Lauwen M, Ossendorp F, Melief CJ, Raposo G, Wubbolts R, Wauben MH, Stoorvogel W (2009) MHC II in dendritic cells is targeted to lysosomes or T cell-induced exosomes via distinct multivesicular body pathways. Traffic 10(10):1528–1542. https://doi.org/10.1111/j.1600-0854.2009.00963.x

Buzas EI (2023) The roles of extracellular vesicles in the immune system. Nat Rev Immunol 23:236–250. https://doi.org/10.1038/s41577-022-00763-8

Cai Z, Yang F, Yu L, Yu Z, Jiang L, Wang Q, Yang Y, Wang L, Cao X, Wang J (2012) Activated T cell exosomes promote tumor invasion via Fas signaling pathway. J Immunol 188(12):5954–5961. https://doi.org/10.4049/jimmunol.1103466

Campana S, De Pasquale C, Carrega P, Ferlazzo G, Bonaccorsi I (2015) Cross-dressing: an alternative mechanism for antigen presentation. Immunol Lett 168(2):349–354. https://doi.org/10.1016/j.imlet.2015.11.002

Carvalho AS, Baeta H, Henriques AFA, Ejtehadifar M, Tranfield EM, Sousa AL, Farinho A, Silva BC, Cabeçadas J, Gameiro P, Silva MGD, Beck HC, Matthiesen R (2021) Proteomic landscape of extracellular vesicles for diffuse large B-cell lymphoma subtyping. Int J Mol Sci 22(20):11004. https://doi.org/10.3390/ijms222011004

Céspedes PF, Jainarayanan A, Fernández-Messina L, Valvo S, Saliba DG, Kurz E, Kvalvaag A, Chen L, Ganskow C, Colin-York H, Fritzsche M, Peng Y, Dong T, Johnson E, Siller-Farfán JA, Dushek O, Sezgin E, Peacock B, Law A, Aubert D et al (2022) T-cell trans-synaptic vesicles are distinct and carry greater effector content than constitutive extracellular vesicles. Nat Commun 13(1):3460. https://doi.org/10.1038/s41467-022-31160-3

Chaplin DD (2010) Overview of the immune response. J Allergy Clin Immunol 125(2 Suppl 2):S3–S23. https://doi.org/10.1016/j.jaci.2009.12.980

Chen G, Huang AC, Zhang W, Zhang G, Wu M, Xu W, Yu Z, Yang J, Wang B, Sun H, Xia H, Man Q, Zhong W, Antelo LF, Wu B, Xiong X, Liu X, Guan L, Li T, Liu S et al (2018) Exosomal PD-L1 contributes to immunosuppression and is associated with anti-PD-1 response. Nature 560(7718):382–386. https://doi.org/10.1038/s41586-018-0392-8

Claesson-Welsh L, Dejana E, McDonald DM (2021) Permeability of the endothelial barrier: identifying and reconciling controversies. Trends Mol Med 27(4):314–331. https://doi.org/10.1016/j.molmed.2020.11.006

Clayton A, Al-Taei S, Webber J, Mason MD, Tabi Z (2011) Cancer exosomes express CD39 and CD73, which suppress T cells through adenosine production. J Immunol 187(2):676–683. https://doi.org/10.4049/jimmunol.1003884

Coakley G, McCaskill JL, Borger JG, Simbari F, Robertson E, Millar M, Harcus Y, McSorley HJ, Maizels RM, Buck AH (2017) Extracellular vesicles from a helminth parasite suppress macrophage activation and constitute an effective vaccine for protective immunity. Cell Rep 19(8):1545–1557

Couch Y, Akbar N, Davis S, Fischer R, Dickens AM, Neuhaus AA, Burgess AI, Rothwell PM, Buchan AM (2017) Inflammatory stroke extracellular vesicles induce macrophage activation. Stroke 48(8):2292–2296. https://doi.org/10.1161/STROKEAHA.117.017236

Cronan MR, Beerman RW, Rosenberg AF, Saelens JW, Johnson MG, Oehlers SH, Sisk DM, Jurcic Smith KL, Medvitz NA, Miller SE, Trinh LA, Fraser SE, Madden JF, Turner J, Stout JE, Lee S, Tobin DM (2016) Macrophage epithelial reprogramming underlies mycobacterial granuloma formation and promotes infection. Immunity 45(4):861–876. https://doi.org/10.1016/j.immuni.2016.09.014

Danesh A, Inglis HC, Jackman RP, Wu S, Deng X, Muench MO, Heitman JW, Norris PJ (2014) Exosomes from red blood cell units bind to monocytes and induce proinflammatory cytokines, boosting T-cell responses in vitro. Blood 123(5):687–696. https://doi.org/10.1182/blood-2013-10-53046

Davoine F, Lacy P (2014) Eosinophil cytokines, chemokines, and growth factors: emerging roles in immunity. Front Immunol 5:570. https://doi.org/10.3389/fimmu.2014.00570

DeRita RM, Zerlanko B, Singh A, Lu H, Iozzo RV, Benovic JL, Languino LR (2017) c-Src, insulin-like growth factor I receptor, G-protein-coupled receptor kinases and focal adhesion kinase are enriched into prostate cancer cell exosomes. J Cell Biochem 118(1):66–73. https://doi.org/10.1002/jcb.25611

Dosil SG, Lopez-Cobo S, Rodriguez-Galan A, Fernandez-Delgado I, Ramirez-Huesca M, Milan-Rois P, Castellanos M, Somoza A, Gómez MJ, Reyburn HT, Vales-Gomez M, Sánchez Madrid

F, Fernandez-Messina L (2022) Natural killer (NK) cell-derived extracellular-vesicle shuttled microRNAs control T cell responses. eLife 11:e76319. https://doi.org/10.7554/eLife.76319
Du H, Bartleson JM, Butenko S, Alonso V, Liu WF, Winer DA, Butte MJ (2023) Tuning immunity through tissue mechanotransduction. Nat Rev Immunol 23(3):174–188. https://doi.org/10.1038/s41577-022-00761-w
Durlanik S, Loyal L, Stark R, Sercan Alp Ö, Hartung A, Radbruch A, von Herrath M, Matzmohr N, Frentsch M, Thiel A (2016) CD40L expression by CD4+ but not CD8+ T cells regulates antiviral immune responses in acute LCMV infection in mice. Eur J Immunol 46(11):2566–2573. https://doi.org/10.1002/eji.201646420
Edrissi H, Schock SC, Hakim AM, Thompson CS (2016) Microparticles generated during chronic cerebral ischemia increase the permeability of microvascular endothelial barriers in vitro. Brain Res 1634:83–93. https://doi.org/10.1016/j.brainres.2015.12.032
Fendl B, Weiss R, Eichhorn T, Linsberger I, Afonyushkin T, Puhm F, Binder CJ, Fischer MB, Weber V (2021) Extracellular vesicles are associated with C-reactive protein in sepsis. Sci Rep 11(1):6996. https://doi.org/10.1038/s41598-021-86489-4
Feng D, Zhao WL, Ye YY, Bai XC, Liu RQ, Chang LF, Zhou Q, Sui SF (2010) Cellular internalization of exosomes occurs through phagocytosis. Traffic 11(5):675–687. https://doi.org/10.1111/j.1600-0854.2010.01041.x
Fernández-Messina L, Rodríguez-Galán A, de Yébenes VG, Gutiérrez-Vázquez C, Tenreiro S, Seabra MC, Ramiro AR, Sánchez-Madrid F (2020) Transfer of extracellular vesicle-microRNA controls germinal center reaction and antibody production. EMBO Rep 21(4):e48925. https://doi.org/10.15252/embr.201948925
Fischer D, Büssow J, Meybohm P, Weber CF, Zacharowski K, Urbschat A, Müller MM, Jennewein C (2017) Microparticles from stored red blood cells enhance procoagulant and proinflammatory activity. Transfusion 57(11):2701–2711. https://doi.org/10.1111/trf.14268
Fitzgerald W, Freeman ML, Lederman MM, Vasilieva E, Romero R, Margolis L (2018) A system of cytokines encapsulated in ExtraCellular vesicles. Sci Rep 8(1):8973. https://doi.org/10.1038/s41598-018-27190-x
French SL, Butov KR, Allaeys I, Canas J, Morad G, Davenport P, Laroche A, Trubina NM, Italiano JE, Moses MA, Sola-Visner M, Boilard E, Panteleev MA, Machlus KR (2020) Platelet-derived extracellular vesicles infiltrate and modify the bone marrow during inflammation. Blood Adv 4(13):3011–3023. https://doi.org/10.1182/bloodadvances.2020001758
Gao Y, Jin H, Tan H, Cai X, Sun Y (2022) Erythrocyte-derived extracellular vesicles aggravate inflammation by promoting the proinflammatory macrophage phenotype through TLR4-MyD88-NF-κB-MAPK pathway. J Leukoc Biol 112(4):693–706. https://doi.org/10.1002/JLB.3A0821-451RR
Gardell JL, Parker DC (2017) CD40L is transferred to antigen-presenting B cells during delivery of T-cell help. Eur J Immunol 47(1):41–50. https://doi.org/10.1002/eji.201646504
Garraud O, Cognasse F (2015) Are platelets cells? And if yes, are they immune cells? Front Immunol 6:70. https://doi.org/10.3389/fimmu.2015.00070
Gasser O, Schifferli JA (2004) Activated polymorphonuclear neutrophils disseminate anti-inflammatory microparticles by ectocytosis. Blood 104(8):2543–2548. https://doi.org/10.1182/blood-2004-01-0361
Groot Kormelink T, Arkesteijn GJ, van de Lest CH, Geerts WJ, Goerdayal SS, Altelaar MA, Redegeld FA, Nolte-'t Hoen EN, Wauben MH (2016) Mast cell degranulation is accompanied by the release of a selective subset of extracellular vesicles that contain mast cell-specific proteases. J Immunol 197(8):3382–3392. https://doi.org/10.4049/jimmunol.1600614
Groot Kormelink T, Mol S, de Jong EC, Wauben MHM (2018) The role of extracellular vesicles when innate meets adaptive. Semin Immunopathol 40(5):439–452. https://doi.org/10.1007/s00281-018-0681-1
Gutiérrez-Vázquez C, Villarroya-Beltri C, Mittelbrunn M, Sánchez-Madrid F (2013) Transfer of extracellular vesicles during immune cell-cell interactions. Immunol Rev 251(1):125–142. https://doi.org/10.1111/imr.12013

Gutknecht MF, Holodick NE, Rothstein TL (2023) B cell extracellular vesicles contain monomeric IgM that binds antigen and enters target cells. iScience 26(9):107526. https://doi.org/10.1016/j.isci.2023.107526

Häder A, Schäuble S, Gehlen J, Thielemann N, Buerfent BC, Schüller V, Hess T, Wolf T, Schröder J, Weber M, Hünniger K, Löffler J, Vylkova S, Panagiotou G, Schumacher J, Kurzai O (2023) Pathogen-specific innate immune response patterns are distinctly affected by genetic diversity. Nat Commun 14(1):3239. https://doi.org/10.1038/s41467-023-38994-5

Hämälistö S, Del Valle Batalla F, Yuseff MI, Mattila PK (2024) Endolysosomal vesicles at the center of B cell activation. J Cell Biol 223(3):e202307047. https://doi.org/10.1083/jcb.202307047

Harrington LE, Hatton RD, Mangan PR, Turner H, Murphy TL, Murphy KM, Weaver CT (2005) Interleukin 17-producing CD4+ effector T cells develop via a lineage distinct from the T helper type 1 and 2 lineages. Nat Immunol 6(11):1123–1132. https://doi.org/10.1038/ni1254

Herrero-Cervera A, Soehnlein O, Kenne E (2022) Neutrophils in chronic inflammatory diseases. Cell Mol Immunol 19(2):177–191. https://doi.org/10.1038/s41423-021-00832-3

Hong CW (2018) Extracellular vesicles of neutrophils. Immune Netw 18(6):e43. https://doi.org/10.4110/in.2018.18.e43

Hood JL, San RS, Wickline SA (2011) Exosomes released by melanoma cells prepare sentinel lymph nodes for tumor metastasis. Cancer Res 71(11):3792–3801. https://doi.org/10.1158/0008-5472.CAN-10-4455

Hwang HS, Kim H, Han G, Lee JW, Kim K, Kwon IC, Yang Y, Kim SH (2021) Extracellular vesicles as potential therapeutics for inflammatory diseases. Int J Mol Sci 22(11):5487. https://doi.org/10.3390/ijms22115487

Hyenne V, Ghoroghi S, Collot M, Bons J, Follain G, Harlepp S, Mary B, Bauer J, Mercier L, Busnelli I, Lefebvre O, Fekonja N, Garcia-Leon MJ, Machado P, Delalande F, López AA, Silva SG, Verweij FJ, van Niel G, Djouad F et al (2019) Studying the fate of tumor extracellular vesicles at high spatiotemporal resolution using the Zebrafish embryo. Dev Cell 48(4):554–572.e7. https://doi.org/10.1016/j.devcel.2019.01.014

Iannotta D, Amruta A, Kijas AW, Rowan A, Wolfram J (2024) Entry and exit of extracellular vesicles to and from the blood circulation. Nat Nanotechnol 19:13–20. https://doi.org/10.1038/s41565-023-01522-z

Iwasaki A, Medzhitov R (2010) Regulation of adaptive immunity by the innate immune system. Science 327(5963):291–295. https://doi.org/10.1126/science.1183021

Jong AY, Wu CH, Li J, Sun J, Fabbri M, Wayne AS, Seeger RC (2017) Large-scale isolation and cytotoxicity of extracellular vesicles derived from activated human natural killer cells. J Extracell Vesicles 6(1):1294368. https://doi.org/10.1080/20013078.2017.1294368

Kalluri R (2016) The biology and function of exosomes in cancer. J Clin Invest 126(4):1208–1215. https://doi.org/10.1172/JCI81135

Kato T, Miyaki S, Ishitobi H, Nakamura Y, Nakasa T, Lotz MK, Ochi M (2014) Exosomes from IL-1β stimulated synovial fibroblasts induce osteoarthritic changes in articular chondrocytes. Arthritis Res Ther 16(4):R163. https://doi.org/10.1186/ar4679

Kolonics F, Kajdácsi E, Farkas VJ, Veres DS, Khamari D, Kittel Á, Merchant ML, McLeish KR, Lőrincz ÁM, Ligeti E (2021) Neutrophils produce proinflammatory or anti-inflammatory extracellular vesicles depending on the environmental conditions. J Leukoc Biol 109(4):793–806. https://doi.org/10.1002/JLB.3A0320-210R

Kranich J, Chlis NK, Rausch L, Latha A, Schifferer M, Kurz T, Foltyn-Arfa Kia A, Simons M, Theis FJ, Brocker T (2020) In vivo identification of apoptotic and extracellular vesicle-bound live cells using image-based deep learning. J Extracell Vesicles 9(1):1792683. https://doi.org/10.1080/20013078.2020.1792683

Kuokkanen E, Šuštar V, Mattila PK (2015) Molecular control of B cell activation and immunological synapse formation. Traffic 16(4):311–326. https://doi.org/10.1111/tra.12257

Lacy P, Stow J (2011) Cytokine release from innate immune cells: association with diverse membrane trafficking pathways. Blood 118(1):9–18. https://doi.org/10.1182/blood-2010-08-265892

Lee H, Park H, Noh GJ, Lee ES (2018) pH-responsive hyaluronate-anchored extracellular vesicles to promote tumor-targeted drug delivery. Carbohydr Polym 202:323–333. https://doi.org/10.1016/j.carbpol.2018.08.141

Li F, Wang Y, Lin L, Wang J, Xiao H, Li J, Peng X, Dai H, Li L (2016) Mast cell-derived exosomes promote Th2 cell differentiation via OX40L-OX40 ligation. J Immunol Res 2016:3623898. https://doi.org/10.1155/2016/3623898

Li Y, Tan J, Miao Y, Zhang Q (2021) MicroRNA in extracellular vesicles regulates inflammation through macrophages under hypoxia. Cell Death Discov 7(1):285. https://doi.org/10.1038/s41420-021-00670-2

Lu H, Bowler N, Harshyne LA, Craig Hooper D, Krishn SR, Kurtoglu S, Fedele C, Liu Q, Tang HY, Kossenkov AV, Kelly WK, Wang K, Kean RB, Weinreb PH, Yu L, Dutta A, Fortina P, Ertel A, Stanczak M, Forsberg F et al (2018) Exosomal αvβ6 integrin is required for monocyte M2 polarization in prostate cancer. Matrix Biol 70:20–35. https://doi.org/10.1016/j.matbio.2018.03.009

Lu M, DiBernardo E, Parks E, Fox H, Zheng SY, Wayne E (2021) The role of extracellular vesicles in the pathogenesis and treatment of autoimmune disorders. Front Immunol 12:566299. https://doi.org/10.3389/fimmu.2021.566299

Luketic L, Delanghe J, Sobol PT, Yang P, Frotten E, Mossman KL, Gauldie J, Bramson J, Wan Y (2007) Antigen presentation by exosomes released from peptide-pulsed dendritic cells is not suppressed by the presence of active CTL. J Immunol 179(8):5024–5032. https://doi.org/10.4049/jimmunol.179.8.5024

Ma SR, Xia HF, Gong P, Yu ZL (2023a) Red blood cell-derived extracellular vesicles: an overview of current research progress, challenges, and opportunities. Biomedicines 11(10):2798. https://doi.org/10.3390/biomedicines11102798

Ma Y, Ma QW, Sun Y, Chen XF (2023b) The emerging role of extracellular vesicles in the testis. Hum Reprod 38(3):334–351. https://doi.org/10.1093/humrep/dead015

Marar C, Starich B, Wirtz D (2021) Extracellular vesicles in immunomodulation and tumor progression. Nat Immunol 22(5):560–570. https://doi.org/10.1038/s41590-021-00899-0

Marshall JS, Warrington R, Watson W et al (2018) An introduction to immunology and immunopathology. Allergy Asthma Clin Immunol 14(Suppl 2):49. https://doi.org/10.1186/s13223-018-0278-1

Martin-Jaular L, Nevo N, Schessner JP, Tkach M, Jouve M, Dingli F, Loew D, Witwer KW, Ostrowski M, Borner GHH, Théry C (2021) Unbiased proteomic profiling of host cell extracellular vesicle composition and dynamics upon HIV-1 infection. EMBO J 40(8):e105492. https://doi.org/10.15252/embj.2020105492

Mause SF, von Hundelshausen P, Zernecke A, Koenen RR, Weber C (2005) Platelet microparticles: a transcellular delivery system for RANTES promoting monocyte recruitment on endothelium. Arterioscler Thromb Vasc Biol 25(7):1512–1518. https://doi.org/10.1161/01.ATV.0000170133.43608.37

Mazzeo C, Cañas JA, Zafra MP, Rojas Marco A, Fernández-Nieto M, Sanz V, Mittelbrunn M, Izquierdo M, Baixaulli F, Sastre J, Del Pozo V (2015) Exosome secretion by eosinophils: a possible role in asthma pathogenesis. J Allergy Clin Immunol 135(6):1603–1613. https://doi.org/10.1016/j.jaci.2014.11.026

McDonald MK, Tian Y, Qureshi RA, Gormley M, Ertel A, Gao R, Aradillas Lopez E, Alexander GM, Sacan A, Fortina P, Ajit SK (2014) Functional significance of macrophage-derived exosomes in inflammation and pain. Pain 155(8):1527–1539. https://doi.org/10.1016/j.pain.2014.04.029

Mesri M, Altieri DC (1998) Endothelial cell activation by leukocyte microparticles. J Immunol 161(8):4382–4387

Miskolci V, Klemm LC, Huttenlocher A (2021) Cell migration guided by cell-cell contacts in innate immunity. Trends Cell Biol 31(2):86–94. https://doi.org/10.1016/j.tcb.2020.11.002

Mitroulis I, Alexaki VI, Kourtzelis I, Ziogas A, Hajishengallis G, Chavakis T (2015) Leukocyte integrins: role in leukocyte recruitment and as therapeutic targets in inflammatory disease. Pharmacol Ther 147:123–135. https://doi.org/10.1016/j.pharmthera.2014.11.008

Mittelbrunn M, Gutiérrez-Vázquez C, Villarroya-Beltri C, González S, Sánchez-Cabo F, González MÁ, Bernad A, Sánchez-Madrid F (2011) Unidirectional transfer of microRNA-loaded exosomes from T cells to antigen-presenting cells. Nat Commun 2:282. https://doi.org/10.1038/ncomms1285

Mobarrez F, Fuzzi E, Gunnarsson I, Larsson A, Eketjäll S, Pisetsky DS, Svenungsson E (2019) Microparticles in the blood of patients with SLE: size, content of mitochondria and role in circulating immune complexes. J Autoimmun 102:142–149. https://doi.org/10.1016/j.jaut.2019.05.003

Molfetta R, Lecce M, Quatrini L, Caracciolo G, Digiacomo L, Masuelli L, Milito ND, Vulpis E, Zingoni A, Galandrini R, Santoni A, Paolini R (2020) Immune complexes exposed on mast cell-derived nanovesicles amplify allergic inflammation. Allergy 75(5):1260–1263. https://doi.org/10.1111/all.14103

Morandi MI, Busko P, Ozer-Partuk E, Khan S, Zarfati G, Elbaz-Alon Y, Abou Karam P, Napso Shogan T, Ginini L, Gil Z, Regev-Rudzki N, Avinoam O (2022) Extracellular vesicle fusion visualized by cryo-electron microscopy. PNAS nexus 1(4):pgac156. https://doi.org/10.1093/pnasnexus/pgac156

Muntasell A, Berger AC, Roche PA (2007) T cell-induced secretion of MHC class II-peptide complexes on B cell exosomes. EMBO J 26(19):4263–4272. https://doi.org/10.1038/sj.emboj.7601842

Németh A, Orgovan N, Sódar BW, Osteikoetxea X, Pálóczi K, Szabó-Taylor KÉ, Vukman KV, Kittel Á, Turiák L, Wiener Z, Tóth S, Drahos L, Vékey K, Horvath R, Buzás EI (2017) Antibiotic-induced release of small extracellular vesicles (exosomes) with surface-associated DNA. Sci Rep 7(1):8202. https://doi.org/10.1038/s41598-017-08392-1

Nolte-'t Hoen EN, Buschow SI, Anderton SM, Stoorvogel W, Wauben MH (2009) Activated T cells recruit exosomes secreted by dendritic cells via LFA-1. Blood 113(9):1977–1981. https://doi.org/10.1182/blood-2008-08-174094

Nouri N, Kleinstein SH (2020) Somatic hypermutation analysis for improved identification of B cell clonal families from next-generation sequencing data. PLoS Comput Biol 16(6):e1007977. https://doi.org/10.1371/journal.pcbi.1007977

Phan HD, Longjohn MN, Gormley DJB, Smith RH, Dang-Lawson M, Matsuuchi L, Gold MR, Christian SL (2021) CD24 and IgM stimulation of B cells triggers transfer of functional B cell receptor to B cell recipients via extracellular vesicles. J Immunol 207(12):3004–3015. https://doi.org/10.4049/jimmunol.2100025

Piccioli D, Sbrana S, Melandri E, Valiante NM (2002) Contact-dependent stimulation and inhibition of dendritic cells by natural killer cells. J Exp Med 195(3):335–341. https://doi.org/10.1084/jem.20010934

Pitt JM, André F, Amigorena S, Soria JC, Eggermont A, Kroemer G, Zitvogel L (2016) Dendritic cell-derived exosomes for cancer therapy. J Clin Invest 126(4):1224–1232. https://doi.org/10.1172/JCI81137

Pizzirani C, Ferrari D, Chiozzi P, Adinolfi E, Sandonà D, Savaglio E, Di Virgilio F (2007) Stimulation of P2 receptors causes release of IL-1beta-loaded microvesicles from human dendritic cells. Blood 109(9):3856–3864. https://doi.org/10.1182/blood-2005-06-031377

Poon KS, Palanisamy K, Chang SS, Sun KT, Chen KB, Li PC, Lin TC, Li CY (2017) Plasma exosomal miR-223 expression regulates inflammatory responses during cardiac surgery with cardiopulmonary bypass. Sci Rep 7(1):10807. https://doi.org/10.1038/s41598-017-09709-w

Pucci F, Garris C, Lai CP, Newton A, Pfirschke C, Engblom C, Alvarez D, Sprachman M, Evavold C, Magnuson A, von Andrian UH, Glatz K, Breakefield XO, Mempel TR, Weissleder R, Pittet MJ (2016) SCS macrophages suppress melanoma by restricting tumor-derived vesicle-B cell interactions. Science 352(6282):242–246. https://doi.org/10.1126/science.aaf1328

Qazi KR, Torregrosa Paredes P, Dahlberg B, Grunewald J, Eklund A, Gabrielsson S (2010) Proinflammatory exosomes in bronchoalveolar lavage fluid of patients with sarcoidosis. Thorax 65(11):1016–1024. https://doi.org/10.1136/thx.2009.132027

Qiu Y, Yang Y, Yang R, Liu C, Hsu JM, Jiang Z, Sun L, Wei Y, Li CW, Yu D, Zhang J, Hung MC (2021) Activated T cell-derived exosomal PD-1 attenuates PD-L1-induced immune dysfunction in triple-negative breast cancer. Oncogene 40(31):4992–5001. https://doi.org/10.1038/s41388-021-01896-1

Raposo G, Nijman HW, Stoorvogel W, Liejendekker R, Harding CV, Melief CJ, Geuze HJ (1996) B lymphocytes secrete antigen-presenting vesicles. J Exp Med 183(3):1161–1172. https://doi.org/10.1084/jem.183.3.1161

Rastogi I, Jeon D, Moseman JE, Muralidhar A, Potluri HK, McNeel DG (2022) Role of B cells as antigen presenting cells. Front Immunol 13:954936. https://doi.org/10.3389/fimmu.2022.954936

Rayamajhi S, Sulthana S, Ferrel C, Shrestha TB, Aryal S (2023) Extracellular vesicles production and proteomic cargo varies with incubation time and temperature. Exp Cell Res 422(2):113454. https://doi.org/10.1016/j.yexcr.2022.113454

Sanwlani R, Gangoda L (2021) Role of extracellular vesicles in cell death and inflammation. Cells 10(10):2663. https://doi.org/10.3390/cells10102663

Satta N, Toti F, Feugeas O, Bohbot A, Dachary-Prigent J, Eschwège V, Hedman H, Freyssinet JM (1994) Monocyte vesiculation is a possible mechanism for dissemination of membrane-associated procoagulant activities and adhesion molecules after stimulation by lipopolysaccharide. J Immunol 153(7):3245–3255

Saunderson SC, Dunn AC, Crocker PR, McLellan AD (2014) CD169 mediates the capture of exosomes in spleen and lymph node. Blood 123(2):208–216. https://doi.org/10.1182/blood-2013-03-489732

Segura E, Guérin C, Hogg N, Amigorena S, Théry C (2007) CD8+ dendritic cells use LFA-1 to capture MHC-peptide complexes from exosomes in vivo. J Immunol 179(3):1489–1496. https://doi.org/10.4049/jimmunol.179.3.1489

Sharonov GV, Serebrovskaya EO, Yuzhakova DV et al (2020) B cells, plasma cells and antibody repertoires in the tumour microenvironment. Nat Rev Immunol 20:294–307. https://doi.org/10.1038/s41577-019-0257-x

Shefler I, Salamon P, Mekori YA (2021) Extracellular vesicles as emerging players in intercellular communication: relevance in mast cell-mediated pathophysiology. Int J Mol Sci 22(17):9176. https://doi.org/10.3390/ijms22179176

Shen Y, Guo D, Weng L, Wang S, Ma Z, Yang Y, Wang P, Wang J, Cai Z (2017) Tumor-derived exosomes educate dendritic cells to promote tumor metastasis via HSP72/HSP105-TLR2/TLR4 pathway. Oncoimmunology 6(12):e1362527. https://doi.org/10.1080/2162402X.2017.1362527

Shitara K, Nishikawa H (2018) Regulatory T cells: a potential target in cancer immunotherapy. Ann N Y Acad Sci 1417(1):104–115. https://doi.org/10.1111/nyas.13625

Sims PJ, Wiedmer T, Esmon CT, Weiss HJ, Shattil SJ (1989) Assembly of the platelet prothrombinase complex is linked to vesiculation of the platelet plasma membrane. Studies in Scott syndrome: an isolated defect in platelet procoagulant activity. J Biol Chem 264(29):17049–17057. https://doi.org/10.1016/S0021-9258(18)71457-9

Singhto N, Thongboonkerd V (2018) Exosomes derived from calcium oxalate-exposed macrophages enhance IL-8 production from renal cells, neutrophil migration and crystal invasion through extracellular matrix. J Proteome 185:64–76. https://doi.org/10.1016/j.jprot.2018.06.015

Slater TW, Finkielsztein A, Mascarenhas LA, Mehl LC, Butin-Israeli V, Sumagin R (2017) Neutrophil microparticles deliver active myeloperoxidase to injured mucosa to inhibit epithelial wound healing. J Immunol 198(7):2886–2897. https://doi.org/10.4049/jimmunol.1601810

Smyth LA, Ratnasothy K, Tsang JY, Boardman D, Warley A, Lechler R, Lombardi G (2013) CD73 expression on extracellular vesicles derived from CD4+ CD25+ Foxp3+ T cells contributes to their regulatory function. Eur J Immunol 43(9):2430–2440. https://doi.org/10.1002/eji.201242909

Straat M, van Hezel ME, Böing A, Tuip-De Boer A, Weber N, Nieuwland R, van Bruggen R, Juffermans NP (2016) Monocyte-mediated activation of endothelial cells occurs only after binding to extracellular vesicles from red blood cell products, a process mediated by β-integrin. Transfusion 56(12):3012–3020. https://doi.org/10.1111/trf.13851

Tang N, Sun B, Gupta A, Rempel H, Pulliam L (2016) Monocyte exosomes induce adhesion molecules and cytokines via activation of NF-κB in endothelial cells. FASEB J 30(9):3097–3106. https://doi.org/10.1096/fj.201600368RR

Thangaraju K, Neerukonda SN, Katneni U, Buehler PW (2020) Extracellular vesicles from red blood cells and their evolving roles in health, coagulopathy and therapy. Int J Mol Sci 22(1):153. https://doi.org/10.3390/ijms22010153

Théry C, Ostrowski M, Segura E (2009) Membrane vesicles as conveyors of immune responses. Nat Rev Immunol 9(8):581–593. https://doi.org/10.1038/nri2567

Torralba D, Baixauli F, Villarroya-Beltri C, Fernández-Delgado I, Latorre-Pellicer A, Acín-Pérez R, Martín-Cófreces NB, Jaso-Tamame ÁL, Iborra S, Jorge I, González-Aseguinolaza G, Garaude J, Vicente-Manzanares M, Enríquez JA, Mittelbrunn M, Sánchez-Madrid F (2018) Priming of dendritic cells by DNAcontaining extracellular vesicles from activated T cells through antigen-driven contacts. Nature communications 9(1):2658. https://doi.org/10.1038/s41467-018-05077-9

Tung SL, Boardman DA, Sen M, Letizia M, Peng Q, Cianci N, Dioni L, Carlin LM, Lechler R, Bollati V, Lombardi G, Smyth LA (2018) Regulatory T cell-derived extracellular vesicles modify dendritic cell function. Sci Rep 8(1):6065. https://doi.org/10.1038/s41598-018-24531-8

van Niel G, D'Angelo G, Raposo G (2018) Shedding light on the cell biology of extracellular vesicles. Nat Rev Mol Cell Biol 19(4):213–228. https://doi.org/10.1038/nrm.2017.125

Vega VL, Rodríguez-Silva M, Frey T, Gehrmann M, Diaz JC, Steinem C, Multhoff G, Arispe N, De Maio A (2008) Hsp70 translocates into the plasma membrane after stress and is released into the extracellular environment in a membrane-associated form that activates macrophages. J Immunol 180(6):4299–4307. https://doi.org/10.4049/jimmunol.180.6.4299

Verweij FJ, Balaj L, Boulanger CM, Carter DRF, Compeer EB, D'Angelo G, El Andaloussi S, Goetz JG, Gross JC, Hyenne V, Krämer-Albers EM, Lai CP, Loyer X, Marki A, Momma S, Nolte-'t Hoen ENM, Pegtel DM, Peinado H, Raposo G, Rilla K et al (2021) The power of imaging to understand extracellular vesicle biology in vivo. Nat Methods 18(9):1013–1026. https://doi.org/10.1038/s41592-021-01206-3

Vojdani A, Koksoy S, Vojdani E, Engelman M, Benzvi C, Lerner A (2024) Natural killer cells and cytotoxic T cells: complementary partners against microorganisms and cancer. Microorganisms 12(1):230. https://doi.org/10.3390/microorganisms12010230

Volgers C, Benedikter BJ, Grauls GE, Savelkoul P, Stassen F (2017) Immunomodulatory role for membrane vesicles released by THP-1 macrophages and respiratory pathogens during macrophage infection. BMC Microbiol 17:216. https://doi.org/10.1186/s12866-017-1122-3

Walker JD, Maier CL, Pober JS (2009) Cytomegalovirus-infected human endothelial cells can stimulate allogeneic CD4+ memory T cells by releasing antigenic exosomes. J Immunol 182(3):1548–1559. https://doi.org/10.4049/jimmunol.182.3.1548

Welsh JA, Goberdhan DCI, O'Driscoll L, Buzas EI, Blenkiron C, Bussolati B, Cai H, Di Vizio D, Driedonks TAP, Erdbrügger U, Falcon-Perez JM, Fu Q-L, Hill AF, Lenassi M, Lim SK, Mahoney MG, Mohanty S, Möller A, Nieuwland R et al (2024) Minimal information for studies of extracellular vesicles (MISEV2023): from basic to advanced approaches. J Extracell Vesicles 13:e12404. https://doi.org/10.1002/jev2.12404

Wiklander OP, Nordin JZ, O'Loughlin A, Gustafsson Y, Corso G, Mäger I, Vader P, Lee Y, Sork H, Seow Y, Heldring N, Alvarez-Erviti L, Smith CI, Le Blanc K, Macchiarini P, Jungebluth P, Wood MJ, Andaloussi SE (2015) Extracellular vesicle in vivo biodistribution is determined by cell source, route of administration and targeting. J Extracell Vesicles 4:26316. https://doi.org/10.3402/jev.v4.26316

Xu Y, Liu Y, Yang C, Kang L, Wang M, Hu J, He H, Song W, Tang H (2016) Macrophages transfer antigens to dendritic cells by releasing exosomes containing dead-cell-associated antigens partially through a ceramide-dependent pathway to enhance CD4(+) T-cell responses. Immunology 149(2):157–171. https://doi.org/10.1111/imm.12630

Yang P, Peng Y, Feng Y, Xu Z, Feng P, Cao J, Chen Y, Chen X, Cao X, Yang Y, Jie J (2021) Immune cell-derived extracellular vesicles - new strategies in cancer immunotherapy. Front Immunol 12:771551. https://doi.org/10.3389/fimmu.2021.771551

Yang M, Walker SA, Díaz A, de León JS, Davidovich I, Broad K, Talmon Y, Borges CR, Wolfram J (2022) Extracellular vesicle glucose transporter-1 and glycan features in monocyte-endothelial inflammatory interactions. Nanomedicine 42:102515. https://doi.org/10.1016/j.nano.2022.102515

Yu X, Huang C, Song B, Xiao Y, Fang M, Feng J, Wang P (2013a) CD4+CD25+ regulatory T cells-derived exosomes prolonged kidney allograft survival in a rat model. Cell Immunol 285(1–2):62–68. https://doi.org/10.1016/j.cellimm.2013.06.010

Yu L, Yang F, Jiang L, Chen Y, Wang K, Xu F, Wei Y, Cao X, Wang J, Cai Z (2013b) Exosomes with membrane-associated TGF-β1 from gene-modified dendritic cells inhibit murine EAE independently of MHC restriction. Eur J Immunol 43(9):2461–2472. https://doi.org/10.1002/eji.201243295

Yuseff MI, Pierobon P, Reversat A, Lennon-Duménil AM (2013) How B cells capture, process and present antigens: a crucial role for cell polarity. Nat Rev Immunol 13(7):475–486. https://doi.org/10.1038/nri3469

Zhang H, Xie Y, Li W, Chibbar R, Xiong S, Xiang J (2011) CD4(+) T cell-released exosomes inhibit CD8(+) cytotoxic T-lymphocyte responses and antitumor immunity. Cell Mol Immunol 8(1):23–30. https://doi.org/10.1038/cmi.2010.59

Chapter 6
Extracellular Vesicles in Synovial Fluid: Their Role in Joint Homeostasis and Pathophysiology

Enrico Ragni and Laura de Girolamo

Abstract Synovial fluid is a crucial component of joints. It serves as lubricant of articular cartilage and nutrient source for articular structures including cartilage and meniscus. The synovial fluid is produced by the synovial membrane and is made of polymers including hyaluronan, lubricin, proteinase, collagenases, and prostaglandins. Being populated by immune cells and floating stromal components and being in contact with the articular tissues, the synovial fluid is populated by different classes of extracellular vesicles (EVs). EVs can be free or entrapped in the network through physical binding with hyaluronan and the other polymers. Thus, their movement and shuttling between cells and tissues is tightly regulated and part of the mechanisms allowing joints component to communicate. For this reason, alteration in joint tissues under stress or pathological conditions may alter EVs composition and communication and these changes may be useful as biomarkers for the disease. Moreover, in recent years, EVs released from mesenchymal stromal cells (MSCs) have been envisioned as potential therapeutic agents, due to the anti-inflammatory and pro-regenerative features mirrored from their releasing cells. An active debate is also undergoing to find the most powerful way of EVs administration, using new-generation scaffolds as vehicle to protect EVs from degradation and therefore to prolong their therapeutic activity. Thus, synovial fluid and EVs, both endogenous and administered, are crucial actors for joint homeostasis and therapy.

6.1 Introduction

The joints form an integral part of the body framework holding segments of the bony skeleton together and allowing motion to take place. There are several different types of joints, with the synovial joint being encountered most often. The

E. Ragni (✉) · L. de Girolamo
IRCCS Ospedale Galeazzi – Sant'Ambrogio, Laboratorio di Biotecnologie Applicate all'Ortopedia, Milan, Italy
e-mail: enrico.ragni@grupposandonato.it; laura.degirolamo@grupposandonato.it

K. Rilla (ed.), *Extracellular Vesicles as Matrix Messengers*, Biology of Extracellular Matrix 15, https://doi.org/10.1007/978-3-031-68811-9_6

synovial joint allows the movement in one or more directions between two or more major segments of the skeleton. The joint provides movement of the body with low friction of the articulation, under both weight-bearing and non-weight-bearing conditions. A synovial joint is made of a few components: the joint capsule, the synovial membrane, the articular cartilage, intra-articular structures (such as menisci in the knee joint), and the synovial fluid (SF) (Juneja et al. 2023).

The SF is made of fluids confined within the joint space with the role of both lubricant of articular cartilage and nutrient source for articular structures including cartilage and meniscus. Synovial fluid is produced from type B fibroblast-like cells of the synovial membrane and is primarily composed of hyaluronan, lubricin, proteinases, collagenases, and prostaglandins. Under trauma, inflammation, and infection, the SF changes in its features such as volume and content (Seidman and Limaiem 2023). Recently, also extracellular vesicles (EVs) have been detected in the synovial fluid, thus playing a crucial role in joint physiology and pathological state (Domenis et al. 2017).

Of note, under disease conditions such as osteoarthritis (OA) or rheumatoid arthritis (RA), it clearly emerged that the molecular content of EVs may account for the pathological differences between healthy individuals and patients (Withrow et al. 2016). This feature resembles the differential state of the major EVs secreting cell types, including chondrocytes and synovial membrane cells. In fact, under degenerative and inflammatory conditions, whole joint released molecules including cytokines and chemokines alter their fingerprint (Molnar et al. 2021; Kondo et al. 2021). Similarly, a modulation in both the concentration and degree of polymerization of hyaluronic acid was reported (Stafford et al. 1964). This is strictly linked with EVs, since EVs can be bound to hyaluronic acid through CD44 receptor (Boere et al. 2016) and are also active transporter of polymer fragments (Mustonen et al. 2022). Thus, an overall change in synovial fluid composition and molecular architecture has a strong impact on associated EVs, altering their distribution and bioavailability.

The aim of this chapter is to discuss the relationship between the different joint components, with a particular focus on synovial fluid, and EVs in the context of healthy and diseased subjects. The role of EVs as biomarkers of disease and possible therapeutic mediators will also be addressed.

6.2 Main Part

6.2.1 Joint Structure

The most common type of joints is the synovial joint, also called diarthrosis (Juneja et al. 2023). It contains a joint cavity filled with the SF and is located between two or more bones. The presence of a synovial membrane gives the name of synovial joints to diarthrodial structures that, regardless of their subtypes, always share seven elements:

- Capsule: connective tissue, binds the joint together
- Cartilage: composed of chondrocytes and matrix, absorbs and dispels compressive forces
- Synovial membrane: composed of Type A and Type B synoviocytes, produces synovial fluid
- Synovial fluid: supplies joint lubrication and nutrition
- Capsular ligaments: connective tissue, limits excessive joint motion
- Blood vessels: provide nutrients
- Sensory nerves: transmit signals of pain and proprioception

Synovial joints can be classified by anatomists into categories regarding their structural features that shape their functional potential. To help in the understanding of the structure and function of most joints within the body, the two following similitudes may be of general help.

Hinge joint: allows motion in only one plane about a single axis of rotation (e.g. humeroulnar joint (elbow), interphalangeal joints (fingers and toes)).

Pivot joint: allows rotation about a single longitudinal axis of rotation (e.g. proximal radioulnar joint, atlantoaxial joint between the first and second cervical vertebrae).

6.2.2 *Capsule*

Dense fibrous connective tissue is the main component of the synovial joint capsule (Ralphs and Benjamin 1994). The connective tissue is lined with synovium and forms a sleeve around the articulating bones to which it is attached. Bundles of collagen fibres compose the capsule and adhere firmly to the bones, with nerves and blood vessels passing through the tissue. As a general rule, capsule thickness and fibre orientation are dependent on the loads applied to the different area. Often, there are 2 or more localized thickenings of the capsule, which form ligaments that may be further supported by accessory ligaments which also restrict motion. Capsule is innervated from nerves commonly derived from those supplying muscles acting on the joint.

6.2.3 *Cartilage*

Articular cartilage is a highly specialized connective tissue of synovial joints that provide a smooth and lubricated surface and facilitate the transmission of loads with a low frictional coefficient (Fox et al. 2009). Articular cartilage is hyaline cartilage, 2–4 mm thick, devoid of blood vessels and innervation, and has a limited capacity for intrinsic healing and repair. Cartilage architecture is based on a dense extracellular matrix (ECM), mainly composed of water, collagen, and proteoglycans. Embedded within ECM, highly specialized cells called chondrocytes (articular chondrocytes, AC) are dispersed and organized in various zones (superficial, middle, deep, and calcified). The collagen fibres of the superficial zone (10–20% of tissue thickness) are packed tightly and aligned parallel to the articular surface, with a high number of flattened chondrocytes. The collagen fibres of the middle zone (40–60% of tissue thickness) are thicker with an oblique organization, with chondrocytes that are spherical and at low density. The deep zone (30% of tissue thickness) confers tissue resistance, being collagen fibrils arranged perpendicular to the articular surface in a radial disposition, with chondrocytes that are arranged in columnar orientation, parallel to the collagen fibres and perpendicular to the joint line. Eventually, the calcified layer secures the cartilage to bone, with a scarce population of hypertrophic chondrocytes.

Chondrocytes are embedded in the extracellular matrix (ECM) which is composed mainly of water (65–80%) (Mow et al. 1992). Collagens and proteoglycans account for the remaining dry weight, 60% and 15%, respectively. The most abundant collagen in healthy cartilage is type II (95%), while the most frequent proteoglycan is aggrecan, able to interact with hyaluronan. Chondrocytes constitute around 2% of the total volume of articular cartilage and differentiate from mesenchymal stromal cells (Alford and Cole 2005). Chondrocytes are responsible for the turnover of the ECM and respond to growth factors, mechanical loads, and hydrostatic pressures (Lin et al. 2006). They are able to secrete several factors, including ADAMTS, Apo, BMPs, CCLs, CHAD, ChM, CILP, CHI3L1, COMP, CRLF, CRTAC, CSF, CTGF, Cx, CXCLs, ECM, ENPP, EMI, HSPA5, HTRA, ILs, LTBP, MGP, MMPs, PCOLCE, PLOD, PRELP, SERPINs, SOD, SPARC, TGFBI, TGF, TIMPs, TNAP, and TSP (Sanchez et al. 2017) and EVs (Song et al. 2021).

6.2.4 Synovial Membrane

The synovial membrane forms the inner side of the joint capsule. This structure produces the synovial fluid lubricating the joint during movement and nourishing the avascular cartilage. The synovial membrane is crucial for maintaining joint homeostasis and a well-functioning articulation (Firestein et al. 2016). Two layers define a highly functionalized and multifunctional structure. The first one, which is thin (<2 mm) and highly cellularized, is composed of synovial macrophages (Type A cells) and synovial fibroblasts (type B cells). Both the ECM network supporting the synovium and the hyaluronic acid and lubricin that form the synovial fluid matrix are produced by the synovial fibroblasts. Their endoplasmic reticulum is crucial for synthesizing ECM molecules in synovial fluid. Synovial macrophages account for 25% of the residential cells in synovial membrane, and their function is presumed to be sentinel of joint homeostasis. They are derived from monocyte lineage and are mainly located in the inner surface of the joint cavity. Synovial macrophages are responsible for removing debris in the synovial fluid through phagocytosis, and a complex and active mitochondrial network is important to fulfil this energy-consuming task. In addition to fibroblasts and macrophages, few other immune cells (lymphocytes, mast cells, and dendritic cells) can be found and localized mainly in perivascular areas of the sublining layer (Singh et al. 2004). The second layer (subintima) is composed of loose connective tissue, with the presence of fibroblasts, sympathetic and sensory nerves, and blood and lymphatic vasculature for oxygen/nutrients supply and immune drainage. It is able to reshape itself to avoid any collision between the muscles and acting as the line-up between the bursae, tendon sheaths, and joints.

6.2.5 Synovial Fluid

The cavities of synovial joints are filled with synovial fluid, which is a viscous non-Newtonian fluid with an albumen-like viscosity. The main role of synovial fluid is to reduce friction between the articular cartilages of synovial joints during motion. Synovial fluid is a component of the transcellular fluid compartment of extracellular fluid that fills up the spaces of chambers that are created or formed from the linings of epithelial cells.

The synovial fluid is an ultrafiltrate from plasma and contains plasma-derived proteins (Keenan 2022). Synovial fluid exhibits non-Newtonian flow characteristics with the viscosity coefficient that is not constant making the fluid not linearly viscous. Synovial fluid has thixotropic properties, meaning that it can change its viscosity when loaded by stress. Thus, under conditions of sudden pressure rise, such as some types of mechanical shock typical of joints, its viscosity increases suddenly. It is produced by cells within joint tissue that are responsible for its homeostasis and turnover. Type A cells remove wear debris while Type B cells produce hyaluronic acid (1–4 mg/ml), a disaccharide polymer composed of D-glucuronic acid and D-N-acetylglucosamine with alternating β-1,4 and β-1,3 glycosidic bonds. Lubricin (PRG4, 0.05–0.35 mg/ml), a secondary lubricating component, is secreted by Type B cells, surface chondrocytes of articular cartilage, and meniscus cells. Eventually, surface-active phospholipids (0.1 mg/ml) are produced by Type B cells (Tamer 2013).

Cartilage surface is covered by a thin layer (50 μm) of synovial fluid that penetrates into the microcavities and irregularities of the articular cartilage surface, filling all empty spaces. The fluid soaking the articular cartilage serves as a supply during exercise and motion. In fact, the retained fluid within cartilage can be mechanically squeezed out, to maintain a thin sheet on the cartilage surface and give a wet lubrication to reduce friction. Together with this function, synovial fluid may (Tamer 2013): (i) absorb shocks; (ii) function as a dilatant fluid that has rheopectic properties and becomes more viscous when pressure is applied; (iii) transport oxygen and nutrients or waste products such as carbon dioxide and metabolic waste from the chondrocytes of the surrounding cartilage; (iv) act as a molecular sieve since the pressure within the joint forces the hyaluronic acid against the synovial membrane, forming a barrier to cells migrating into and fluid from the joint space; v) help regulate synovial cell growth; and vi) contain phagocytic cells that remove microbes and debris from normal wear and tear in joints.

6.2.6 Hyaluronic Acid

Hyaluronic acid (HA), also called hyaluronan, is the main component of synovial fluid. HA is a biopolymer composed of repeating disaccharide units formed by N-acetyl-d-glucosamine and d-glucuronic acid. HA reaches up to several million Daltons and is synthesized at the plasma membrane, where sulfated glycosaminoglycans are added to protein cores (Itano and Kimata 2002). At high concentrations, HA has an extremely high but shear-dependent viscosity, being jelly at 1%, and traps approximately 1000 times its weight in water (Cowman and Matsuoka 2005). Notably, HA and other polysaccharides are responsible for partitioning plasma proteins between the vascular and extravascular spaces, thus affecting the solubility of other macromolecules, changing chemical equilibria, and stabilizing the structure of collagen fibres in the soaked cartilage. The degradation of high-molecular-weight HA occurring under inflammation and/or oxidative stress is accompanied by impairment and loss of its viscoelastic properties (Soltés et al. 2005). Low-molecular-weight HA chains and fragments of 25–50 disaccharide units are immune-stimulatory, inflammatory, and highly angiogenic. HA fragments act as endogenous signals for danger, reflecting tissue and joint damage (Stern et al., 2007).

6.2.7 Joint Disorders

Joint disorders affect the health and function of a joint. Joint disorders may be distinguished into two principal categories: (i) those in which inflammation is the principal set of signs or symptoms (e.g. inflammatory arthritis, like rheumatoid arthritis, and collagen diseases) and those called non-inflammatory, in which inflammation may be present to some degree (as after an injury) but is not the essential feature (e.g. traumatic and degenerative arthritis, like osteoarthritis). The two most studied pathologies are:

- Rheumatoid arthritis (RA): RA is a chronic and inflammatory joint disease of autoimmune nature characterized by autoantibodies to immunoglobulin G and citrullinated proteins (Smolen et al. 2018). RA is characterized by synovial inflammation based on mononuclear cell infiltration, mainly $CD4^+$ T cells and macrophages, with high expression of matrix-degrading enzymes (such as matrix metalloproteinases (MMPs)) in the synovial intimal lining by Type B cells (Bartok and Firestein 2010). Infiltrated Type A macrophages produce a variety of pro-inflammatory cytokines, including IL-1, IL-6, tumor necrosis factor, and others. Together with inflammation, damage to cartilage and bone due to synovial invasion into adjacent articular structures is a distinctive sign of RA. Matrix destruction is promoted mainly by Type B synoviocytes, which under RA acquire an aggressive phenotype and are able to invade cartilage (Müller-Ladner et al. 1996). This phenotype is activated by immune synovial cells, producing cytokines that act in a paracrine or autocrine fashion and can enhance and perpetuate inflammation in RA. For example, Type A cells produce cytokines that activate adjacent FLS, T cells, and dendritic cells that produce additional cytokines activating other cells in the joint environment and promoting inflammation and cartilage degradation. This increase in production of inflammatory cytokines and mediators is mirrored by their presence in the synovial fluid from patients (Hampel et al. 2013). Moreover, RA synovial fluid can contain large quantities of immune complexes that are able to activate neutrophils that discharge large quantities of toxic products, overwhelming the antioxidant and antiprotease shields of synovial fluid (Ossanna et al. 1986). Eventually, the intrinsic viscosity of HA in synovial fluid decreases significatively in RA (around 30–40%) with concomitant reduction in hyaluronic acid concentration (Kofoed and Barceló 1978).

- Osteoarthritis (OA): OA is the most common joint disorder involving structural alterations in the hyaline articular cartilage, subchondral bone, ligaments, capsule, synovium, and periarticular muscles (Hunter and Bierma-Zeinstra 2019). Infiltration of the synovial membrane with activated B cells and T lymphocytes and overexpression of pro-inflammatory mediators is common in OA with production of high levels of IL-1, tumor necrosis factor, vascular endothelial growth factor (VEGF), and intercellular adhesion molecules (Benito et al. 2005). These factors, together with Th1 cytokines produced by activated T cells, are initially produced by the synovial membrane and diffuse into the cartilage via the synovial fluid (Pelletier et al. 2001). These molecules stimulate chondrocytes to produce other soluble and catabolic factors such as MMPs, which degrade matrix proteins and collagen and can promote catabolic activation, aberrant, hypertrophy-like differentiation, and apoptosis. Further, inflamed chondrocytes produce IL-1, which then stimulates MMP production on a feedback loop model. Moreover, a typical landmark of OA is synovial angiogenesis that exacerbates inflammation. This is promoted by VEGFA production by Type A macrophages, present in abundance in chronically inflamed OA synovium (Haywood et al. 2003). VEGFA produced in the synovial membrane and conveyed by synovial fluid to cartilage can promote vascular penetration from the subchondral bone (Goldring 2012). In this frame, synovial fluid plays a crucial role in OA pathogenesis and progression. The concentration of HA and lubricin is decreased in OA synovial fluid, partly owing to changes in synovial membrane permeability due to alteration in HA size (Temple-Wong et al. 2016). Also, synovial fluid viscosity is decreased in OA and may be related to joint pain (Balazs 2004). Eventually, the synovial fluid being the collector of molecules produced by Type A and B cells and chondrocytes, OA synovial fluid was found enriched in cytokines, chemokines, and growth factors. These molecules are able to accelerate chondrocyte dedifferentiation, increase chondrocyte fibrosis, elevate the inflammatory response, produce extracellular matrix-degrading enzymes, and enable chondrocyte proliferation (Housmans et al. 2023).

6.2.8 Extracellular Vesicles (EVs) in Joint Diseases

EVs reflect the physiological and pathological state of parent cells and can be released into the extracellular space and transferred to recipient cells, for tissue–tissue and cell–cell communication in homeostasis and disease. Under this paradigm, synovial fluid is the collector of all EVs secreted by synovial membrane and cartilage cells and orchestrates their diffusion to target tissues. Alterations in synovial fluid viscosity, volume, or concentration of polymers can therefore alter these mechanisms and link disease with EVs biological activity. Thus, EVs in the synovial fluid, both physiological and exogenously administered, can be described under a dual perspective: markers of disease or vehicle for therapeutics.

6.2.9 EVs Interaction with Synovial Fluid

EVs can be isolated directly from synovial fluid. Several methods have been used showing that it is possible to collect EVs, and this is of particular relevance when analysing EVs as physiological and active components of joints and/or markers of altered tissue functions during diseases (Domenis et al. 2017). Of note, EVs in the synovial fluid can be present in two states: floating free or interacting with HA (Boere et al. 2016). In fact, depending on the secreting cells, EVs can be coated with different surface molecules. One of these proteins is CD44, an HA receptor and present on synovial fibroblasts, which are important EV producers in SF. Treating synovial fluid with hyaluronidase before EVs isolation allowed a higher recovery of CD44-coated EVs. The lower yield of $CD44^+$ EVs from non-hyaluronidase-treated samples indicates the presence of complexes of these EVs with the large chains of non-degraded HA and that hyaluronidase degrades HA and liberates EVs that are CD44-bound, resulting in increased EV recovery.

The physical interaction of $CD44^+$ EVs with HA might also mirror a functional role in the joint for this subset of EVs, in both healthy and diseased states, and be a fair example for other EV subpopulations in the synovial fluid. In fact, as previously discussed, the degree of HA polymerization, length of chains, and concentration are influenced by pathologies of the joint (Murai 2015). As a consequence, interaction of CD44-EVs with HA network might be altered by pathology, leading to alteration of functional effects due to increased EVs motility within the meshes of the net and eventual availability at level of target cells in the synovial membrane or the cartilage. On the other side of the coin, a reduced or increased CD44-HA binding can also modify EVs-facilitated shuttling within the joint space of free HA fragments, altering their availability to target cells of synovial membrane or cartilage. In fact, free HA can be transported by EVs to synovial cells and chondrocytes and the eventual interaction results in an anabolic (and anti-inflammatory) cascade by inhibiting TNF- or IL-1-induced matrix MMP production and aggrecanase expression (Masuko et al. 2009). Also, the interaction between HA and CD44 can be used by CD44-coated EVs to bind the HA "shield" present on chondrocytes and synoviocytes plasma membrane (Rilla et al. 2008; Ragni et al. 2019) and further facilitate EVs uptake or free HA bound stimulation of target cells. Thus, the CD44+ subset of EVs can be a guide for how synovial fluid state and polymer size and concentration may alter specific EVs populations availability and functionality. A striking example for synovial fluid is the effect of inflammation with consequent infiltration of interstitial fluid resulting in the reduction of synovial fluid viscosity and the increased production of low-molecular-weight HA that might end up in increased trapping of $CD44^+$ EVs into the extracellular matrix (Murai 2015).

6.2.10 EVs as Biomarkers: RA

EVs can store data from their provenance cell, giving them potential as disease biomarkers due to their surface fingerprint and condition-specific cargo. EVs in RA synovial fluid are highly heterogeneous in size and frequently express the integrin CD41, consistent with platelet origin, and display autoantigens vimentin and fibrinogen. Functionally, platelet EVs are highly pro-inflammatory, eliciting leukotriene production by neutrophils (Cloutier et al. 2013). Annexin V^+, leukocyte-derived, granulocyte-derived, monocyte-derived, and T cell-derived EVs were found at high levels in RA synovial fluid (Michael et al. 2019). Other studies showed a difference between RA and OA as EVs presence in the synovial fluid. RA samples contained more $CD61^+$ (Michael et al. 2018), $CD45^+$ (Michael et al. 2019), $CD3^+$ (György et al. 2012), $CD4^+$ (Michael et al. 2019), $CD8^+$ (Michael et al. 2019; György et al. 2012), $CD14^+$ (Michael et al. 2019), and $CD66^+$ (Michael et al. 2019) EVs than OA samples. Also, thrombin-generating capacity (factor VIIa) is high for RA synovial fluid EVs (Berckmans et al. 2002).

Specific synovial fluid EVs presence and abundance might also be correlated with disease severity or progression. RA synovial fluid $CD41^+$ EVs levels are correlated with disease duration (György et al. 2012) while $CD66b^+$ EVs are more abundant in established RA than early RA (Michael et al. 2019). A positive association was found between rheumatoid factor and synovial fluid $CD3^+$, $CD4^+$, and $CD8^+$ (György et al. 2012). A weak correlation was observed between anti-citrullinated peptides antibodies titer, $CD4^+$ EVs, and annexin $V^+/CD45^+$ EVs (Michael et al. 2019).

6.2.11 EVs as Biomarkers: OA

It is well established that synovial fluid contains various cytokines, whose amount may greatly change during disease. Interestingly, most of these factors are not only in free form, but are also enriched in EVs that, compared with the cytokine of synovial fluid, are enriched in these proteins in patients with end-stage OA (Carlson et al. 2018). Zhao and colleagues demonstrated that the expression level of synovial fluid-derived EVs lncRNAs is increased in the primary stage of OA; in particular lncRNA prostate-specific transcript 1 (PCGEM1) showed appreciable differences in different stages of OA and was significantly elevated in advanced OA compared with primary OA as well as advanced OA compared with healthy controls (Zhao and Xu 2018). Moreover, authors found that lncRNA PCGEM1 exhibited a positive relationship with the WOMAC Index (the Western Ontario and McMaster Universities Osteoarthritis Index). A similar role as biomarker was also demonstrated for circular RNAs, since EVs-associated hsa-circ-0104595 in synovial fluid has the potential to be transported via bodily fluids and could serve as diagnostic biomarker (Zhang et al. 2021).

Gender can also influence markers associated with EVs of OA synovial fluid. In the synovial fluid EVs of female OA, the levels of haptoglobin, orosomucoid, and ceruloplasmin are upregulated, while the level of apolipoprotein is downregulated. Conversely, in the synovial fluid EVs of male OA, the levels of β-2-glycoprotein and complement component five protein are significantly upregulated, while the level of Spt-Ada-Gcn5 acetyltransferase-related factor 29 is downregulated. The same was observed for miRNA-associated EVs, since there is significant gender-specific differential expression of EVs miRNAs in SF from OA patients. In the male group, 69 miRNAs were downregulated while 45 miRNAs were downregulated. In the female group, 91 miRNAs were decreased while 53 miRNAs were increased (Kolhe et al. 2017). Interestingly, female OA-specific miRNAs are estrogen responsive and target toll-like receptor signalling pathways.

As for RA, EVs in synovial fluid could be envisioned in the future to understand progression of the pathology after treatment. miRNA and protein content of EVs isolated from synovial fluid in OA were similar to those of senescent chondrocytes, suggesting that EVs can be used not only to diagnose OA but also to evaluate the effects of treatment (Jeon et al. 2019).

6.2.12 EVs Biological Effect: RA

EVs found in synovial fluid are released mainly by chondrocytes, Type A and B cells of the synovial membrane, and immune cells. Very few reports described the activity of specific subpopulations of EVs isolated from synovial fluid. Therefore, the knowledge about the biological/pathological role of EVs released in the synovial fluid and acting on other joint tissues and cell types very often derives from in vitro studies with population-specific EVs.

Regarding the effects of RA synovial fluid EVs, it was demonstrated that these particles are able to induce B cell-activating factor, thymic stroma lymphopoietin, and secretory leukocyte protease inhibitor release by rheumatoid arthritis fibroblast-like synoviocytes (Messer et al. 2009). This is of importance for joint homeostasis and inflammatory status, since EVs appear to be potent inducers of pro-inflammatory factors as well as B-cell survival and promote the release of activation factors such as BAFF or TSLP by RA Type B cells. Thus, synovial fluid EVs might be involved in the control of humoral B-cell responses in RA synovium.

Moreover, synovial fluid EVs from RA patients have high osteoclastogenic potential (Eun Song et al. 2021), which may partly explain the characteristic bony erosion observed in inflamed joints of RA patients, not frequently observed in other inflammatory arthritis like OA. Intriguingly, synovial fluid EVs of RA patients contain high levels of RANKL, essential for osteoclast differentiation, which could effectively contribute to characteristic bony erosion in RA (Eun Song et al. 2021). Consistently, high levels of miR-574-5p were detected in EVs from the synovial fluid of RA patients, and high levels of this miRNA were reported to significantly increase osteoclastogenesis and elevate IL-23 and IFN-α mRNA levels in CD14+ monocytes via TLR7/8 (Hegewald et al. 2020). This is of importance since TLRs trigger an immune response and inflammation involved in osteoclast-mediated bone erosion and joint vascularization in RA.

Among the totality of synovial fluid EVs, synovial membrane-derived EVs play a role in cartilage failure and joint impairment. Accordingly, RA-Type B-cell EVs reduce osteoblast proliferation, mineralization, and differentiation (Chen et al. 2020). Moreover, RA-Type B-cell EVs suppress the proliferation and migration of chondrocytes and promote their apoptosis (Liu et al. 2020a, b). On the other hand, release of pro-inflammatory mediators, chemokines, and matrix-degrading enzymes in RA-Type B cells, thereby aggravating angiogenesis, matrix degradation, and cartilage damage in RA, is triggered by EVs from T cells and monocytes (Bao et al. 2021). Also, RA Type B-cell EVs cause increased resistance to apoptosis and prolong the survival time in autoimmune T cells (Zhang et al. 2006). Similarly, miR-424 in EVs derived from RA-Type B cells induced Th17 differentiation and inhibited Treg cells differentiation under hypoxic conditions (Ding et al. 2020). Thus, there is a cross-talk between synovial membrane cells and immune cells of the synovial fluid that, at least in part, is mediated by EVs of different origins that are released in the joint synovial fluid.

6.2.13 EVs Biological Effect: OA

As well as in RA, EVs released in the synovial fluid by different joint cell types are involved in OA pathology. Again, synovial Type B cells play a crucial role. Their EVs induce macrophages to release an array of chemokines and pro-inflammatory cytokines, such as IL-1β, CCL8, CCL15, and CCL20, while the secretion of IL-16, IL-8, and CCL7 is downregulated, overall promoting cartilage destruction and local inflammation in joints (Domenis et al. 2017). Consistently, EVs from OA-Type B cells stimulated with IL-1β induce M1 pro-inflammatory macrophages to secrete MMP12 and MMP7 and to suppress MMP8, thus contributing to the degradation of the extracellular matrix. Therefore, Type B cells during OA can trigger and contribute to the extension of the inflammatory process via the stimulation of M1 macrophages to secrete critical molecules functioning in the inflammatory process and cartilage damage. Moreover, inflamed synovial Type B cells are also implicated in promoting angiogenic activities (Kato et al. 2014), which facilitate inflammation and contribute to structural disease progression in osteoarthritis.

Together with Type B cells also other joint cell types can influence EVs through the release of EVs able to interact with other joint tissues and cells. Regarding cartilage, abnormal biomechanical loading typical of the pathology leads to the increased release of chondrocyte-derived EVs, in which promoters of calcification increased and inhibitors decreased, resulting in accelerating abnormal cartilage calcification in OA (Liu et al. 2022). Chondrocytes are also affected in their capability to defend against oxidative stress by EVs released by vascular endothelial cells by downregulating autophagy and p21 expression, thus increasing the cellular ROS content and contributing to apoptosis (Yang et al. 2021).

Intriguingly, EVs released in the synovial fluid can alter joint tissue homeostasis via the shuttling of several molecules, including non-coding RNAs. Levels of 50 miRNAs differ between exosomes from IL-1 stimulated and non-stimulated synovial type B cells (Kato et al. 2014). As an example of EV-miRNA role, miR-126-3p, a transcriptional modulator of chronic inflammation and innate immunity, is significantly reduced in EVs from Type B cells (Zhou et al. 2021), and its supplementation exerts antiapoptotic and anti-inflammatory effects on articular cartilage in vivo. Chondrocyte–synovial Type B cells inter-communication may also rely on other types of nucleic acids. One example is the role of long non-coding RNAs. EVs isolated from OA-Type B cells have high levels of lncRNA-PCGEM1, involved in chondrocyte apoptosis and cartilage matrix degradation, as observed in OA patient cartilage (Zeng et al. 2022). Of note, a feed-back loop between chondrocytes and synoviocytes is also present. When EVs from IL-1-treated chondrocytes are applied to OA synovium, a marked increase in MMP13, IL-1, TNF, and cyclooxygenase-2 (COX-2) emerges (Nakasa et al. 2016).

Alongside Type B cells and chondrocytes, EVs-mediated cross-talk also involves immune cells of the synovium and synovial fluid. EVs isolated from the synovial fluid of OA patients induce an abundant release of IL-1 and chemokines in GM-CSF-primed M1 macrophages (Domenis et al. 2017). Similarly, being part of the EVs pool of the synovial fluid, EVs originating from OA chondrocytes can be internalized by LPS-primed macrophages and further enhance inflammasome activation and subsequent IL-1 processing (Ni et al. 2019). On the contrary, macrophages can influence chondrocytes. IL-4- or Il-13-treated macrophages release EVs enriched in

SOX9 mRNA and protein that can be transferred to chondrocytes, where they can promote the secretion of ECM components (Bai et al. 2020).

6.2.14 EVs Therapeutic Strategy: RA

EVs-based vehicle delivery systems represent advanced therapeutic strategies that have recently attracted increasing attention. When envisioning EVs to be administered to a diseased joint, synovial fluid plays a crucial role since injected particles directly interact with and diffuse through the joint space.

Among the different cell types as EVs source for therapeutic approaches, mesenchymal stromal cells (MSCs) and their EVs represent the most promising option due to their anti-inflammatory and regenerative properties (Dabrowska et al. 2021). Bone marrow MSCs (BMSCs)-EVs showed their efficacy in alleviating experimental RA by inhibiting T and B lymphocyte proliferation, as well as by inducing Treg and IL-10-expressing regulatory B cells in a dose-dependent manner (Cosenza et al. 2018). BMSC-EVs have also been shown to have a therapeutic effect on synovial Type B cells, especially through miRNA transfer. miR-34 in BMSC-EVs can reduce RA-Type B cells proliferation and inflammation (Wu et al. 2021), miR-320 inhibits the activation, migration, and invasion (Meng et al. 2020), and miR-150-5p inhibits angiogenesis by downregulating the levels of matrix metallopeptidase MMP14 and vascular endothelial growth factor (Chen et al. 2018). Regarding adipose MSCs (ASCs)-EVs, results on RA-independent models suggest their potency for RA treatment. In fact, ASC-EVs can drive both M2 macrophage polarization, thus reducing the ability of macrophages to evoke inflammatory responses (Hu et al. 2021), and an immunosuppressive effect on T cells (Heidari et al. 2021; Blazquez et al. 2014). Eventually, umbilical cord MSCs (UCMSCs)-EVs significantly promote the proliferation of chondrocytes and the M2 polarization of macrophages, in addition to regulating inflammation of the joint cavity (Jiang et al. 2021).

6.2.15 EVs Therapeutic Strategy: OA

EVs derived from MSCs are also actively explored for their therapeutic potential in OA at several levels. OA is characterized by changes in ECM composition and structure, with loss of collagen type II and aggrecan (major ECM components) and increase of collagenases, metalloproteinases, and aggrecanases (Maldonado and Nam 2013). EVs from MSCs were able to reverse cartilage destructive processes. In fact, BMSC-EVs promote cartilage matrix regeneration by promoting aggrecan expression and by supplementing mitochondrial-related proteins, thus rescuing the imbalance in energy metabolism, another trait of OA cartilage (Chen et al. 2019). Together with aggrecan, BMSC-EVs upregulate the expression of type II collagen while decreasing the expression levels of MMP-13, a disintegrin and metalloproteinase with thrombospondin motifs 5 (ADAMTS5) and iNOS (Cosenza et al. 2018). Intriguingly, a comparative study using EVs from three MSC sources, bone marrow, adipose tissue, and synovium, showed that ASC-EVs have the highest potential for promoting chondrogenicity by inducing focal adhesion and ECM–receptor interaction, with ASC-EVs having highest abundance of related proteins (Li et al. 2021). Accordingly, ASC-EVs are enriched by both versican, a chondroitin sulphate proteoglycan that promotes chondrogenesis and joint morphogenesis (Giannasi et al. 2020), and tissue inhibitors of metalloproteinase-1 and -2 (TIMP-1 and TIMP-2), which counteracts MMP activities (Woo et al. 2020).

Together with ECM destruction, inflammation is another trigger of OA progression. ASC-EVs not only promote collagen II deposition but alongside impair the production of pro-inflammatory mediators in osteoarthritic chondrocytes and increase the levels of anti-inflammatory cytokine IL-10 (Tofiño-Vian et al. 2018). The anti-inflammatory features were also confirmed for BMSC-EVs, able to reduce the expression of pro-inflammatory interleukins, such as IL-1A, IL-1B, IL-6, IL-8, and IL-17, in OA chondrocytes (Vonk et al. 2018; Liu et al. 2020a, b). As for RA, also for OA the effects of MSC-EVs on macrophage polarization are of crucial relevance. This was shown to be due to several EVs-encapsulated miRNAs, in both ASCs and BMSCs (Ragni et al. 2020a, 2022). This is of relevance since synovial pro-inflammatory M1 macrophages, which are reduced by MSC-EVs miRNAs, secrete pro-inflammatory factors, while anti-inflammatory M2 macrophages, which are promoted by EVs miRNAs, ensure cartilage graft survival and support healthy cartilage (Fahy et al. 2014).

6.2.16 Therapeutic EVs Interaction with Joint Tissues

As discussed in the previous paragraphs, EVs from MSCs used for therapeutic approaches showed high potential for tissue healing and inflammation reduction. A crucial question still to be fully addressed is how these particles interact with cells of synovial and cartilage tissues or with immune cells of the synovial fluid. Amniotic membrane MSC (hAMSC)-EVs showed a differential incorporation among the different immune populations, with CD14+ cells, representative of the monocyte compartment, being more prone to interaction in comparison to the lymphocyte compartment given by T and B lymphocytes (Papait et al. 2022). The reason for this difference between immune cells is at the moment not clear but may be crucial for envisioning a targeted approach for specific cell types involved in the pathology. Regarding synoviocytes, a quick uptake starting from few minutes after administration and continuing up to 24 h was reported both in 2D and in 3D cultures (Ragni et al. 2019, 2020b). For synoviocytes, in 2D cultures, EVs-cell binding was shown to be dependent on the interaction between the hyaluronan coating surrounding the cells and the CD44 receptor for hyaluronan present on EVs surface (Ragni et al. 2019). Again, for a target therapy, the possibility of an increase in the amount of CD44 on EVs would be an efficient strategy to improve their uptake. For chondrocytes, a fast interaction was observed in both 2D and 3D cultures, with EVs able to penetrate cartilage ECM (Mortati et al. 2020). Of note, EVs not only interact with ECM but are able to be incorporated by chondrocytes, thus explaining the results observed in in vivo models after injection in the joint.

6.2.17 EVs and Scaffold: A New Option

The ability of EVs to interact with natural ECM or ECM-like structures opened the way to couple EVs with scaffolds to position the particles at the site of injury and to improve their bioavailability over time. A possibility is combining EVs with a suitable biomaterial, such as a hydrogel or scaffold. One example is given by a hydrogel glue (comprising o-nitrobenzyl alcohol moieties, modified hyaluronic acids, and gelatine) that was able to retain over 90% of embedded hiPSC-derived MSC-EVs (Liu et al. 2017). Importantly, the EVs-hydrogel could form a seamless interface in the rabbit articular cartilage defect model. Of note, synthetic scaffolds can be combined with natural structures. A decellularized cartilage ECM was associated with gelatine methacrylate (GelMA) to biofabricate an ECM/GelMA/EXO complex using BMSC-EVs, and EVs enhanced the positive effect of ECM on chondrocyte migration in vitro and showed the best therapeutic activity promoting cartilage regeneration in a rabbit osteochondral defect model (Chen et al. 2019).

An advantage of using biomaterials is their ease of positioning at the site of injury while a limitation is their poor strength. To overcome the latter, nanomaterials may improve the mechanical characteristics of a hydrogel (Fan et al. 2022). As an example, the addition of nanoclay in a gelatine methacrylate hydrogel produced a potential material for cartilage regeneration which promoted the sustained release of EVs from umbilical cord MSC that promoted cartilage regeneration in a rat cartilage defect model (Hu et al. 2020). On the other hand, having a scaffold able to be injected can be a valuable option to prevent the immediate washout of the active molecules and prolong the general therapeutic exposure when the positioning is not crucial. A mixture of Poloxamer-407 and 188 is a functional injectable carrier to deliver platelet-rich plasma-derived EVs that remained functional, demonstrating positive effects on chondrocytes in vitro and protective activity against inflammation-mediated apoptosis and degeneration of chondrocytes in vivo (Zhang et al. 2022).

6.3 Conclusion and Future Perspectives

The synovial fluid is a specialized structure that is essential for lubricating joints and providing nutrients to tissues and cells. Its particular structure made by polymers allows for creating a network where EVs from either joint tissues and cells or exogenously administered can be entrapped, shuttled, and delivered (Fig. 6.1). This allows synovial fluid to be crucial for both diagnosis and therapy. To date, there is

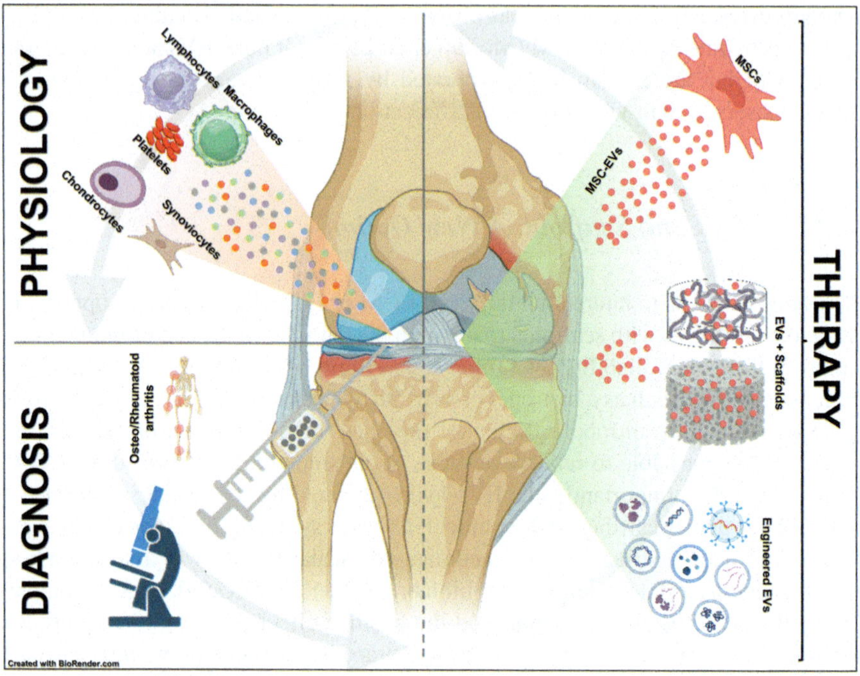

Fig. 6.1 EVs role in joint physiology, diagnosis, and therapy. EVs in synovial fluid may be used to understand joint homeostasis and contribution of the different cells and tissues (physiology), to identify disease-specific markers to drive personalized treatments (diagnosis), and eventually to counteract pathology features (therapy) both directly and coupled with next-generation scaffolds or engineering approaches

still a lack of markers for joint pathologies. With future screening it is possible that, with a simple synovial fluid collection, disease and stage-specific EVs and markers will be tested to drive the most effective cure. On this side, EVs-based therapies are gaining interest and they will be the future for joint pathologies. What remains to be deciphered is the way EVs can be conveyed in the joint to facilitate interaction with target tissue and to prolong their stability and availability over time. The use of scaffolds will be the next step for EVs therapeutics for joint pathologies.

Acknowledgements The authors want to thank all the clinicians, operating room staff, and researchers of IRCCS Ospedale Galeazzi—Sant'Ambrogio for their exceptional work during the pandemic. The authors want to thank all authors of the works cited in the present chapter; without their data it would have been impossible to complete this chapter.

Compliance with Ethical Standards The authors declare that they have no conflict of interest. The study did not receive funding from any source. This article does not contain any studies involving animals performed by any of the authors. This article does not contain any studies involving human participants performed by any of the authors.

References

Alford JW, Cole BJ (2005) Cartilage restoration, part 1: basic science, historical perspective, patient evaluation, and treatment options. Am J Sports Med 33:295–306. https://doi.org/10.1177/0363546504273510

Bai J, Zhang Y, Zheng X et al (2020) LncRNA MM2P-induced, exosome-mediated transfer of Sox9 from monocyte-derived cells modulates primary chondrocytes. Cell Death Dis 11:763. https://doi.org/10.1038/s41419-020-02945-5

Balazs EA (2004) Viscosupplementation for treatment of osteoarthritis: from initial discovery to current status and results. Surg Technol Int 12:278–289

Bao X, Ma L, He C et al (2021) MicroRNA-23a-5p regulates cell proliferation, migration and inflammation of TNF-α-stimulated human fibroblast-like MH7A synoviocytes by targeting TLR4 in rheumatoid arthritis. Exp Ther Med 21:479. https://doi.org/10.3892/etm.2021.9910

Bartok B, Firestein GS (2010) Fibroblast-like synoviocytes: key effector cells in rheumatoid arthritis. Immunol Rev 233:233–255. https://doi.org/10.1111/j.0105-2896.2009.00859.x

Benito MJ, Veale DJ, FitzGerald O et al (2005) Synovial tissue inflammation in early and late osteoarthritis. Ann Rheum Dis 64:1263–1267. https://doi.org/10.1136/ard.2004.025270

Berckmans RJ, Nieuwland R, Tak PP et al (2002) Cell-derived microparticles in synovial fluid from inflamed arthritic joints support coagulation exclusively via a factor VII-dependent mechanism. Arthritis Rheum 46:2857–2866. https://doi.org/10.1002/art.10587

Blazquez R, Sanchez-Margallo FM, de la Rosa O et al (2014) Immunomodulatory potential of human adipose mesenchymal stem cells derived exosomes on in vitro stimulated T cells. Front Immunol 5:556. https://doi.org/10.3389/fimmu.2014.00556

Boere J, van de Lest CHA, Libregts SFWM et al (2016) Synovial fluid pretreatment with hyaluronidase facilitates isolation of CD44+ extracellular vesicles. J Extracell Vesicles 5:31751. https://doi.org/10.3402/jev.v5.31751

Carlson AK, Rawle RA, Adams E et al (2018) Application of global metabolomic profiling of synovial fluid for osteoarthritis biomarkers. Biochem Biophys Res Commun 499:182–188. https://doi.org/10.1016/j.bbrc.2018.03.117

Chen Z, Wang H, Xia Y (2018) Therapeutic potential of mesenchymal cell-derived miRNA-150-5p-expressing exosomes in rheumatoid arthritis mediated by the modulation of MMP14 and VEGF. J Immunol 201:2472–2482. https://doi.org/10.4049/jimmunol.1800304

Chen P, Zheng L, Wang Y et al (2019) Desktop-stereolithography 3D printing of a radially oriented extracellular matrix/mesenchymal stem cell exosome bioink for osteochondral defect regeneration. Theranostics 9:2439–2459. https://doi.org/10.7150/thno.31017

Chen J, Liu M, Luo X et al (2020) Exosomal miRNA-486-5p derived from rheumatoid arthritis fibroblast-like synoviocytes induces osteoblast differentiation through the Tob1/BMP/Smad pathway. Biomater Sci 8:3430–3442. https://doi.org/10.1039/c9bm01761e

Cloutier N, Tan S, Boudreau LH et al (2013) The exposure of autoantigens by microparticles underlies the formation of potent inflammatory components: the microparticle-associated immune complexes. EMBO Mol Med 5:235–249. https://doi.org/10.1002/emmm.201201846

Cosenza S, Toupet K, Maumus M et al (2018) Mesenchymal stem cells-derived exosomes are more immunosuppressive than microparticles in inflammatory arthritis. Theranostics 8:1399–1410. https://doi.org/10.7150/thno.21072

Cowman MK, Matsuoka S (2005) Experimental approaches to hyaluronan structure. Carbohydr Res 40:791–809. https://doi.org/10.1016/j.carres.2005.01.022

Dabrowska S, Andrzejewska A, Janowski M et al (2021) Immunomodulatory and regenerative effects of mesenchymal stem cells and extracellular vesicles: therapeutic outlook for inflammatory and degenerative diseases. Front Immunol 11:591065. https://doi.org/10.3389/fimmu.2020.591065

Ding Y, Wang L, Wu H et al (2020) Exosomes derived from synovial fibroblasts under hypoxia aggravate rheumatoid arthritis by regulating Treg/Th17 balance. Exp Biol Med (Maywood) 245:1177–1186. https://doi.org/10.1177/1535370220934736

Domenis R, Zanutel R, Caponnetto F et al (2017) Characterization of the proinflammatory profile of synovial fluid-derived exosomes of patients with osteoarthritis. Mediat Inflamm 2017:4814987. https://doi.org/10.1155/2017/4814987

Eun Song J, Soo Kim J, Hye Shin J et al (2021) Role of synovial exosomes in osteoclast differentiation in inflammatory. Arthritis Cells 10:120. https://doi.org/10.3390/cells10010120

Fahy N, de Vries-van Melle ML, Lehmann J et al (2014) Human osteoarthritic synovium impacts chondrogenic differentiation of mesenchymal stem cells via macrophage polarisation state. Osteoarthr Cartil 22:1167–1175. https://doi.org/10.1016/j.joca.2014.05.021

Fan WJ, Liu D, Pan LY, Wang WY, Ding YL, Zhang YY, Ye RX, Zhou Y, An SB, Xiao WF (2022 Jul) Exosomes in osteoarthritis: updated insights on pathogenesis, diagnosis, and treatment. Front Cell Dev Biol 26(10):949690. https://doi.org/10.3389/fcell.2022.949690

Firestein GS, Gabriel SE, McInnes IB et al (2016) Kelley and Firestein's textbook of rheumatology, 10th edn. Elsevier, Amsterdam

Fox AJS, Bedi A, Rodeo SA (2009) The basic science of articular cartilage: structure, composition, and function. Sports Health 1:461–468. https://doi.org/10.1177/1941738109350438

Giannasi C, Niada S, Magagnotti C et al (2020) Comparison of two ASC-derived therapeutics in an in vitro OA model: secretome versus extracellular vesicles. Stem Cell Res Ther 11:521. https://doi.org/10.1186/s13287-020-02035-5

Goldring MB (2012) Articular cartilage degradation in osteoarthritis. HSS J 8:7–9. https://doi.org/10.1007/s11420-011-9250-z

György B, Szabó TG, Turiák L et al (2012) Improved flow cytometric assessment reveals distinct microvesicle (cell-derived microparticle) signatures in joint diseases. PLoS One 7:e49726. https://doi.org/10.1371/journal.pone.0049726

Hampel U, Sesselmann S, Iserovich P et al (2013) Chemokine and cytokine levels in osteoarthritis and rheumatoid arthritis synovial fluid. J Immunol Methods 396:134–139. https://doi.org/10.1016/j.jim.2013.08.007

Haywood L, McWilliams DF, Pearson CI et al (2003) Walsh inflammation and angiogenesis in osteoarthritis. Arthritis Rheum 48:2173–2177. https://doi.org/10.1002/art.11094

Hegewald AB, Breitwieser K, Ottinger SM et al (2020) Extracellular miR-574-5p induces osteoclast differentiation via TLR 7/8 in rheumatoid arthritis. Front Immunol 11:585282. https://doi.org/10.3389/fimmu.2020.585282

Heidari N, Abbasi-Kenarsari H, Namaki S et al (2021) Adipose-derived mesenchymal stem cell-secreted exosome alleviates dextran sulfate sodium-induced acute colitis by Treg cell induction and inflammatory cytokine reduction. J Cell Physiol 236:5906–5920. https://doi.org/10.1002/jcp.30275

Housmans BAC, van den Akker GGH, Neefjes M et al (2023) Direct comparison of non-osteoarthritic and osteoarthritic synovial fluid-induced intracellular chondrocyte signaling and phenotype changes. Osteoarthr Cartil 31:60–71. https://doi.org/10.1016/j.joca.2022.09.004

Hu H, Dong L, Bu Z et al (2020) miR-23a-3p-abundant small extracellular vesicles released from Gelma/nanoclay hydrogel for cartilage regeneration. J Extracell Vesicles 9:1778883. https://doi.org/10.1080/20013078.2020.1778883

Hu Q, Lyon CJ, Fletcher JK et al (2021) Extracellular vesicle activities regulating macrophage- and tissue-mediated injury and repair responses. Acta Pharm Sin B 11:1493–1512. https://doi.org/10.1016/j.apsb.2020.12.014

Hunter DJ, Bierma-Zeinstra S (2019) Osteoarthritis. Lancet 393:1745–1759. https://doi.org/10.1016/S0140-6736(19)30417-9

Itano N, Kimata K (2002) Mammalian hyaluronan synthases. IUBMB Life 54:195–199. https://doi.org/10.1080/15216540214929

Jeon OH, Wilson DR, Clement CC et al (2019) Senescence cell-associated extracellular vesicles serve as osteoarthritis disease and therapeutic markers. JCI Insight 4:e125019. https://doi.org/10.1172/jci.insight.125019

Jiang S, Tian G, Yang Z et al (2021) Enhancement of acellular cartilage matrix scaffold by Wharton's jelly mesenchymal stem cell-derived exosomes to promote osteochondral regeneration. Bioact Mater 6:2711–2728. https://doi.org/10.1016/j.bioactmat.2021.01.031

Juneja P, Munjal A, Hubbard JB (2023) Anatomy, joints. In: StatPearls [internet]. StatPearls Publishing, Treasure Island, FL. https://www.ncbi.nlm.nih.gov/books/NBK507893/. Accessed 20 Dec 2023

Kato T, Miyaki S, Ishitobi H et al (2014) Exosomes from IL-1β stimulated synovial fibroblasts induce osteoarthritic changes in articular chondrocytes. Arthritis Res Ther 16:R163. https://doi.org/10.1186/ar4679

Keenan RT (2022) Biochemical composition of synovial fluid in health and disease. In: Mandell BF (ed) Synovial fluid analysis and the evaluation of patients with arthritis. Springer, Cham. https://doi.org/10.1007/978-3-030-99612-3

Kofoed JA, Barceló AC (1978) The synovial fluid hyaluronic acid in rheumatoid arthritis. Experientia 34:1545–1546. https://doi.org/10.1007/BF02034662

Kolhe R, Hunter M, Liu S et al (2017) Gender-specific differential expression of exosomal miRNA in synovial fluid of patients with osteoarthritis. Sci Rep 7:2029. https://doi.org/10.1038/s41598-017-01905-y

Kondo N, Kuroda T, Kobayashi D (2021) Cytokine networks in the pathogenesis of rheumatoid arthritis. Int J Mol Sci 22:10922. https://doi.org/10.3390/ijms222010922

Li Q, Yu H, Sun M et al (2021) The tissue origin effect of extracellular vesicles on cartilage and bone regeneration. Acta Biomater 125:253–266. https://doi.org/10.1016/j.actbio.2021.02.039

Lin Z, Willers C, Xu J et al (2006) The chondrocyte: biology and clinical application. Tissue Eng 12:1971–1984. https://doi.org/10.1089/ten.2006.12.1971

Liu X, Yang Y, Li Y et al (2017) Integration of stem cell-derived exosomes with in situ hydrogel glue as a promising tissue patch for articular cartilage regeneration. Nanoscale 9:4430–4438. https://doi.org/10.1039/c7nr00352h

Liu D, Fang Y, Rao Y et al (2020a) Synovial fibroblast-derived exosomal microRNA-106b suppresses chondrocyte proliferation and migration in rheumatoid arthritis via down-regulation of PDK4. J Mol Med (Berl) 98:409–423. https://doi.org/10.1007/s00109-020-01882-2

Liu C, Li Y, Yang Z et al (2020b) Kartogenin enhances the therapeutic effect of bone marrow mesenchymal stem cells derived exosomes in cartilage repair. Nanomedicine (Lond) 15(3):273–288. https://doi.org/10.2217/nnm-2019-0208

Liu Q, Wang R, Hou S et al (2022) Chondrocyte-derived exosomes promote cartilage calcification in temporomandibular joint osteoarthritis. Arthritis Res Ther 24:44. https://doi.org/10.1186/s13075-022-02738-5

Maldonado M, Nam J (2013) The role of changes in extracellular matrix of cartilage in the presence of inflammation on the pathology of osteoarthritis. Biomed Res Int 2013:284873. https://doi.org/10.1155/2013/284873

Masuko K, Murata M, Yudoh K et al (2009) Anti-inflammatory effects of hyaluronan in arthritis therapy: not just for viscosity. Int J Gen Med 2:77–81. https://doi.org/10.2147/ijgm.s5495

Meng Q, Qiu B et al (2020) Exosomal microRNA-320a derived from mesenchymal stem cells regulates rheumatoid arthritis fibroblast-like synoviocyte activation by suppressing CXCL9 expression. Front Physiol 11:441. https://doi.org/10.3389/fphys.2020.00441

Messer L, Alsaleh G, Freyssinet J (2009) Microparticle-induced release of B-lymphocyte regulators by rheumatoid synoviocytes. Arthritis Res Ther 11:R40. https://doi.org/10.1186/ar2648

Michael BNR, Durga Prasanna M, Chengappa KG et al (2018) Relevance of elevated microparticles in peripheral blood and synovial fluid of patients with rheumatoid arthritis. Indian J Rheumatol 13:222–228. https://doi.org/10.4103/injr.injr_101_18

Michael BNR, Kommoju V, Ganapathy CK et al (2019) Characterization of cell-derived microparticles in synovial fluid and plasma of patients with rheumatoid arthritis. Rheumatol Int 39:1377–1387. https://doi.org/10.1007/s00296-019-04337-1

Molnar V, Matišić V, Kodvanj I (2021) Cytokines and chemokines involved in osteoarthritis pathogenesis. Int J Mol Sci 22:9208. https://doi.org/10.3390/ijms22179208

Mortati L, de Girolamo L, Perucca Orfei C et al (2020) In vitro study of extracellular vesicles migration in cartilage-derived osteoarthritis samples using real-time quantitative multimodal nonlinear optics imaging. Pharmaceutics 12:734. https://doi.org/10.3390/pharmaceutics12080734

Mow VC, Ratcliffe A, Poole AR (1992) Cartilage and diarthrodial joints as paradigms for hierarchical materials and structures. Biomaterials 13:67–97. https://doi.org/10.1016/0142-9612(92)90001-5

Müller-Ladner U, Kriegsmann J, Franklin BN et al (1996) Synovial fibroblasts of patients with rheumatoid arthritis attach to and invade normal human cartilage when engrafted into SCID mice. Am J Pathol 149:1607–1615

Murai T (2015) Lipid raft-mediated regulation of hyaluronan-CD44 interactions in inflammation and cancer. Front Immunol 6:420. https://doi.org/10.3389/fimmu.2015.00420

Mustonen A, Lehmonen N, Oikari S et al (2022) Counts of hyaluronic acid-containing extracellular vesicles decrease in naturally occurring equine osteoarthritis. Sci Rep 12:17550. https://doi.org/10.1038/s41598-022-21398-8

Nakasa T, Miyaki S, Kato T et al (2016) Exosome derived from osteoarthritis cartilage induces catabolic factor gene expressions in synovium. Trans Orth Res Soc 2012:abstract 708

Ni Z, Kuang L, Chen H et al (2019) The exosome-like vesicles from osteoarthritic chondrocyte enhanced mature IL-1β production of macrophages and aggravated synovitis in osteoarthritis. Cell Death Dis 10:522. https://doi.org/10.1038/s41419-019-1739-2

Ossanna PJ, Test ST, Matheson NR et al (1986) Oxidative regulation of neutrophil elastase-alpha-1-proteinase inhibitor interactions. J Clin Invest 77:1939–1951. https://doi.org/10.1172/JCI112523

Papait A, Ragni E, Cargnoni A et al (2022) Comparison of EV-free fraction, EVs, and total secretome of amniotic mesenchymal stromal cells for their immunomodulatory potential: a translational perspective. Front Immunol 13:960909. https://doi.org/10.3389/fimmu.2022.960909

Pelletier JP, Martel-Pelletier J, Abramson SB (2001) Osteoarthritis, an inflammatory disease: potential implication for the selection of new therapeutic targets. Arthritis Rheum 44:1237–1247. https://doi.org/10.1002/1529-0131(200106)44:6<1237::AID-ART214>3.0.CO;2-F

Ragni E, Perucca Orfei C, De Luca P et al (2019) Interaction with hyaluronan matrix and miRNA cargo as contributors for in vitro potential of mesenchymal stem cell-derived extracellular vesicles in a model of human osteoarthritic synoviocytes. Stem Cell Res Ther 10:109. https://doi.org/10.1186/s13287-019-1215-z

Ragni E, Perucca Orfei C, De Luca P et al (2020a) Inflammatory priming enhances mesenchymal stromal cell secretome potential as a clinical product for regenerative medicine approaches through secreted factors and EV-miRNAs: the example of joint disease. Stem Cell Res Ther 11:165. https://doi.org/10.1186/s13287-020-01677-9

Ragni E, Palombella S, Lopa S et al (2020b) Innovative visualization and quantification of extracellular vesicles interaction with and incorporation in target cells in 3D microenvironments. Cells 9:1180. https://doi.org/10.3390/cells9051180

Ragni E, Perucca Orfei C, de Girolamo L (2022) Secreted factors and extracellular vesicles account for the immunomodulatory and tissue regenerative properties of bone-marrow-derived mesenchymal stromal cells for osteoarthritis. Cells 11:3501. https://doi.org/10.3390/cells11213501

Ralphs JR, Benjamin M (1994) The joint capsule: structure, composition, ageing and disease. J Anat 184:503–509

Rilla K, Tiihonen R, Kultti A et al (2008) Pericellular hyaluronan coat visualized in live cells with a fluorescent probe is scaffolded by plasma membrane protrusions. J Histochem Cytochem 56:901–910. https://doi.org/10.1369/jhc.2008.951665

Sanchez C, Bay-Jensen A, Pap T et al (2017) Chondrocyte secretome: a source of novel insights and exploratory biomarkers of osteoarthritis. Osteoarthr Cartil 25:1199–1209. https://doi.org/10.1016/j.joca.2017.02.797
Seidman AJ, Limaiem F (2023) Synovial fluid analysis. In: StatPearls [internet]. StatPearls Publishing, Treasure Island, FL. https://www.ncbi.nlm.nih.gov/books/NBK537114/. Accessed 20 Dec 2023
Singh JA, Arayssi T, Duray P et al (2004) Immunohistochemistry of normal human knee synovium: a quantitative study. Ann Rheum Dis 63:785–790. https://doi.org/10.1136/ard.2003.013383
Smolen JS, Aletaha D, Barton A et al (2018) Rheumatoid arthritis. Nat Rev Dis Primers 4:18001. https://doi.org/10.1038/nrdp.2018.1
Soltés L, Stankovská M, Kogan G et al (2005) Contribution of oxidative-reductive reactions to high-molecular-weight hyaluronan catabolism. Chem Biodivers 2:1242–1245. https://doi.org/10.1002/cbdv.200590094
Song H, Zhao J, Cheng J et al (2021) Extracellular vesicles in chondrogenesis and cartilage regeneration. J Cell Mol Med 25:4883–4892. https://doi.org/10.1111/jcmm.16290
Stafford CT, Niedermeier W, Holley L et al (1964) Studies on the concentration and intrinsic viscosity of hyaluronic acid in synovial fluids of patients with rheumatic diseases. Ann Rheum Dis 23:152–157. https://doi.org/10.1136/ard.23.2.152
Stern R, Kogan G, Jedrzejas MJ et al (2007) The many ways to cleave hyaluronan. Biotechnol Adv 25:537–557. https://doi.org/10.1016/j.biotechadv.2007.07.001
Tamer TM (2013) Hyaluronan and synovial joint: function, distribution and healing. Interdiscip Toxicol 6:111–125. https://doi.org/10.2478/intox-2013-0019
Temple-Wong MM, Ren S, Quach P et al (2016) Hyaluronan concentration and size distribution in human knee synovial fluid: variations with age and cartilage degeneration. Arthritis Res Ther 18:18. https://doi.org/10.1186/s13075-016-0922-4
Tofiño-Vian M, Guillén MI, Del Caz MDP et al (2018) Microvesicles from human adipose tissue-derived mesenchymal stem cells as a new protective strategy in osteoarthritic chondrocytes cell. Physiol Biochem 47:11–25. https://doi.org/10.1159/000489739
Vonk LA, van Dooremalen SFJ, Liv N et al (2018) Mesenchymal stromal/stem cell-derived extracellular vesicles promote human cartilage regeneration in vitro. Theranostics 8:906–920. https://doi.org/10.7150/thno.20746
Withrow J, Murphy C, Liu Y et al (2016) Extracellular vesicles in the pathogenesis of rheumatoid arthritis and osteoarthritis. Arthritis Res Ther 18:286. https://doi.org/10.1186/s13075-016-1178-8
Woo CH, Kim HK, Jung GY et al (2020) Small extracellular vesicles from human adipose-derived stem cells attenuate cartilage degeneration. J Extracell Vesicles 9:1735249. https://doi.org/10.1080/20013078.2020.1735249
Wu H, Zhou X, Wang X et al (2021) miR-34a in extracellular vesicles from bone marrow mesenchymal stem cells reduces rheumatoid arthritis inflammation via the cyclin I/ATM/ATR/p53 axis. J Cell Mol Med 25:1896–1910. https://doi.org/10.1111/jcmm.15857
Yang R, Zheng H, Xu W et al (2021) Vascular endothelial cell-secreted exosomes facilitate osteoarthritis pathogenesis by promoting chondrocyte apoptosis. Aging (Albany NY) 13:4647–4662. https://doi.org/10.18632/aging.202506
Zeng G, Deng G, Xiao S et al (2022) Fibroblast-like synoviocytes-derived exosomal PCGEM1 accelerates IL-1β-induced apoptosis and cartilage matrix degradation by miR-142-5p/RUNX2 in chondrocytes. Immunol Investig 51:1284–1301. https://doi.org/10.1080/08820139.2021.1936010
Zhang H, Liu C, Su K et al (2006) A membrane form of TNF-alpha presented by exosomes delays T cell activation-induced cell death. J Immunol 176:7385–7393. https://doi.org/10.4049/jimmunol.176.12.7385
Zhang W, Qi L, Chen R (2021) Circular RNAs in osteoarthritis: indispensable regulators and novel strategies in clinical implications. Arthritis Res Ther 23:23. https://doi.org/10.1186/s13075-021-02420-2

Zhang Y, Wang X, Chen J et al (2022) Exosomes derived from platelet-rich plasma administration in site mediate cartilage protection in subtalar osteoarthritis. J Nanobiotechnol 20:56. https://doi.org/10.1186/s12951-022-01245-8

Zhao Y, Xu J (2018) Synovial fluid-derived exosomal lncRNA PCGEM1 as biomarker for the different stages of osteoarthritis. Int Orthop 42:2865–2872. https://doi.org/10.1007/s00264-018-4093-6

Zhou Y, Ming J, Li Y et al (2021) Exosomes derived from miR-126-3p-overexpressing synovial fibroblasts suppress chondrocyte inflammation and cartilage degradation in a rat model of osteoarthritis. Cell Death Discov 7:37. https://doi.org/10.1038/s41420-021-00418-y

Chapter 7
Extracellular Vesicles from Obese Adipose Tissue and Their Relationship with the Development of Comorbidities

Camino Tamara, Lago-Baameiro Nerea, and Pardo Maria

Abstract Extracellular vesicles (EVs) have emerged as key players in intercellular communication and metabolic regulation, with potential implications for obesity and related health problems. This chapter provides a comprehensive overview of current research on EVs released from adipose tissue, including their composition and molecular content. It highlights the dynamic nature of EVs secretion in response to pathophysiological states and discusses the potential impact of EVs on metabolic regulation, obesity-associated inflammatory diseases, and the extracellular matrix (ECM), which is a key component in the development of obesity and its comorbidities due to its remodelling and degradation. Furthermore, it highlights the potential therapeutic applications of EVs in tissue regeneration through ECM modulation and the need for further studies on the effect of EVs on the ECM during physiological processes such as tissue regeneration. Therefore, this chapter is a valuable resource for researchers and practitioners interested in understanding the broader implications of EVs in metabolic disorders, their interaction with the ECM, and their potential therapeutic applications.

7.1 Introduction

Today, sedentary lifestyles and the consumption of processed foods are increasing the rate of obesity to alarmingly pandemic proportions worldwide. In particular, the World Health Organisation (WHO) estimates that 39% of adults over the age of 18

C. Tamara (✉) · L.-B. Nerea
Grupo Obesidómica, Instituto de Investigación Sanitaria de Santiago (IDIS), Complexo Hospitalario Universitario de Santiago/SERGAS, Santiago de Compostela, Spain

P. Maria
Grupo Obesidómica, Instituto de Investigación Sanitaria de Santiago (IDIS), Complexo Hospitalario Universitario de Santiago/SERGAS, Santiago de Compostela, Spain

CIBER de Fisiopatología de la Obesidad y la Nutrición, Instituto de Salud Carlos III, Madrid, Spain

K. Rilla (ed.), *Extracellular Vesicles as Matrix Messengers*, Biology of Extracellular Matrix 15, https://doi.org/10.1007/978-3-031-68811-9_7

are overweight or obese (Marques et al. 2018; Ruíz-Roso et al. 2020). Moreover, abdominal obesity, which is associated with an increased risk of other pathologies, is increasing morbidity and mortality in populations in developing countries, at a high economic cost to global health systems and individual health (Apovian 2016; Andolfi and Fisichella 2018).

Obesity, which is a metabolic disease of multifactorial origin, is triggered by an imbalance in energy homeostasis, causing excess energy to be stored in the form of triglycerides in adipocytes. This causes adipose tissue to undergo a process of hypertrophy (increased size) and hyperplasia (increased cell production), promoting inflammation and fibrosis of this tissue (Fantuzzi 2005; Balistreri et al. 2010; Saltiel and Olefsky 2017). These changes affect not only adipocytes, but also the other cellular components of adipose tissue, such as stromal vascular fraction (preadipocytes, fibroblasts, vascular endothelial cells, pericytes, stromal cells, and immune cells), and their interaction with the extracellular matrix (ECM) (Giordano et al. 2014). This matrix is a complex structure consisting of different molecules such as proteins, proteoglycans, and polysaccharides. It provides an anchorage site for cells by regulating biological processes such as cell adhesion, migration, tissue repair and regeneration, survival, and development (Lin et al. 2016; Ruiz-Ojeda et al. 2019; Al Halawani et al. 2022). Therefore, adipose tissue expansion requires ECM remodelling and reorganisation to provide sufficient space during the process of cell hypertrophy and hyperplasia (Ruiz-Ojeda et al. 2019).

Moreover, fat tissue is not merely a storage site, but it is an endocrine organ that secretes different molecules capable of interacting both locally and peripherally with other tissues or organs, and even centrally with appetite control areas in the hypothalamus (Trayhurn and Beattie 2001; McGown et al. 2014). This endocrine organ is also involved in the regulation of physiological processes such as sexual dimorphism, immunity, reproduction, angiogenesis, adipogenesis, extracellular matrix restructuring, steroid metabolism, and body temperature control, among others (Newbold et al. 2008). Thus, during obesity there is an altered secretion of proteins shed by adipose tissue (adipokines) and other molecules, which are involved in obesity-associated comorbidities such as insulin resistance, inflammation, hypertension, cardiovascular disease, type 2 diabetes, non-alcoholic fatty liver disease (NAFLD), metabolic disorders, increased risk of infectious disease complications (COVID-19), and cancer and thus reduced life expectancy (Apovian 2016; Andolfi and Fisichella 2018; Reiterer et al. 2021). However, it is still unclear how adipokines and cytokines are involved in these comorbidities and how these signalling molecules vary depending on the excess location of different adipose tissue depots. In this context, different alternative communication pathways have recently emerged for adipose tissue, such as the non-classical secretion of proteins (Tanaka et al. 2007), the release of extracellular vesicles (EVs) (Colombo et al. 2014), or the release of long non-coding RNAs (Greenhill 2017) and microRNAs (Zhang et al. 2015) that can travel freely or within EVs.

Therefore, it is necessary that the adipose tissue has an adequate secretion and that these adipokines interact correctly with the ECM and other target organs to achieve the physiological functioning of the adipose tissue. For this reason, it is essential to better understand the molecular mechanisms involved in this disease, as well as to deepen our knowledge of the new signals secreted by adipose tissue and their relationship with the ECM and other distant tissues in both physiological and pathological conditions.

7.2 Extracellular Vesicles (EVs), Biogenesis, Composition, and Biological Functions

Extracellular vesicles (EVs) are small, lipid membrane-bounded, mobile spheres of varying size (30–5000 nm) that are released into the extracellular space (Théry et al. 2006). Initially, these particles were thought to be mere cellular debris, until they were first described in 1981 as enzymes secreted into the extracellular space (ecto-enzymes) (Trams et al. 1981). These particles can be classified according to their size and biogenesis and can be identified in three major groups: microvesicles (MVs) (100 nm–1 μm), derived from plasma membrane evagination; exosomes (30–100 nm), assembled into multivesicular endosomes (MBVs) that are secreted by exocytosis; and larger vesicles (50–5000 nm), comprising apoptotic bodies released by cells preceding apoptosis (Fig. 7.1a) (Raposo and Stoorvogel 2013).

Currently, because there is no international consensus on specific markers for the different EV subtypes, the assignment of an EV to a specific biogenesis pathway remains complex. In case no specific markers of subcellular origin can be established, ISEV (International Society for Extracellular Vesicles) recommends classification according to Théry et al. (2018):

- Physical characteristics, such as size: small EVs (sEVs: small Extracellular Vesicles) that are smaller than 200 nm, and medium/large EVs (m/lEVs medium/large Extracellular Vesicles) that are larger than 200 nm.
- Biochemical composition (EVs positive/negative for CD63/CD81 markers, etc.).
- Description of conditions or cells of origin (hypoxic EVs, large oncosomes, apoptotic bodies, etc.).

7.2.1 *Biogenesis of EVs*

The formation of these vesicles occurs mainly at two subcellular sites: at the plasma membrane, as in the case of MVs; or from the endosomal pathway in exosomes (Fig. 7.1a) (Abels and Breakefield 2016). We can establish a series of biomarkers according to this place of origin. For example, tetraspanins, which are a superfamily of ubiquitous transmembrane molecules that mediate the formation of specialised microdomains in plasma and intracellular membranes (this is the case for CD9, CD61, CD63, and CD81); ESCRT proteins, endosomal sorting complexes required for the transport machinery, and associated proteins such as TSG101 (tumour

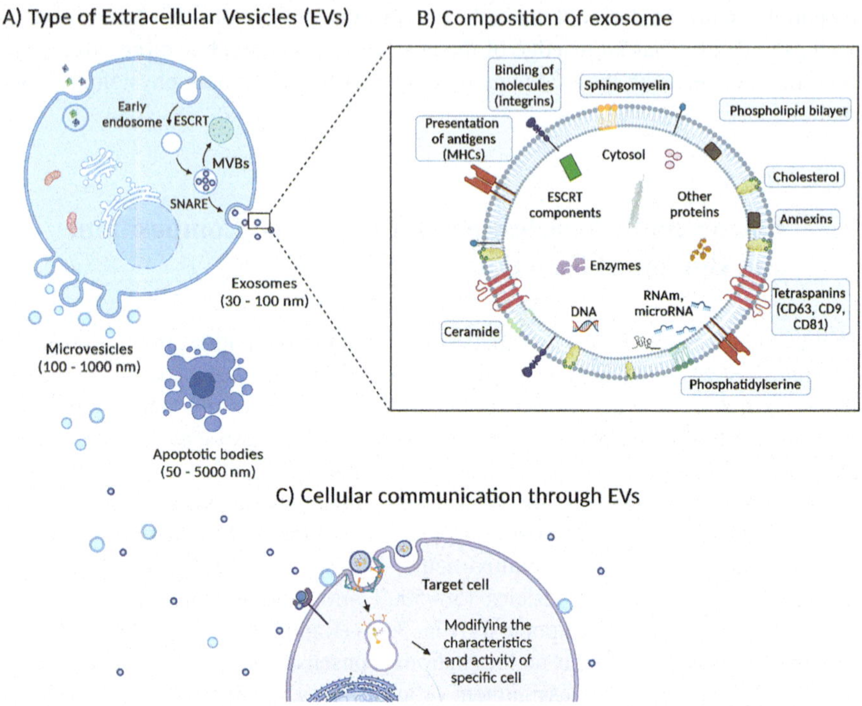

Fig. 7.1 Biogenesis and characteristics of extracellular vesicles (EVs). EVs can be classified according to their biogenesis into microvesicles (from the plasma membrane), exosomes (from the formation of multivesicular curls that are shed by exocytosis), and apoptotic bodies (from cellular apoptosis) (**a**). Due to their biogenesis and cell of origin, the composition of EVs is dynamic, but highlights transmembrane proteins such as receptors, endosomal markers such as tetraspanins, cytosolic proteins and enzymes, as well as lipids and genetic material (RNA, mRNA, and DNA, among others) (**b**). Upon release, EVs pass through the extracellular matrix to interact with target cells or are transported in biological fluids. Interactions between EVs and target cells occur through several mechanisms. Upon interaction with target cells, EV cargo can cause post-translational changes in cellular mRNA, resulting in positive or negative functional changes (**c**). Created by BioRender

susceptibility gene 101 protein), Alix (protein interacting with programmed cell death protein 6), as well as syntenin, flotillin, and heat shock proteins, are characteristic markers of EVs derived from the endosomal pathway (exosomes) (Colombo et al. 2013; Toribio and Yáñez-Mó 2022). In contrast, arrestin domain-containing protein 1 and CD73 are suggested as markers of EVs released from the plasma membrane (Colombo et al. 2014). Moreover, since exosomes are formed in multivesicular bodies/endosomes (MVBs) and MVs originate by direct budding from the plasma membrane, the cellular machinery involved in their formation and release is likely to differ, although the mechanistic elements may be shared (Colombo et al. 2014). However, since EVs biogenesis is dynamic, it should be noted that vesicle-specific markers may vary depending on the cell type and the metabolic and functional state of these cells; indeed, vesicle levels may vary over time and after physiological and pathological alteration (Abels and Breakefield 2016; Camino et al. 2020b).

7.2.2 Composition and Functionality of EVs

Interestingly, EVs contain different bioactive material such as proteins, lipids, different types of RNA and DNA (Fig. 7.1b) (Doyle and Wang 2019). Most studies on the biochemical composition of EVs involve the analysis of heterogeneous vesicle populations due to the difficulties of different isolation methods to obtain pure vesicle subtypes (Konoshenko et al. 2018; Sidhom et al. 2020). Therefore, the actual composition of each EVs subtype is generally unknown even though it is recognised to be relevant playing a key role in cell communication and cell transfer (Kalluri and LeBleu 2020). Specifically, small particles (exosome) are composed of several proteins: transmembrane proteins such as tetraspanins (CD63, CD81), antigen presenting molecules (MHC), glycoproteins, and adhesion molecules; and proteins in the exosome lumen such as heat shock proteins (HSP), cytoskeleton proteins, components of the ESCRT complex, transport proteins (SNARE), fusion proteins, growth factors, and cytokines (Doyle and Wang 2019). In addition, those EVs are composed of multiple lipids such as cholesterol, ceramides, sphingomyelin, among others (Vickers and Remaley 2012).

Also, a diverse composition of genetic material is found in EVs. In a small number of cases, DNA has been found, including genomic and mitochondrial DNA (Doyle and Wang 2019). But in general, EVs are mainly enriched with mRNAs and microRNAs, small RNAs, many of them derived from 18S and 28S ribosomal rRNAs, tRNAs, long-chain non-coding RNAs (lncRNAs), nuclear small RNAs (snRNAs), and piwi-interacting RNAs (Abels and Breakefield 2016; Greenhill 2017; Pathan et al. 2019). In particular, microRNAs play an important role in target cell gene regulation and in various biological processes such as exocytosis, haematopoiesis, and cell communication (Abels and Breakefield 2016).

This composition is highly dynamic and is conditioned by the type of biogenesis as well as by the type and physiological state of the cell of origin, whereby these vesicles reflect the state and pathophysiological changes of the cell of origin (Campello et al. 2016). Interestingly, EVs are released by all cells in the body and can therefore be detected both in the extracellular environment and in virtually all body fluids (blood, urine, cerebrospinal fluid, saliva, milk, among others) (Théry et al. 2006; Russell et al. 2019). Because of this, these particles have recently emerged as potentially good biomarkers of disease with potential use for monitoring pathology or even the efficacy of therapeutic treatments (Properzi et al. 2013; Barile and Vassalli 2017). In addition, EVs, unlike other classical biomarkers, allow minimally invasive analysis without the need for biopsies, permitting sequential sampling (Simeone et al. 2020; Camino et al. 2020b).

Moreover, most interestingly, due to their composition, EVs can exert stimulatory or inhibitory functional effects, such as cell proliferation, apoptosis, cytokine production, immune modulation, or metastasis, inducing cellular transduction signals or genetic and epigenetic changes, and thus modifying the characteristics and activity of specific cells and tissues (Nazarenko et al. 2013). In general, vesicles exert their intercellular signalling function in two ways: by transmitting information to recipient cells by direct contact via their surface ligands; or they can transfer proteins and nucleic acids, including RNA and DNA, to target cells by direct membrane fusion, phagocytosis, or clathrin-dependent endocytosis, among others (Fig. 7.1c) (Mathivanan et al. 2010; Tual-Chalot et al. 2011; Meldolesi 2018).

7.3 Extracellular Vesicles and Obesity

Moreover, extracellular vesicles (EVs) have been postulated as key factors in several physiological processes, such as development, ageing, homeostasis, metabolic regulation, as well as in non-infectious (cancer, metabolic disorders) and infectious (malaria, among others) diseases (Tual-Chalot et al. 2011; Kalluri and LeBleu 2020). In this context, several studies on the biological role of EVs in various human pathologies have emerged, providing a new approach in biomedical research (Tual-Chalot et al. 2011). Especially, in metabolic diseases, several studies suggest that EVs are key players in metabolic organ communication in energy homeostasis at both paracrine and endocrine levels, probably participating in metabolic alterations related to obesity and its comorbidities (Pardo et al. 2018). It should be noted that the functionality of EVs is mainly determined by their cargo, so it is necessary to know their composition in order to discern changes or alterations associated with the physiology of the cell of origin (Kalluri and LeBleu 2020). Unlike other tissues, there are few reports on the composition of EVs released from adipose tissue under both physiological and pathological conditions.

7.3.1 White Adipose Tissue Vesicular Biomarkers in Obesity

White adipose tissue (WAT) is characterised by the fact that it is composed of the most abundant adipocytes in the body: white adipocytes. These are large round cells between 20 and 200 μm in size and unilocular. They store triglycerides in the form of large lipid droplets in a lipid vacuole which represents 90% of the cell volume (Giordano et al. 2014). When energy supply to the body is excessive and/or energy expenditure decreases, excess energy is effectively deposited in the form of triacylglycerols in the WAT (Giordano et al. 2014). On the other hand, when energy intake is inadequate and/or energy expenditure increases, the WAT mobilises lipid storages to release fatty acids and glycerol, which are transported by the blood to the tissues, where they are oxidised for energy (Scheja and Heeren 2019). Thus, between meals and during fasting, cells break down the fat stored in these droplets to provide energy, thus providing a source of energy when food is not available. It also has the function of insulating and providing mechanical protection to some vital organs (Giordano et al. 2014).

In addition to this, white adipocytes have an important endocrine function as they release adipokines (peptides), such as leptin and adiponectin, and lipokines (lipid-like in nature) that can autocrine/paracrine and/or endocrine regulate energy homeostasis (Fantuzzi 2005; Wasim et al. 2016; Engin and Engin 2017). These proteins have a wide versatility of functions such as proinflammatory cytokines, lipid metabolism and transport, steroid metabolism enzymes, or immunity, among others (Fantuzzi 2005). For this reason, the loss of adipocyte function during obesity triggers an imbalance or disruption in adipokine production.

In obesity, there is an increase in white adipose tissue, since, due to the plasticity of this tissue, adipocytes have the capacity to buffer energy surplus because they can expand by hyperplasia and/or hypertrophy (Giordano et al. 2014). These hypertrophied adipocytes can release an increased concentration of fatty acids and proinflammatory adipokines (Apovian 2016). This process is accompanied by changes in tissue vascularisation leading to a fibrotic state with subsequent cell death (McGown et al. 2014).

White fat is distributed throughout the body (Fig. 7.2) and can be classified anatomically as follows: subcutaneous (SAT) and visceral (VAT) adipose tissue (Ibrahim 2010). In healthy humans, SAT comprises about 80% of body fat, whereas VAT accounts for 10–20% (Ibrahim 2010). Thus, SAT is mainly located in the buttocks, thighs, and abdomen, under the skin, surrounded by the connective capsule and innervated by peripheral blood vessels and nerves. In contrast, VAT is mainly found inside the abdomen around the umbilicus (omental), intestine (mesenteric) and perirenal areas (retroperitoneal), and in the pericardial area (Alvarez-Llamas et al. 2007; Ibrahim 2010). Visceral fat is known to consist of a greater number of immune cells and adipocytes, which are metabolically more active, as they take up a greater amount of glucose and lipids and have a greater innervation; they also have a smaller number of preadipocytes than SAT (Roca-Rivada et al. 2015).

These two types of white fat have differences at both the genetic and metabolic level. The expression of gene families such as *HOXA5* (Homeobox A5) and *TBX15* (transcription factor T-Box 15) is known to predominate in VAT, whereas *GPC4* (glypican 4) is more strongly expressed in SAT (Ibrahim 2010). In visceral fat, *GPC4* and *HOXA5* correlate positively with BMI (Body Mass Index) and WC (Waist Circumference), whereas *TBX15* expression shows a negative correlation. However, in subcutaneous fat, *HOXA5* expression levels correlate positively with BMI and WC, whereas *GPC4* has a negative correlation with both parameters, with the expression levels being extremely low for BMI values above 30. Therefore, it has been suggested that the expression of these genes represents an excellent marker of visceral fat accumulation (Ibrahim 2010). Furthermore, under conditions of caloric excess, the SAT expands by hypertrophy and hyperplasia, whereas the VAT expands predominantly by hypertrophy of adipocytes (Curat et al. 2006). Hypertrophy of these VAT adipocytes, but not of SAT adipocytes, together with

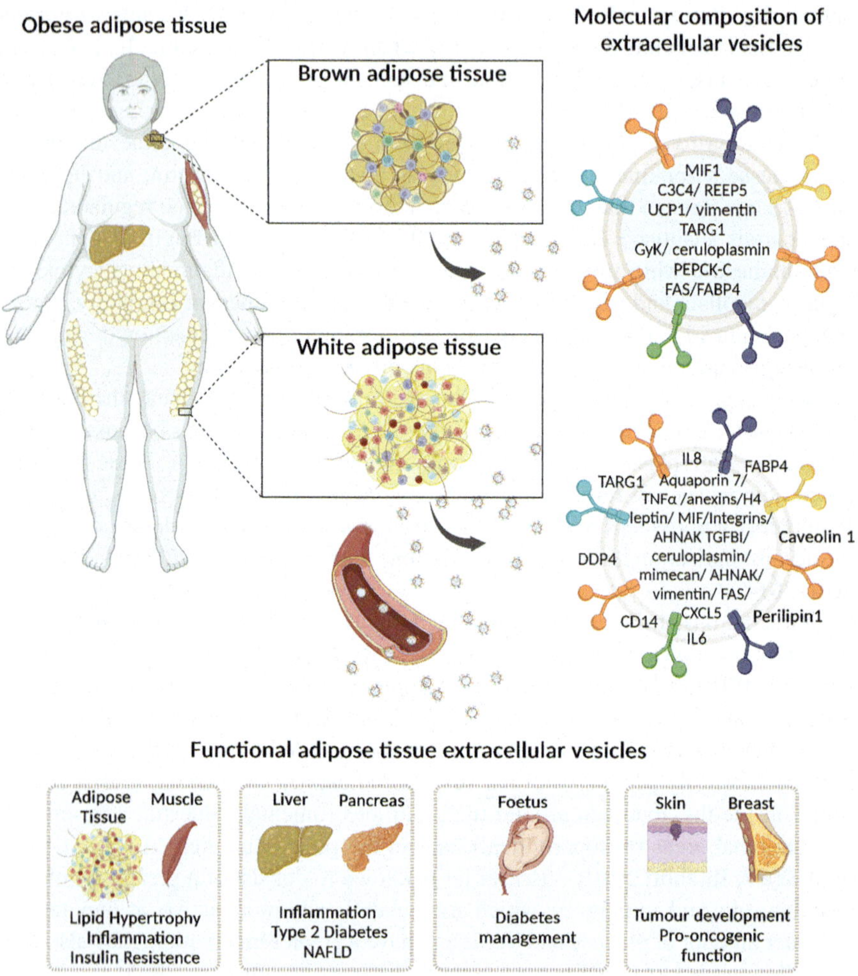

Fig. 7.2 Extracellular vesicles (EVs) as carriers of adipokines and their role in metabolic diseases. The secretion of EVs by brown and white adipose tissue is an alternative pathway of communication at the paracrine and endocrine level that reflects the characteristics of the metabolic state of the cell of origin, thus representing a rich source of disease biomarkers and exerting functions involved in obesity-associated comorbidities. The protein composition of EVs secreted by brown and white adipose tissue and their possible involvement in obesity-associated pathologies are shown. Created by BioRender

secretion of inflammation-related adipokines and infiltration of macrophages and immune cells, is associated with metabolic dysfunction such as insulin resistance (IR), hypertension, type 2 diabetes mellitus (DM2), and cardiovascular disease (Apovian 2020; Camino et al. 2022a). It should also be noted that during obesity, visceral fat accumulates ectopically in tissues other than adipose tissue, in atopic locations such as the liver, pancreas, skeletal muscle, and heart, leading to malfunctioning of these organs. For example, lipid accumulation in the liver (steatosis) leads to oxidative stress, inflammation, and insulin resistance (IR) (Dorairaj et al. 2020). In the pancreas, this fat accumulation causes β-cell malfunction, which also leads to a state of IR (Khushman et al. 2017). Therefore, the localisation of white fat has genetic, metabolic, and endocrine differences, with VAT being a risk factor for the development of metabolic diseases in obesity, whereas SAT plays a more protective role (Verboven et al. 2018). Therefore, due to the important endocrine role of this tissue and the differences depending on its location, the study of the composition of EVs released by WAT will be essential to better understand the metabolic imbalances and diseases associated with obesity (Fig. 7.2).

One of the first published works in this field was by Kranendonk et al. in which they describe the protein content of EVs isolated from human adipocytes of Simpson-Golabi-Behmel Syndrome (SGBS) (Kranendonk et al. 2014a). These vesicles were shown to carry FABP4 (adipocyte fatty acid binding protein 4), TNF-α (tumour necrosis factor-alpha), adiponectin, MCSF (macrophage colony-stimulating factor), RBP-4 (retinol binding protein 4), and MIF (macrophage migration inhibitory factor) (Kranendonk et al. 2014a). In addition, another study identified a variety of proteins involved in angiogenesis, blood coagulation, apoptosis, extracellular matrix remodelling, and regulation of inflammation in EVs isolated from porcine tissue-derived mesenchymal stem cells (PBSCs) (Koeck et al. 2014).

In a research work focusing on the relationship of obesity with cancer, Lazar et al. performed a proteomic analysis of EVs released from the murine adipocyte cell line 3T3-F442A differentiated in vitro in which they identified 324 proteins of which the majority were related to lipid metabolism (Lazar et al. 2016). In addition, another proteomic study showed that lEVs were enriched in membrane components, organelles, and cellular parts in contrast to sEVs that contained mainly ECM components (Durcin et al. 2017). At the pathological level, the study by Lee et al. showed a higher proportion of aquaporin 7, caveolin, and lipoprotein lipase in EVs isolated from primary adipocytes of obese diabetic rats compared to non-diabetic rats (Lee et al. 2015). In addition, recent studies have also identified novel adipokines (nucleoplasmin-3, DAD-1, and STEAP3 metalloreductase) and observed an increase in these in EVs isolated from rat (Sprague-Dawley) groin tissue with obesity compared to lean animals (Zhang et al. 2020).

Of particular interest are our recent studies that demonstrated variation in the protein composition of EVs released from adipose tissue depending on the metabolic state of the cell/tissue of origin (obesity, normal weight) and its anatomical location (target: subcutaneous, visceral; or brown) (Camino et al. 2020b, 2023). These proteomic studies have been conducted on the composition of EVs released from murine pathological adipocyte cell models including lipid hypertrophy and insulin-resistant adipocytes, as well as whole explants of subcutaneous and visceral adipose tissue from obese patients, and white and brown adipose tissue from obese and lean rats (Camino et al. 2020a, 2022a, b, 2023). These studies describe the first reference proteomic maps of EVs released from human obese adipose tissue according to their anatomical location, showing that vesicles released from obese adipose tissue share structural vesicular proteins, but also contain components specific to the location and metabolic state of the depot (Camino et al. 2022a, b). In these studies (Camino et al. 2022a), common proteins identified in the EVs released by adipose tissue under different conditions were cytoskeletal proteins such as annexin A1/A6, which modulate anti-inflammatory processes that may be altered in obesity-associated DM2 (type 2 diabetes) (Krautbauer et al. 2017; Pietrani et al. 2018), as well as histones (H4), whose epigenetic changes such as acetylation may trigger metabolic disorders (Pessoa Rodrigues et al. 2021). In addition, interestingly, other known proteins related to fatty acid transport or synthesis, or adiposity, were also identified in all obese EVs analysed, such as perilipins (perilipin-1), caveolin-1 (Cav-1), FABP4, or fatty acid synthase (FAS) (Camino et al. 2022a, 2023). In particular, Cav-1 and perilipin-1 proteins have already been proposed in the literature as vesicular biomarkers of adipose deposition (Tai and Ordovas 2007; Camino et al. 2023), especially perilipin-1, which has been identified as a biomarker of adipose tissue at the circulating level (Tai and Ordovas 2007; Yllmaz et al. 2020). Notably, perilipin-1 protein is one of the most abundant adipocyte proteins, which plays a key role in the regulation of lipid and glucose metabolism, being also associated with body weight and related complications of obesity (Tansey et al. 2004). On the other hand, EVs carrying caveolin-1 are known to be involved in communication between different cell types within adipose tissue, such as endothelial cells and adipocytes (Fernández-Real et al. 2010), as well as in tumour development or progression by controlling cell metabolism (Nwosu et al. 2016).

Furthermore, it has been observed that FABP4 protein was detected as elevated in EVs released from visceral adipose tissue under obese conditions in contrast to normal weight conditions (Camino et al. 2022b). This finding is consistent with circulating levels of this protein correlating positively with the incidence of metabolic disease, which may suggest that vesicles carrying FABP4 may be contributing to the levels of this protein detected in peripheral blood (Prentice et al. 2019). Another protein identified in vesicles released from the fat depot under obese conditions was FAS (Camino et al. 2020a, 2022a, b, 2023). Notably, high levels of this protein have previously been associated with impaired insulin sensitivity and adipose tissue malfunction in obesity (Blüher et al. 2014).

Furthermore, it has been reported that EVs secreted by visceral adipose tissue from obese individuals showed a greater variety of proteins compared to subcutaneous, and more interestingly, the functional classification of these proteins revealed that many were previously related to adipose tissue, obesity, and obesity-associated inflammation (Camino et al. 2022a, 2023). These findings may be explained by the fact that VAT of obese individuals has been reported to be more inflamed than subcutaneous (Ibrahim 2010), thus, it is invaded by macrophages and other immune cells, which also release EVs that can modulate the type, amount, and dynamic release of these particles. In this case, some of the proteins identified were leptin (Wasim 2015), IL-6 (interleukin-6) (Sadeghabadi et al. 2019), CXCL5 (chemosyn 5 with C-X-C motif) (Cereijo et al. 2018), DDP-4 (dipeptidyl-peptidase) (Lamers et al. 2011), and among others (Camino et al. 2023).

On the other hand, novel adipokines such as mimecan and ceruloplasmin were also observed in EVs of visceral origin compared to those of subcutaneous origin. EVs from visceral adipose tissue of patients with obesity have been shown to contain a higher amount of the mimecan protein than EVs isolated from SAT or adipocytes without metabolic disturbances (Camino et al. 2020a, 2022a, 2023). This protein is considered an adipokine involved in the control of satiety because it inhibits food intake by inducing IL-1 and IL-6 expression in the hypothalamus, independent of leptin signalling (Cao et al. 2015). Therefore, mimecan, like leptin, at the vesicular level may be released in high concentrations as compensation for the development of obesity (Camino et al. 2023). In the same context, circulating ceruloplasmin protein has also been identified in elevated concentrations in obese individuals compared to healthy individuals (Kim et al. 2011). Thus, vesicular ceruloplasmin is postulated as a biomarker of adiposity (Camino et al. 2023).

Interestingly, one of the most remarkable findings of our work was the identification of vesicular TFGBI (growth-induced ig-h3 protein) associated with visceral fat deposition in obese conditions, and almost exclusively present in EVs released from adipocytes exhibiting insulin resistance (Camino et al. 2020a, 2022a, 2023). In particular, the gene encoding this protein, *TGFBI*, has been described as a diabetes risk gene in both mice and humans (Han et al. 2014). Furthermore, this vesicular protein could be detected at the circulating level to be present at a higher ratio in those plasma vesicles from obese patients with diabetes compared to vesicles from donors without diabetes (Han et al. 2014). Therefore, it is predicted that vesicular TGFBI protein may be a good biomarker of early insulin resistance and progression of this pathology (Camino et al. 2023).

In addition, other proteins that were observed in a higher ratio in visceral tissue EVs from patients with obesity compared to subcutaneous fat vesicles were AHNAK, CD14, and vimentin (Camino et al. 2023). AHNAK protein plays a crucial role in body fat accumulation by regulating adipose tissue development and being involved in metabolic homeostasis (Shin et al. 2015). In turn, CD14 protein has previously been associated with macrophage infiltration of adipose tissue leading to an increased inflammatory state of adipose tissue (Duque and Descoteaux 2014; Ali et al. 2022). On the other hand, the protein vimentin plays a key role in adipose tissue plasticity and has also been found to be involved in obesity and type 2 diabetes (Kim et al. 2021).

In summary, adipose tissue under conditions of obesity undergoes changes not only at the morphological level, but also at its secretory profile, and in this process, the role of EVs is crucial. Thus, different studies show that these particles contain a high number of proteins related to obesity comorbidities such as diabetes, cardiovascular problems, cancer, fatty liver; therefore, these molecules may be good candidates for biomarkers of the development and monitoring of these pathologies (Fig. 7.2).

7.3.2 Brown Adipose Tissue Vesicular Biomarkers in Obesity

Brown adipose tissue (BAT) is characterised by brown adipocytes, which are smaller in size and have a polygonal shape with multiple small lipid vacuoles (multilocular) than white adipocytes (Cannon and Nedergaard 2004). In addition, this tissue is more vascularised than WAT, and its adipocytes have a high density of mitochondria in which UCP-1 protein (mitochondrial respiratory chain uncoupling protein) is present (Perdikari et al. 2018). This enzyme is key in heat production by uncoupling the respiratory chain from ATP synthase, using fatty acid oxidation as a substrate, which allows protons to pass back into the mitochondrial matrix, thus dissipating the energy gradient, preventing ATP synthesis, and releasing excess energy in the form of heat (Villarroya et al. 2019; Lim et al. 2020). In other words, it allows thermogenesis to take place in these adipocytes without shivering, which is controlled both centrally and peripherally by the sympathetic nervous system (SNS) (Perdikari et al. 2018).

In contrast to WAT, brown adipose tissue is only found in mammals in superficial (interscapular, cervical, axillary regions) and deep (perirenal, inguinal regions) areas (Fig. 7.2) (Doyle and Wang 2019; Cannon and Nedergaard 2004). Furthermore, similar to white adipocytes, brown adipocytes have been shown to secrete a variety of adipokines, called batokines, such as neuregulin 4 (NRG-4), fibroblast growth factor 21 (FGF-21), vascular endothelial growth factor (VEGF), chemokine 14 with C-X-C motif (C-X-C motif chemokine 14), bone morphogenetic protein 8B (BMP-8B), and the homologous Slit homologue 2 protein (Slit-2), among others (Villarroya et al. 2019). Therefore, this has motivated the study of the role of brown adipose tissue in recent years as a therapeutic target for the treatment of obesity and diabetes through a pharmacological approach to stimulate growth, differentiation, and/or activation of WAT (Cypess and Kahn 2010). Under this context, EVs emerge as potential batokines still to be explored being essential to getter knowledge about the composition of EVs secreted by brown adipose tissue in the prevention of obesity and its associated pathologies.

Recent studies have characterised and compared the protein composition of EVs released from whole explants of brown adipose tissue isolated from obese and control animals (Camino et al. 2022b, 2023). Interestingly, EVs from obese and lean brown adipose tissues were found to have a significant mitochondrial component (40–45%) characterised by proteins of the beta-oxidation pathway compared to white depots (Camino et al. 2022b). Also relevant is the fact that not only was UCP1 protein identified for the first time at the vesicular level, but it was observed that this protein was elevated in EVs isolated from brown adipose tissue of obese animals compared to control animals (Camino et al. 2022b, 2023). This result seems contradictory to previous research, because although the brown fat depot is more resistant to inflammation than the white fat depot (Villarroya et al. 2018), the secretion of proinflammatory cytokines associated with obesity is known to alter the thermogenic function of adipose tissue by reducing the expression levels of UCP1 and other markers of thermogenesis under obese conditions (Cereijo et al. 2018). However, some studies of diet-induced obesity (DIO), such as the work of Alcalá M. et al., indicate that UCP1 levels are up-regulated in obese mice after a 20-week exposure to a high-fat diet (HFD) (Alcalá et al. 2017). Therefore, EVs released from brown fat are predicted to represent the adaptive state of this depot, which triggers an early adaptive thermogenic response induced by an HFD diet, as reflected by an increase in UCP1 protein levels (Camino et al. 2023). Moreover, this is in agreement with previous studies in which the presence of mitochondrial proteins in EVs was demonstrated (Rosina et al. 2022; Zhou et al. 2023). For example, Rosina et al. identified PDHE1-o (pyruvate dehydrogenase beta, mitochondrial) in vesicles isolated from primary cultures of brown adipocytes extracted from mice subjected to thermogenic stress (Rosina et al. 2022). It is noteworthy that studies have observed that mitochondria can travel whole in EVs (Zhou et al. 2023).

On the other hand, other enzymes and proteins related to the insulin signalling pathway and inflammation have also been identified, such as C3/C4, AOC3 (amine oxidase copper containing 3), MIF1 (macrophage migration inhibitor 1), REEP5 (receptor enhancer protein 5 expression), PEPCK-C (phosphoenolipoylpyruvate carboxykinase), TARG1 (regulator of GLUT4 trafficking 1), and Gyk (glucose glucolide carboxykinase) in EVs isolated from brown tissue under obese conditions (Camino et al. 2022b) (Fig. 7.2). In this pathological context, numerous reports have shown an increase in circulating AOC3 protein in obese and diabetic conditions in rodents (Camino et al. 2023). Also proteins of the same family as REEP5, such as REEP6, are known to regulate cold-induced thermogenesis, demonstrating that a deficiency of this protein leads to an obesity-prone phenotype (Son et al. 2022). Furthermore, overexpression of PEPCK-C, a glycoerongenesis-regulating enzyme in adipose tissue, has been reported to lead to a high susceptibility to insulin resistance and diet-induced obesity (Franckhauser et al. 2006). TARG1 protein is also known to positively regulate GLUT4 trafficking and insulin sensitivity in adipocytes (Duan et al. 2022). Furthermore, it has been postulated that Gyk protein stimulates UCP1 expression through activation of the AAR-cAMP-CREB pathway and regulation of lipid metabolism (Sanchez-Gurmaches et al. 2018). On the other hand, it is worth highlighting that BAT in conditions of obesity, like WAT, secretes EVs that carry proteins associated with obesity and lipid accumulation such as perilipins, vimentin, ceruloplasmin, FABP4, FAS, among others which suggest that brown adipose tissue also suffers from obesity deleterious effects (Camino et al. 2022b).

However, brown adipose tissue EVs under normal weight conditions have been observed to contain up-regulated proteins such as ACLY (ATP citrate synthetase), Rab 14 and ACC (acetyl-CoA carboxylase), compared to brown depot EVs under obese conditions (Camino et al. 2022b, 2023). ACLY is a protein involved in lipogenesis, and it is known that during the development of obesity there is a decrease in this protein (Sanchez-Gurmaches et al. 2018). On the other hand, Rab 14 protein is involved in the transit of GLUT4, internalising it from early endosomes to the Golgi complex (Reed et al. 2013). In addition, ACC, anabolic de novo lipogenesis gene, are known to be activated under thermogenic conditions that correlate with UCP1 expression (Feldmann et al. 2009). Therefore, brown adipose tissue secretes EVs, which differ in composition according to metabolic state that dynamically reflects changes in the cell of origin; that is, these EVs carry biomarkers of brown adipose tissue-specific thermogenic activity, oxidative stress, and inflammation during the development of obesity (Fig. 7.2) (Camino et al. 2023).

7.3.3 Functional Role of Adipose Tissue EVs in Obesity

At present, knowledge on the role of EVs in the development and maintenance of metabolic diseases remains scarce (Nazarenko et al. 2013). There is increasing evidence for the involvement of EVs in obesity-associated metabolic disorders and, more specifically, in local and systemic inflammation related to liver and adipose tissues (Ferrante et al. 2015). In this context, EVs released from adipose tissue have been proposed to be involved in the interaction between adipocytes and macrophages, affecting insulin signalling and the expression of genes related to the control of energy homeostasis in muscle and liver cells, leading to metabolic diseases (Ferrante et al. 2015). Since EV biogenesis is a dynamic process and depends on the pathophysiological state of the cell of origin, the functional role of EVs in obesity will differ depending on the molecular composition of these particles (Fig. 7.2) (Colombo et al. 2014).

Different functional assays have been performed in which EVs released from adipose tissue interact with metabolic cells or tissues involved in the development of obesity (Camino et al. 2023). Thus, EVs from adipocytes of the C3H10T1/2 cell line with lipid hypertrophy and insulin resistance have been reported to induce and stimulate the differentiation and lipid hypertrophy of neighbouring healthy adipocytes (Camino et al. 2020a). This is consistent with previous studies in which EVs isolated from adipose tissue internalise and, in turn, induce adipogenesis in adipose-derived stem cells (ASCs) (Crewe et al. 2018). Moreover, EVs released from hypertrophied, insulin-resistant adipocytes were also found to be able to promote insulin resistance in these same neighbouring adipocytes by inhibiting the phosphorylation of several proteins involved in the insulin signalling pathway (Camino et al. 2020a). In this context, Kranendonk et al. observed that EVs released from subcutaneous adipose tissue of some individuals inhibit Akt phosphorylation in liver cells (HepG2) and myotubes (C2C12) (Kranendonk et al. 2014b). In addition, others have reported that vesicles released from fat depots induce insulin resistance in macrophages (Deng et al. 2009). The study by Ferrante et al. has shown that EVs released from adipocytes in adipose tissue of obese individuals transfer microRNAs that target genes involved in inflammatory signalling pathways (Ferrante et al. 2015). Similarly, Gesmundo et al. observed that EVs released from healthy 3T3-L1 adipocyte cultures increased survival and proliferation of pancreatic INS-1E cells and also of human pancreatic islets, thereby improving insulin sensitivity (Gesmundo et al. 2021). In addition, the study by Jayabalan et al. found that EVs isolated from omental tissue of pregnant women with gestational diabetes promoted the expression of genes associated with glycolysis and gluconeogenesis in human placental cells compared to EVs from pregnant women without gestational diabetes (Jayabalan et al. 2019). It is also noteworthy that adipocyte EVs are able to cross the blood–brain barrier and regulate POMC (pro-opiomelanocortin arcuate) expression through hypothalamic mTOR signalling in vivo and in vitro, thus affecting body energy intake (Gao et al. 2020). Therefore, several studies demonstrate that EVs released by adipocytes under obese conditions can alter insulin sensitivity by promoting a state of insulin resistance in different cells/tissues, such as adipose tissue itself, pancreas, muscle, and liver cells, among others (Camino et al. 2023).

On the other hand, EVs secreted by C3H10T1/2 adipocytes with hypertrophy have also been found to induce inflammation in healthy, non-inflamed macrophages (Camino et al. 2020a). In the same context, other studies have shown that EVs derived from adipose tissue, especially visceral fat, have a proinflammatory effect on macrophages (Kranendonk et al. 2014a). Deng et al. have shown that EVs isolated from adipocytes of obese mice increased the production of macrophage colony-stimulating factor (MCSF), IL-6, and TNF-o in primary murine macrophages (Deng et al. 2009). Also, EVs secreted by macrophages from mouse adipose tissue have been shown to suppress insulin secretion when it inoculated into lean mice (Ying et al. 2017). Interestingly, in the study by De Silva et al., primary adipocytes isolated from omental tissue treated with EVs from macrophages (THP-1) inflamed with lipopolysaccharide (LPS) expressed a greater number of genes related to inflammatory pathways than untreated adipocytes (De Silva et al. 2018).

Otherwise, Zhao et al. observed that EVs released from activated murine brown adipose tissue played a cardioprotective role in the heart (Zhao et al. 2022). Furthermore, the Kariba et al. study shows that EVs released from brown adipose tissue containing mir-132-3p suppress Srebf1 expression in a primary culture of murine hepatocytes, thereby decreasing the development of fatty liver (Kariba et al. 2020). In this context, the study by Zhou et al. shows that brown adipocyte EVs mitigated metabolic syndrome in animals fed a high-fat diet (Zhou et al. 2020). This research shows how treatment with these vesicles is able to reduce body weight, blood glucose, lipid accumulation, promote cellular oxygen consumption, and restore cardiac function in HFD mice independently of food intake.

Regarding the association of obesity with an increased risk of developing cancer, several studies have shown that EVs secreted by human obese adipocytes may be pro-oncogenic, as they promote the proliferation of MCF7 tumour breast cells and other breast tumour cell lines (Lazar et al. 2016; Clement et al. 2020). In addition, the study by Lin et al. shows that EVs released from human adipose tissue mesenchymal stem cells isolated by liposuction promote MCF7 cell proliferation and migration (Lin et al. 2013). Furthermore, other in vitro studies suggest that adipocyte EVs can promote migration and proliferation of hepatocarcinoma cells (Liu et al. 2019). Other studies have also observed that EVs released from obese adipocytes have a pro-oncogenic effect and promote the aggressiveness of cutaneous melanoma (Ramos-Andrade et al. 2020).

7.4 Extracellular Matrix and Its Interaction with Extracellular Vesicles in Obesity

The extracellular matrix (ECM) is a network of different proteins and proteoglycans that controls differentiation, migration, repair, survival, and development, and it appears that its remodelling is necessary for healthy tissue expansion, such as adipose tissue (Mariman and Wang 2010; Lin et al. 2016). As obesity is characterised by a massive expansion of adipose tissue, it is essential that remodelling and reorganisation of the ECM takes place (Ruiz-Ojeda et al. 2019). In adipose tissue, the ECM is mainly composed of collagens, fibronectin, and laminin (Lin et al. 2016; Ruiz-Ojeda et al. 2019). Several studies show that animals with metabolic disease contain excessive collagen deposition in adipose tissue (Mariman and Wang 2010). Furthermore, collagen V expression is known to be increased in adipose tissue in patients with obesity and correlates with fewer capillaries (Spencer et al. 2011).

However, several components, such as A disintegrin and metalloproteinase-containing protein domains (ADAM), osteopontin (OPN), hyaluronan (HA), thrombospondins (THBS1), matrix metalloproteins (MMP), and tissue inhibitor of metalloproteins (TIMP), play an important role in ECM remodelling and adipose tissue function (Lin et al. 2016). OPN, also known as secretory phosphoprotein 1, is an ECM glycoprotein that is expressed in several cell types and tissues, including adipose tissue (Kahles et al. 2014). OPN expression is dramatically increased in adipose tissues of obese mice, as well as in obese humans, and it has been functionally implicated in the pathogenesis of adipose tissue inflammation and insulin resistance (IR) in obesity (Chapman et al. 2010). On the other hand, HA is a linear glycosaminoglycan consisting of chemically unmodified repeating disaccharide units of D-glucuronic acid and N-acetyl-D-glucosamine (Lin et al. 2016). HA binds to cell surface receptors (HA-mediated motility receptor) and influences cellular responses such as proliferation and migration (Mariman and Wang 2010). Chronic treatment of obese mice with a recombinant human HA has been shown to decrease adiposity and inflammation by preventing IR (Ji et al. 2014). On the other hand, THBS1, an ECM glycoprotein, is predominantly expressed in visceral adipose tissue and is highly expressed in obese and IR humans (Li et al. 2011; Inoue et al. 2013). Also, MMPs, a family of calcium-dependent, zinc-containing endopeptidases, are responsible for the degradation of virtually all ECM proteins. MMPs play an essential role in the regulation of ECM remodelling in both normal physiology and pathology (van Hul and Lijnen 2011). Elevated circulating concentrations of MMP2 and 9 have been seen in obese and diabetic patients (Derosa et al. 2008). In this context, TIMPs play a key role in the regulation of MMPs. Circulating levels of TIMP1 and 2 are increased in patients with metabolic syndrome and diabetes (Lin et al. 2016). Notably, hypothalamic TIMP-1 expression is regulated by lectin and genetic deletion of TIMP-1 causes increased food intake and obesity in mice (Hopps et al. 2013).

In addition, the role of ECM in cell adhesion and signalling is known to be mediated by proteins such as integrins, which transduce signals across the plasma membrane to activate intracellular signalling (Hamidi and Ivaska 2018). Integrins are heterodimeric transmembrane receptors that ensure communication between ECM and the intracellular environment (Fernanda Altei et al. 2021). Integrin 1 has been shown to be critical in regulating obesity-induced IR in skeletal muscles; however, its role in adipose tissues has not been studied (Hamidi and Ivaska 2018). Importantly, this remodelling process allows the formation of new blood vessels which is crucial for the healthy expansion of adipose tissue (Bonnans et al. 2014). However, under obese conditions, this process is dysfunctional, resulting in adipocyte necrosis and hypoxia, triggering chronic inflammation and fibrosis, which is a major player in adipose tissue dysfunction and insulin resistance (IR) (Lin et al. 2016). Indeed, ECM and integrins are important regulators of insulin action and may be a novel therapeutic target to treat the underlying IR associated with diabetes (Ruiz-Ojeda et al. 2019).

Together, the components of the ECM and the mechanical characteristics of the matrix are essential for the health and function of resident cells. Thus, EVs released from healthy and pathological cells can act as ECM mediators, either through direct EV—ECM interactions or by influencing cell–cell interactions (Mariman and Wang 2010). Importantly, EVs are transported through the ECM of the target tissue to reach the recipient cells and thus release their contents to trigger a cellular response (Lewin et al. 2020). Therefore, EVs play a key role in the structure of the ECM and in the modulation and interaction of sender and recipient cells.

However, to date there are no studies on how this interaction looks like and how the composition of EVs affects this process. A recent study by Lenzini et al. on how stress affects ECM relaxation showed that the mechanical properties of the ECM allow EVs to diffuse through the extracellular space, despite their diameter compared to the ECM mesh size (50 nm) (Lenzini et al. 2020). Furthermore, they observed that under these stress conditions there was an increased diffusion of EVs through the matrix. In the same context, deficiencies in the membrane protein aquoporin1 were found to increase ECM stiffness, which reduced the diffusion coefficient of EVs (Clarke-Bland et al. 2022). Therefore, the mechanical properties of the matrix play a role in EV transport.

Furthermore, EVs have also been found to indirectly alter the structural components of the host cell matrix (Patel et al. 2023). In particular, EVs released by tumour cells modulate the host cell matrix to create a tumour niche (Bonnans et al. 2014). Several studies indicate that EVs from cancer cells contain enzymes that degrade the matrix and promote cancer cell migration and growth (Bonnans et al. 2014). In addition to interacting directly with the ECM, EVs secreted into the tumour microenvironment also contain proteins such as MMPs and MMP regulators that are transferred to recipient cells and induce ECM remodelling (Shimoda 2019). For example, EVs released from myeloma patient cells contain high concentrations of MMP9 that promote endothelial cell invasion and reprogramming of healthy cells towards a more cancerous phenotype (Purushothaman et al. 2016). Another MMP regulator that has been associated with poor prognosis in multiple cancer types is TIMP-1 (Song et al. 2016). Overexpression of TIMP-1 in human lung adenocarcinoma cells increased the expression and loading of pro-tumorigenic microRNAs in secreted EVs (Song et al. 2016). These particles induced angiogenic tube formation in human epithelial cells and promoted in vitro migration of lung adenocarcinoma cells, both processes involving ECM remodelling (Dang and Myers 2015). Therefore, these studies indicate that EVs from tumour cells promote recipient cells to adopt a cancerous phenotype involving ECM degradation.

Unlike EVs released by tumour cells, there are no studies on the interaction of EVs released by adipose tissue under obese conditions with the ECM. However, recent proteomic studies have revealed that EVs from obese adipose tissue contain proteins related to the alteration of the ECM and, therefore, to the possible development of comorbidities associated with this pathology (Camino et al. 2020a, 2022a, 2023). Specifically, human vesicles from morbidly obese adipose tissue have been shown to contain ECM components (collagen VI and XV, THBS1, biglycan, osteopontin), ECM modifiers (MMP-1, 2, 9, 14, TIMP-1), and ECM receptors (integrin 1, CD44, CD36) previously associated with obesity and insulin resistance (Table 7.1) (Camino et al. 2022a). Biglycan is a non-fibrillar component of the ECM, released from the matrix in response to tissue stress; the soluble form of biglycan binds to the Toll-like 2/4 receptor on macrophages, triggering the secretion of proinflammatory cytokines (Mariman and Wang 2010). Obese patients have been observed to have a higher concentration of biglycan mRNA in subcutaneous and visceral adipose tissue, and this value was positively correlated with adiposity indices and insulin resistance parameters (Kim et al. 2016). Furthermore, MMP-14 has been found to be overexpressed in obese adipose tissue and is involved in abnormal lipid metabolism and insulin resistance (Li et al. 2020). In the same context, expression of CD44 in adipose tissue, a multifunctional membrane receptor for ECM components (HA and OPN), correlates positively with adipose inflammation and an index of insulin resistance in both obese individuals and mice (Liu et al. 2015). In contrast, treatment with a CD44 monoclonal antibody suppresses visceral adipose tissue inflammation and reduces fasting blood glucose levels, weight gain, hepatic steatosis, and insulin resistance in an obese mouse model (Kodama et al. 2012). Furthermore, CD36, a receptor that binds many ligands, such as collagen, THBS, lipoproteins, and fatty acids, is known to participate in the paracrine inflammatory loop between adipocytes and macrophages that facilitates chronic inflammation and contributes to IR (Kennedy et al. 2011; Cai et al. 2012). This therefore reflects the interaction of the various components of the adipose tissue ECM with EVs released not only by adipocytes, but also by a wide range of inflammatory and endothelial cells that are jointly involved in the development of obesity-associated inflammation and insulin resistance.

Table 7.1 Overview of extracellular matrix (ECM)-related proteins identified in extracellular vesicles (EVs) released from adipose tissue and their involvement in obesity (Camino et al. 2022a)

EVs biomarkers	Relationship with obesity
Collagen VI	Inflammation of adipose tissue (Pasarica et al. 2009)
THBS1	Expressed in visceral adipose tissue and associated with insulin resistance (Li et al. 2011; Inoue et al. 2013)
Byglycan	Inflammation of adipose tissue (Mariman and Wang 2010)
Osteopontin	Inflammation of adipose tissue and insulin resistance (Chapman et al. 2010)
Matrix metalloproteins (MMP 1, 2, 9, 14)	Involved in abnormal lipid metabolism and insulin resistance (Derosa et al. 2008)
TIMP1	Pro-tumorigenic and metabolic syndrome (Song et al. 2016)
Integrin 1	Insulin resistance (Bonnans et al. 2014)
CD44	Inflammation of adipose tissue and insulin resistance (Kodama et al. 2012)
CD36	Inflammation of adipose tissue (Kennedy et al. 2011)

On the other hand, EVs released from mesenchymal stem cells (MSCs) have been shown to have therapeutic capacity by enhancing EMC remodelling and regeneration (Huleihel et al. 2016). For example, EVs released from adipose tissue MSCs (AT-MSCs) decreased scar formation and improved wound healing in an in vivo model (Wang et al. 2017). Specifically in this study, EVs from AT-MSC were administered intravenously to a murine model with a dorsal wound, and this treatment increased the ratio of collagen type 3 and 1 and MMP-3 to TIMP-1 in fibroblasts, thus aiding in matrix remodelling (Wang et al. 2017). Interestingly, fibroblasts isolated from women with stress urinary incontinence had low expression of collagen 1, however, after treatment with EVs isolated from AT-MSC increased expression of type 1 collagen and TIMP-1/3 (Liu et al. 2018). Thus, increased collagen deposition together with inhibition and decreased expression of MMPs led to restoration of ECM stiffness loss during stress urinary incontinence (Liu et al. 2018). Importantly, these studies suggest that stem cell derived EVs are good therapeutic targets for tissue regeneration through modulation of the ECM; thus, studies on the ability of these EVs in the control and progression of obesity are needed.

7.5 Conclusions

In recent years, extracellular vesicles have been of exponential interest due to their role in signalling and as adipokine transporters and biomarkers of the cell and tissue of origin, as they dynamically reflect pathological alterations; therefore, they also provide new therapeutic options. Although studies on the characterisation and functional role of EVs released by adipose tissue took some time to develop, in contrast to other cellular systems, as in the case of pathologies such as cancer, in recent years, the number of investigations on these particles released in the context of metabolic regulation has increased. Consequently, the potential role of EVs in adipose tissue communication at local and peripheral/central levels has been demonstrated. Thus, adipose tissue can send signals at both autocrine/paracrine and endocrine levels, interacting with distant tissues capable of altering the gene expression of target cells due to their content.

However, there are hardly any studies on how these EVs interact with the ECM which is a key component in the development of obesity and its comorbidities due to its remodelling and degradation. Anyhow, some studies have shown that EVs released by adipose tissue contain components characteristic of the ECM such as MMP2, TIMP1, or different types of collagenases or integrins, among others, that are related to the inflammatory process and the development of insulin resistance during obesity.

Therefore, future studies will have to consider the possibility of using the functional effects of adipose EVs for therapeutic purposes, which may represent important curative options to treat obesity. As well as, characterising the effect of EVs on the ECM during physiological processes such as tissue regeneration presents the opportunity to create novel current therapeutic approaches for metabolic diseases.

Acknowledgments CIBERobn (CB06/03) is an initiative from ISCIII. Funding bodies: ISCIII PI19/00305, PI22/00196, and the Axencia Galega de Innovación IN607D 2022/01. Camino T was funded by FPU program (Ministerio de Educación, Cultura y Deporte, Spain); Lago-Baameiro is funded by a 2021predoctoral grant from Xunta de Galicia.

Conflicts of Interest The authors declare no conflict of interest. The funders had no role in the design of the study; in the collection, analyses, or interpretation of data; in the writing of the manuscript; or in the decision to publish the results.

References

Abels ER, Breakefield XO (2016) Introduction to extracellular vesicles: biogenesis, RNA cargo selection, content, release, and uptake. Cell Mol Neurobiol 36:301. https://doi.org/10.1007/S10571-016-0366-Z

Al Halawani A, Mithieux SM, Yeo GC et al (2022) Extracellular vesicles: interplay with the extracellular matrix and modulated cell responses. Int J Mol Sci 23. https://doi.org/10.3390/ijms23063389

Alcalá M, Calderon-Dominguez M, Bustos E et al (2017) Increased inflammation, oxidative stress and mitochondrial respiration in brown adipose tissue from obese mice. Sci Rep 7:1–12. https://doi.org/10.1038/s41598-017-16463-6

Ali MM, Mirza I, Naquiallah D et al (2022) CD147 levels in blood and adipose tissues correlate with vascular dysfunction in obese diabetic adults. J Cardiovasc Dev Dis 9. https://doi.org/10.3390/jcdd9010007

Alvarez-Llamas G, Szalowska E, de Vries MP et al (2007) Characterization of the human visceral adipose tissue secretome. Mol Cell Proteomics 6:589–600. https://doi.org/10.1074/MCP.M600265-MCP200

Andolfi C, Fisichella PM (2018) Epidemiology of obesity and associated comorbidities. J Laparoendosc Adv Surg Tech A 28:919–924. https://doi.org/10.1089/lap.2018.0380

Apovian CM (2016) Obesity: definition, comorbidities, causes, and burden. Am J Manag Care 22:s176–s185

Apovian CM (2020) Obesity: definition, comorbidities, causes, and burden. Am J Manag Care 22:s176–s185

Balistreri CR, Caruso C, Candore G (2010) The role of adipose tissue and adipokines in obesity-related inflammatory diseases. Mediat Inflamm 2010. https://doi.org/10.1155/2010/802078

Barile L, Vassalli G (2017) Exosomes: therapy delivery tools and biomarkers of diseases

Blüher M, Klöting N, Wueest S et al (2014) Fas and FasL expression in human adipose tissue is related to obesity, insulin resistance, and type 2 diabetes. J Clin Endocrinol Metab 99:E36–E44. https://doi.org/10.1210/JC.2013-2488

Bonnans C, Chou J, Werb Z (2014) Remodelling the extracellular matrix in development and disease. Nat Rev Mol Cell Biol 15:786–801. https://doi.org/10.1038/NRM3904

Cai L, Wang Z, Ji A et al (2012) Scavenger receptor CD36 expression contributes to adipose tissue inflammation and cell death in diet-induced obesity. PLoS One 7. https://doi.org/10.1371/journal.pone.0036785

Camino T, Lago-Baameiro N, Bravo SB et al (2020a) Vesicles shed by pathological murine adipocytes spread pathology: characterization and functional role of insulin resistant/hypertrophied adiposomes. Int J Mol Sci 21:2252. https://doi.org/10.3390/ijms21062252

Camino T, Lago-Baameiro N, Martis-Sueiro A et al (2020b) Deciphering adipose tissue extracellular vesicles protein cargo and its role in obesity. Int J Mol Sci 21:1–16. https://doi.org/10.3390/IJMS21249366

Camino T, Lago-Baameiro N, Bravo SB et al (2022a) Human obese white adipose tissue sheds depot-specific extracellular vesicles and reveals candidate biomarkers for monitoring obesity and its comorbidities. Transl Res 239:85–102. https://doi.org/10.1016/j.trsl.2021.01.006

Camino T, Lago-Baameiro N, Sueiro A et al (2022b) Brown adipose tissue sheds extracellular vesicles that carry potential biomarkers of metabolic and thermogenesis activity which are affected by high fat diet intervention. Int J Mol Sci 23. https://doi.org/10.3390/IJMS231810826

Camino T, Lago-Baameiro N, Pardo M (2023) Extracellular vesicles as carriers of adipokines and their role in obesity. Biomedicines 11. https://doi.org/10.3390/BIOMEDICINES11020422

Campello E, Zabeo E, Radu CM et al (2016) Dynamics of circulating microparticles in obesity after weight loss. Intern Emerg Med 11:695–702. https://doi.org/10.1007/s11739-016-1397-7

Cannon B, Nedergaard J (2004) Brown adipose tissue: function and physiological significance. Physiol Rev 84:277–359. https://doi.org/10.1152/PHYSREV.00015.2003

Cao HM, Ye XP, Ma JH et al (2015) Mimecan, a hormone abundantly expressed in adipose tissue, reduced food intake independently of leptin signaling. EBioMedicine 2:1718–1724. https://doi.org/10.1016/j.ebiom.2015.09.044

Cereijo R, Gavaldà-Navarro A, Cairó M et al (2018) CXCL14, a brown adipokine that mediates brown-fat-to-macrophage communication in thermogenic adaptation. Cell Metab 28:750–763.e6. https://doi.org/10.1016/J.CMET.2018.07.015

Chapman J, Miles PD, Ofrecio JM et al (2010) Osteopontin is required for the early onset of high fat diet-induced insulin resistance in mice. PLoS One 5. https://doi.org/10.1371/JOURNAL.PONE.0013959

Clarke-Bland CE, Bill RM, Devitt A (2022) Emerging roles for AQP in mammalian extracellular vesicles. Biochim Biophys Acta Biomembr 1864. https://doi.org/10.1016/J.BBAMEM.2021.183826

Clement E, Lazar I, Attané C et al (2020) Adipocyte extracellular vesicles carry enzymes and fatty acids that stimulate mitochondrial metabolism and remodeling in tumor cells. EMBO J 39. https://doi.org/10.15252/EMBJ.2019102525

Colombo M, Moita C, Van Niel G et al (2013) Analysis of ESCRT functions in exosome biogenesis, composition and secretion highlights the heterogeneity of extracellular vesicles. J Cell Sci 126:5553–5565. https://doi.org/10.1242/jcs.128868

Colombo M, Raposo G, Théry C (2014) Biogenesis, secretion, and intercellular interactions of exosomes and other extracellular vesicles. Annu Rev Cell Dev Biol 30:255–289. https://doi.org/10.1146/ANNUREV-CELLBIO-101512-122326

Crewe C, Joffin N, Rutkowski JM et al (2018) An endothelial-to-adipocyte extracellular vesicle axis governed by metabolic state. Cell 175:695–708.e13. https://doi.org/10.1016/J.CELL.2018.09.005

Curat CA, Wegner V, Sengenès C et al (2006) Macrophages in human visceral adipose tissue: increased accumulation in obesity and a source of resistin and visfatin. Diabetologia 49:744–747. https://doi.org/10.1007/S00125-006-0173-Z

Cypess AM, Kahn CR (2010) Brown fat as a therapy for obesity and diabetes. Curr Opin Endocrinol Diabetes Obes 17:143–149. https://doi.org/10.1097/MED.0B013E328337A81F

Dang K, Myers KA (2015) The role of hypoxia-induced miR-210 in cancer progression. Int J Mol Sci 16:6353–6372. https://doi.org/10.3390/IJMS16036353

De Silva N, Samblas M, Martínez JA, Milagro FI (2018) Effects of exosomes from LPS-activated macrophages on adipocyte gene expression, differentiation, and insulin-dependent glucose uptake. J Physiol Biochem 74:559–568. https://doi.org/10.1007/S13105-018-0622-4

Derosa G, Ferrari I, D'Angelo A et al (2008) Matrix metalloproteinase-2 and -9 levels in obese patients. Endothelium 15:219–224. https://doi.org/10.1080/10623320802228815

Deng Z, Poliakov A, Hardy RW et al (2009) Adipose tissue exosome-like vesicles mediate activation of macrophage-induced insulin resistance. Diabetes 58:2498–2505. https://doi.org/10.2337/DB09-0216

Dorairaj V, Sulaiman SA, Abu N, Murad NAA (2020) Extracellular vesicles in the development of the non-alcoholic fatty liver disease: an update. Biomolecules 10:1–24. https://doi.org/10.3390/BIOM10111494

Doyle LM, Wang MZ (2019) Overview of extracellular vesicles, their origin, composition, purpose, and methods for exosome isolation and analysis. Cells 8

Duan X, Norris DM, Humphrey SJ et al (2022) Trafficking regulator of GLUT4-1 (TRARG1) is a GSK3 substrate. Biochem J 479:1237–1256. https://doi.org/10.1042/BCJ20220153

Duque GA, Descoteaux A (2014) Macrophage cytokines: involvement in immunity and infectious diseases. Front Immunol 5

Durcin M, Fleury A, Taillebois E et al (2017) Characterisation of adipocyte-derived extracellular vesicle subtypes identifies distinct protein and lipid signatures for large and small extracellular vesicles. J Extracell Vesicles:6. https://doi.org/10.1080/20013078.2017.1305677

Engin A, Engin A (2017) Adiponectin-resistance in obesity. Adv Exp Med Biol 960:415–441. https://doi.org/10.1007/978-3-319-48382-5_18

Fantuzzi G (2005) Adipose tissue, adipokines, and inflammation. J Allergy Clin Immunol 115:911–919

Feldmann HM, Golozoubova V, Cannon B, Nedergaard J (2009) UCP1 ablation induces obesity and abolishes diet-induced thermogenesis in mice exempt from thermal stress by living at thermoneutrality. Cell Metab 9:203–209. https://doi.org/10.1016/j.cmet.2008.12.014

Fernanda Altei W, Cruz Pachane B, Karina dos Santos P et al (2021) Extracellular vesicles and integrins: partners in cancer progression. In: Role of exosomes in biological communication systems. Springer, pp 293–310. https://doi.org/10.1007/978-981-15-6599-1_14

Fernández-Real JM, Catalán V, Moreno-Navarrete JM et al (2010) Study of caveolin-1 gene expression in whole adipose tissue and its subfractions and during differentiation of human adipocytes. Nutr Metab (Lond) 7:20. https://doi.org/10.1186/1743-7075-7-20

Ferrante SC, Nadler EP, Pillai DK et al (2015) Adipocyte-derived exosomal miRNAs: a novel mechanism for obesity-related disease. Pediatr Res 77:447–454. https://doi.org/10.1038/PR.2014.202

Franckhauser S, Muñoz S, Elias I et al (2006) Adipose overexpression of phosphoenolpyruvate carboxykinase leads to high susceptibility to diet-induced insulin resistance and obesity. Diabetes 55:273–280. https://doi.org/10.2337/DIABETES.55.02.06.DB05-0482

Gao J, Li X, Wang Y et al (2020) Adipocyte-derived extracellular vesicles modulate appetite and weight through mTOR signalling in the hypothalamus. Acta Physiol 228. https://doi.org/10.1111/apha.13339

Gesmundo I, Pardini B, Gargantini E et al (2021) Adipocyte-derived extracellular vesicles regulate survival and function of pancreatic β cells. JCI Insight 6. https://doi.org/10.1172/JCI.INSIGHT.141962

Giordano A, Smorlesi A, Frontini A et al (2014) White, brown and pink adipocytes: the extraordinary plasticity of the adipose organ. Eur J Endocrinol 170. https://doi.org/10.1530/EJE-13-0945

Greenhill C (2017) Non-coding RNA: exosomal microRNAs as novel adipokines. Nat Rev Genet 18:212. https://doi.org/10.1038/NRG.2017.14

Hamidi H, Ivaska J (2018) Every step of the way: integrins in cancer progression and metastasis. Nat Rev Cancer 18:533–548. https://doi.org/10.1038/S41568-018-0038-Z

Han B, Luo H, Raelson J et al (2014) TGFBI (βIG-H3) is a diabetes-risk gene based on mouse and human genetic studies. Hum Mol Genet 23:4597–4611. https://doi.org/10.1093/HMG/DDU173

Hopps E, Lo PR, Montana M et al (2013) Gelatinases and their tissue inhibitors in a group of subjects with metabolic syndrome. J Investig Med 61:978–983. https://doi.org/10.2310/JIM.0B013E318294E9DA

Huleihel L, Hussey GS, Naranjo JD et al (2016) Matrix-bound nanovesicles within ECM bioscaffolds. Sci Adv 2. https://doi.org/10.1126/SCIADV.1600502

Ibrahim MM (2010) Subcutaneous and visceral adipose tissue: structural and functional differences. Obes Rev 11:11–18. https://doi.org/10.1111/j.1467-789X.2009.00623.x

Inoue M, Jiang Y, Barnes RH et al (2013) Thrombospondin 1 mediates high-fat diet-induced muscle fibrosis and insulin resistance in male mice. Endocrinology 154:4548–4559. https://doi.org/10.1210/EN.2013-1587

Jayabalan N, Lai A, Ormazabal V et al (2019) Adipose tissue exosomal proteomic profile reveals a role on placenta glucose metabolism in gestational diabetes mellitus. J Clin Endocrinol Metab 104:1735–1752. https://doi.org/10.1210/JC.2018-01599

Ji E, Jung MY, Park JH et al (2014) Inhibition of adipogenesis in 3T3-L1 cells and suppression of abdominal fat accumulation in high-fat diet-feeding C57BL/6J mice after downregulation of hyaluronic acid. Int J Obes 38:1035–1043. https://doi.org/10.1038/IJO.2013.202

Kahles F, Findeisen HM, Bruemmer D (2014) Osteopontin: a novel regulator at the cross roads of inflammation, obesity and diabetes. Mol Metab 3:384–393. https://doi.org/10.1016/J.MOLMET.2014.03.004

Kalluri R, LeBleu VS (2020) The biology, function, and biomedical applications of exosomes. Science 1979:367

Kariba Y, Yoshizawa T, Sato Y et al (2020) Brown adipocyte-derived exosomal miR-132-3p suppress hepatic Srebf1 expression and thereby attenuate expression of lipogenic genes. Biochem Biophys Res Commun 530:500–507. https://doi.org/10.1016/J.BBRC.2020.05.090

Kennedy DJ, Kuchibhotla S, Westfall KM et al (2011) A CD36-dependent pathway enhances macrophage and adipose tissue inflammation and impairs insulin signalling. Cardiovasc Res 89:604–613. https://doi.org/10.1093/cvr/cvq360

Khushman M, Bhardwaj A, Patel GK et al (2017) Exosomal markers (CD63 and CD9) expression pattern using immunohistochemistry in resected malignant and nonmalignant pancreatic specimens. In: Pancreas. Lippincott Williams and Wilkins, pp 782–788

Kim OY, Shin MJ, Moon J, Chung JH (2011) Plasma ceruloplasmin as a biomarker for obesity: a proteomic approach. Clin Biochem 44:351–356. https://doi.org/10.1016/j.clinbiochem.2011.01.014

Kim J, Lee SK, Shin JM et al (2016) Enhanced biglycan gene expression in the adipose tissues of obese women and its association with obesity-related genes and metabolic parameters. Sci Rep 6. https://doi.org/10.1038/SREP30609

Kim SY, Kim I, Cho W et al (2021) Vimentin deficiency prevents high-fat diet-induced obesity and insulin resistance in mice. Diabetes Metab J 45:97–108. https://doi.org/10.4093/DMJ.2019.0198

Kodama K, Horikoshi M, Toda K et al (2012) Expression-based genome-wide association study links the receptor CD44 in adipose tissue with type 2 diabetes. Proc Natl Acad Sci USA 109:7049–7054. https://doi.org/10.1073/PNAS.1114513109/-/DCSUPPLEMENTAL/PNAS.201114513SI.PDF

Koeck ES, Sevilla S, Qureshi FG et al (2014) Adipocyte exosomes induce transforming growth factor Beta pathway dysregulation in hepatocytes: a novel paradigm for obesity-related liver disease. J Surg Res 186:573. https://doi.org/10.1016/j.jss.2013.11.500

Konoshenko MY, Lekchnov EA, Vlassov AV, Laktionov PP (2018) Isolation of extracellular vesicles: general methodologies and latest trends. Biomed Res Int 2018. https://doi.org/10.1155/2018/8545347

Kranendonk MEG, Visseren FLJ, Van Balkom BWM et al (2014a) Human adipocyte extracellular vesicles in reciprocal signaling between adipocytes and macrophages. Obesity 22:1296–1308. https://doi.org/10.1002/OBY.20679

Kranendonk MEG, Visseren FLJ, Van Herwaarden JA et al (2014b) Effect of extracellular vesicles of human adipose tissue on insulin signaling in liver and muscle cells. Obesity 22:2216–2223. https://doi.org/10.1002/OBY.20847

Krautbauer S, Haberl EM, Eisinger K et al (2017) Annexin A6 regulates adipocyte lipid storage and adiponectin release. Mol Cell Endocrinol 439:419–430. https://doi.org/10.1016/J.MCE.2016.09.033

Lamers D, Famulla S, Wronkowitz N et al (2011) Dipeptidyl peptidase 4 is a novel adipokine potentially linking obesity to the metabolic syndrome. Diabetes 60:1917–1925. https://doi.org/10.2337/db10-1707

Lazar I, Clement E, Dauvillier S et al (2016) Adipocyte exosomes promote melanoma aggressiveness through fatty acid oxidation: a novel mechanism linking obesity and cancer. Cancer Res 76:4051–4057. https://doi.org/10.1158/0008-5472.CAN-16-0651

Lee JE, Moon PG, Lee IK, Baek MC (2015) Proteomic analysis of extracellular vesicles released by adipocytes of Otsuka Long-Evans Tokushima fatty (OLETF) rats. Protein J 34:220–235. https://doi.org/10.1007/S10930-015-9616-Z

Lenzini S, Bargi R, Chung G, Shin JW (2020) Matrix mechanics and water permeation regulate extracellular vesicle transport. Nat Nanotechnol 15:217–223. https://doi.org/10.1038/S41565-020-0636-2

Lewin S, Hunt S, Lambert DW (2020) Extracellular vesicles and the extracellular matrix: a new paradigm or old news? Biochem Soc Trans 48:2335–2345. https://doi.org/10.1042/BST20200717

Li Y, Tong X, Rumala C et al (2011) Thrombospondin1 deficiency reduces obesity-associated inflammation and improves insulin sensitivity in a diet-induced obese mouse model. PLoS One 6. https://doi.org/10.1371/JOURNAL.PONE.0026656

Li X, Zhao Y, Chen C et al (2020) Critical role of matrix metalloproteinase 14 in adipose tissue remodeling during obesity. Mol Cell Biol 40. https://doi.org/10.1128/mcb.00564-19

Lim J, Park HS, Kim J et al (2020) Depot-specific UCP1 expression in human white adipose tissue and its association with obesity-related markers. Int J Obes 44:697–706. https://doi.org/10.1038/S41366-020-0528-4

Lin R, Wang S, Zhao RC (2013) Exosomes from human adipose-derived mesenchymal stem cells promote migration through Wnt signaling pathway in a breast cancer cell model. Mol Cell Biochem 383:13–20. https://doi.org/10.1007/S11010-013-1746-Z

Lin D, Chun TH, Kang L (2016) Adipose extracellular matrix remodelling in obesity and insulin resistance. Biochem Pharmacol 119:8–16. https://doi.org/10.1016/J.BCP.2016.05.005

Liu LF, Kodama K, Wei K et al (2015) The receptor CD44 is associated with systemic insulin resistance and proinflammatory macrophages in human adipose tissue. Diabetologia 58:1579–1586. https://doi.org/10.1007/S00125-015-3603-Y

Liu X, Wang S, Wu S et al (2018) Exosomes secreted by adipose-derived mesenchymal stem cells regulate type I collagen metabolism in fibroblasts from women with stress urinary incontinence. Stem Cell Res Ther 9. https://doi.org/10.1186/S13287-018-0899-9

Liu Y, Tan J, Ou S et al (2019) Adipose-derived exosomes deliver miR-23a/b to regulate tumor growth in hepatocellular cancer by targeting the VHL/HIF axis. J Physiol Biochem 75:391–401. https://doi.org/10.1007/s13105-019-00692-6

Mariman ECM, Wang P (2010) Adipocyte extracellular matrix composition, dynamics and role in obesity. Cell Mol Life Sci 67:1277–1292. https://doi.org/10.1007/S00018-010-0263-4

Marques A, Peralta M, Naia A et al (2018) Prevalence of adult overweight and obesity in 20 European countries, 2014. Eur J Pub Health 28:295–300. https://doi.org/10.1093/eurpub/ckx143

Mathivanan S, Ji H, Simpson RJ (2010) Exosomes: extracellular organelles important in intercellular communication. J Proteome 73:1907–1920. https://doi.org/10.1016/j.jprot.2010.06.006

McGown C, Birerdinc A, Younossi ZM (2014) Adipose tissue as an endocrine organ. Clin Liver Dis 18:41–58

Meldolesi J (2018) Exosomes and ectosomes in intercellular communication. Curr Biol 28:R435–R444. https://doi.org/10.1016/J.CUB.2018.01.059

Nazarenko I, Rupp AK, Altevogt P (2013) Exosomes as a potential tool for a specific delivery of functional molecules. Methods Mol Biol 1049:495–511. https://doi.org/10.1007/978-1-62703-547-7_37

Newbold RR, Padilla-Banks E, Jefferson WN, Heindel JJ (2008) Effects of endocrine disruptors on obesity. Int J Androl 31:201–208. https://doi.org/10.1111/j.1365-2605.2007.00858.x

Nwosu ZC, Ebert MP, Dooley S, Meyer C (2016) Caveolin-1 in the regulation of cell metabolism: a cancer perspective. Mol Cancer 15. https://doi.org/10.1186/S12943-016-0558-7

Pardo F, Villalobos-Labra R, Sobrevia B et al (2018) Extracellular vesicles in obesity and diabetes mellitus. Mol Asp Med 60:81–91. https://doi.org/10.1016/J.MAM.2017.11.010
Pasarica M, Gowronska-Kozak B, Burk D et al (2009) Adipose tissue collagen VI in obesity. J Clin Endocrinol Metab 94:5155–5162. https://doi.org/10.1210/JC.2009-0947
Patel NJ, Ashraf A, Chung EJ (2023) Extracellular vesicles as regulators of the extracellular matrix. Bioengineering (Basel) 10. https://doi.org/10.3390/BIOENGINEERING10020136
Pathan M, Fonseka P, Chitti SV et al (2019) Vesiclepedia 2019: a compendium of RNA, proteins, lipids and metabolites in extracellular vesicles. Nucleic Acids Res 47:D516–D519. https://doi.org/10.1093/nar/gky1029
Perdikari A, Leparc GG, Balaz M et al (2018) BATLAS: deconvoluting brown adipose tissue. Cell Rep 25:784–797.e4. https://doi.org/10.1016/J.CELREP.2018.09.044
Pessoa Rodrigues C, Chatterjee A, Wiese M et al (2021) Histone H4 lysine 16 acetylation controls central carbon metabolism and diet-induced obesity in mice. Nat Commun 12. https://doi.org/10.1038/S41467-021-26277-W
Pietrani NT, Ferreira CN, Rodrigues KF et al (2018) Proresolving protein Annexin A1: the role in type 2 diabetes mellitus and obesity. Biomed Pharmacother 103:482–489. https://doi.org/10.1016/j.biopha.2018.04.024
Prentice KJ, Saksi J, Hotamisligil GS (2019) Adipokine FABP4 integrates energy stores and counterregulatory metabolic responses. J Lipid Res 60:734–740. https://doi.org/10.1194/JLR.S091793
Properzi F, Logozzi M, Fais S (2013) Exosomes: the future of biomarkers in medicine. Biomark Med 7:769–778
Purushothaman A, Bandari SK, Liu J et al (2016) Fibronectin on the surface of myeloma cell-derived exosomes mediates exosome-cell interactions. J Biol Chem 291:1652–1663. https://doi.org/10.1074/JBC.M115.686295
Ramos-Andrade I, Moraes J, Brandão-Costa RM et al (2020) Obese adipose tissue extracellular vesicles raise breast cancer cell malignancy. Endocr Relat Cancer 27:571–582. https://doi.org/10.1530/ERC-19-0507
Raposo G, Stoorvogel W (2013) Extracellular vesicles: exosomes, microvesicles, and friends. J Cell Biol 200:373–383. https://doi.org/10.1083/JCB.201211138
Reed SE, Hodgson LR, Song S et al (2013) A role for Rab14 in the endocytic trafficking of GLUT4 in 3T3-L1 adipocytes. J Cell Sci 126:1931–1941. https://doi.org/10.1242/JCS.104307/-/DC1
Reiterer M, Rajan M, Gómez-Banoy N et al (2021) Hyperglycemia in acute COVID-19 is characterized by insulin resistance and adipose tissue infectivity by SARS-CoV-2. Cell Metab 33:2174. https://doi.org/10.1016/J.CMET.2021.09.009
Roca-Rivada A, Bravo SB, Pérez-Sotelo D et al (2015) CILAIR-based secretome analysis of obese visceral and subcutaneous adipose tissues reveals distinctive ECM remodeling and inflammation mediators. Sci Rep 5. https://doi.org/10.1038/srep12214
Rosina M, Ceci V, Turchi R et al (2022) Ejection of damaged mitochondria and their removal by macrophages ensure efficient thermogenesis in brown adipose tissue. Cell Metab 34:533. https://doi.org/10.1016/J.CMET.2022.02.016
Ruiz-Ojeda FJ, Méndez-Gutiérrez A, Aguilera CM, Plaza-Díaz J (2019) Extracellular matrix remodeling of adipose tissue in obesity and metabolic diseases. Int J Mol Sci 20. https://doi.org/10.3390/ijms20194888
Ruíz-Roso MB, de Carvalho PP, Matilla-Escalante DC et al (2020) Changes of physical activity and ultra-processed food consumption in adolescents from different countries during Covid-19 pandemic: an observational study. Nutrients 12:2289. https://doi.org/10.3390/NU12082289
Russell AE, Sneider A, Witwer KW et al (2019) Biological membranes in EV biogenesis, stability, uptake, and cargo transfer: an ISEV position paper arising from the ISEV membranes and EVs workshop. J Extracell Vesicles 8:1684862. https://doi.org/10.1080/20013078.2019.1684862
Sadeghabadi ZA, Ziamajidi N, Abbasalipourkabir R et al (2019) Palmitate-induced IL6 expression ameliorated by chicoric acid through AMPK and SIRT1-mediated pathway in the PBMCs of

newly diagnosed type 2 diabetes patients and healthy subjects. Cytokine 116:106–114. https://doi.org/10.1016/j.cyto.2018.12.012
Saltiel AR, Olefsky JM (2017) Inflammatory mechanisms linking obesity and metabolic disease. J Clin Invest 127:1–4
Sanchez-Gurmaches J, Tang Y, Jespersen NZ et al (2018) Brown fat AKT2 is a cold-induced kinase that stimulates ChREBP-mediated de novo lipogenesis to optimize fuel storage and thermogenesis. Cell Metab 27:195–209.e6. https://doi.org/10.1016/J.CMET.2017.10.008
Scheja L, Heeren J (2019) The endocrine function of adipose tissues in health and cardiometabolic disease. Nat Rev Endocrinol 15:507–524
Shimoda M (2019) Extracellular vesicle-associated MMPs: a modulator of the tissue microenvironment. Adv Clin Chem 88:35–66. https://doi.org/10.1016/bs.acc.2018.10.006
Shin JH, Kim IY, Kim YN et al (2015) Obesity resistance and enhanced insulin sensitivity in Ahnak-/- mice fed a high fat diet are related to impaired adipogenesis and increased energy expenditure. PLoS One 10. https://doi.org/10.1371/journal.pone.0139720
Sidhom K, Obi PO, Saleem A (2020) A review of exosomal isolation methods: is size exclusion chromatography the best option? Int J Mol Sci 21:1–19
Simeone P, Bologna G, Lanuti P et al (2020) Extracellular vesicles as signaling mediators and disease biomarkers across biological barriers. Int J Mol Sci 21
Son Y, Choi C, Saha A et al (2022) REEP6 knockout leads to defective β-adrenergic signaling in adipocytes and promotes obesity-related metabolic dysfunction. Metabolism 130:155159. https://doi.org/10.1016/J.METABOL.2022.155159
Song G, Xu S, Zhang H et al (2016) TIMP1 is a prognostic marker for the progression and metastasis of colon cancer through FAK-PI3K/AKT and MAPK pathway. J Exp Clin Cancer Res 35. https://doi.org/10.1186/S13046-016-0427-7
Spencer M, Unal R, Zhu B et al (2011) Adipose tissue extracellular matrix and vascular abnormalities in obesity and insulin resistance. J Clin Endocrinol Metab 96. https://doi.org/10.1210/JC.2011-1567
Tai ES, Ordovas JM (2007) The role of perilipin in human obesity and insulin resistance. Curr Opin Lipidol 18:152–156. https://doi.org/10.1097/MOL.0B013E328086AEAB
Tanaka M, Nozaki M, Fukuhara A et al (2007) Visfatin is released from 3T3-L1 adipocytes via a non-classical pathway. Biochem Biophys Res Commun 359:194–201. https://doi.org/10.1016/J.BBRC.2007.05.096
Tansey JT, Sztalryd C, Hlavin EM et al (2004) The central role of perilipin A in lipid metabolism and adipocyte lipolysis. IUBMB Life 56:379–385
Théry C, Amigorena S, Raposo G, Clayton A (2006) Isolation and characterization of exosomes from cell culture supernatants and biological fluids. Curr Protoc Cell Biol 30. https://doi.org/10.1002/0471143030.CB0322S30
Théry C, Witwer KW, Aikawa E et al (2018) Minimal information for studies of extracellular vesicles 2018 (MISEV2018): a position statement of the International Society for Extracellular Vesicles and update of the MISEV2014 guidelines. J Extracell Vesicles 7:1535750. https://doi.org/10.1080/20013078.2018.1535750
Toribio V, Yáñez-Mó M (2022) Tetraspanins interweave EV secretion, endosomal network dynamics and cellular metabolism. Eur J Cell Biol 101:151229. https://doi.org/10.1016/j.ejcb.2022.151229
Trams EG, Lauter CJ, Norman Salem J, Heine U (1981) Exfoliation of membrane ecto-enzymes in the form of micro-vesicles. Biochim Biophys Acta 645:63–70. https://doi.org/10.1016/0005-2736(81)90512-5
Trayhurn P, Beattie JH (2001) Physiological role of adipose tissue: white adipose tissue as an endocrine and secretory organ. Proc Nutr Soc 60:329–339. https://doi.org/10.1079/pns200194
Tual-Chalot S, Leonetti D, Andriantsitohaina R, Martínez MC (2011) Microvesicles: intercellular vectors of biological messages. Mol Interv 11:88–94. https://doi.org/10.1124/mi.11.2.5

van Hul M, Lijnen HR (2011) Matrix metalloproteinase inhibition impairs murine adipose tissue development independently of leptin. Endocr J 58:101–107. https://doi.org/10.1507/ENDOCRJ.K10E-267

Verboven K, Wouters K, Gaens K et al (2018) Abdominal subcutaneous and visceral adipocyte size, lipolysis and inflammation relate to insulin resistance in male obese humans. Sci Rep 8. https://doi.org/10.1038/s41598-018-22962-x

Vickers KC, Remaley AT (2012) Lipid-based carriers of microRNAs and intercellular communication. Curr Opin Lipidol 23:91–97

Villarroya F, Cereijo R, Gavaldà-Navarro A et al (2018) Inflammation of brown/beige adipose tissues in obesity and metabolic disease. J Intern Med 284:492–504. https://doi.org/10.1111/JOIM.12803

Villarroya F, Gavaldà-Navarro A, Peyrou M et al (2019) Brown adipokines. In: Handbook of experimental pharmacology. Springer, New York LLC, pp 239–256

Wang L, Hu L, Zhou X et al (2017) Exosomes secreted by human adipose mesenchymal stem cells promote scarless cutaneous repair by regulating extracellular matrix remodelling. Sci Rep 7. https://doi.org/10.1038/S41598-017-12919-X

Wasim M (2015) Role of leptin in obesity. J Obes Weight Loss Ther 05:12–14. https://doi.org/10.4172/2165-7904.1000258

Wasim M, Awan FR, Najam SS et al (2016) Role of leptin deficiency, inefficiency, and leptin receptors in obesity. Biochem Genet 54:565–572. https://doi.org/10.1007/s10528-016-9751-z

Ying W, Riopel M, Bandyopadhyay G et al (2017) Adipose tissue macrophage-derived exosomal miRNAs can modulate in vivo and in vitro insulin sensitivity. Cell 171:372–384.e12. https://doi.org/10.1016/J.CELL.2017.08.035

Yllmaz SG, Bozkurt H, Ndadza A et al (2020) Childhood obesity risk in relationship to Perilipin 1 (PLIN1) gene regulation by circulating microRNAs. OMICS 24:43–50. https://doi.org/10.1089/OMI.2019.0150

Zhang J, Li S, Li L et al (2015) Exosome and exosomal microRNA: trafficking, sorting, and function. Genomics Proteomics Bioinformatics 13:17–24

Zhang Y, Yu M, Dong J et al (2020) Identification of novel adipokines through proteomic profiling of small extracellular vesicles derived from adipose tissue. J Proteome Res 19:3130–3142. https://doi.org/10.1021/ACS.JPROTEOME.0C00131

Zhao H, Chen X, Hu G et al (2022) Small extracellular vesicles from brown adipose tissue mediate exercise cardioprotection. Circ Res 130:1490–1506. https://doi.org/10.1161/CIRCRESAHA.121.320458

Zhou X, Li Z, Qi M et al (2020) Brown adipose tissue-derived exosomes mitigate the metabolic syndrome in high fat diet mice. Theranostics 10:8197–8210. https://doi.org/10.7150/THNO.43968

Zhou X, Liu S, Lu Y et al (2023) MitoEVs: a new player in multiple disease pathology and treatment. J Extracell Vesicles 12. https://doi.org/10.1002/JEV2.12320

Chapter 8
Potential of Extracellular Vesicles as Therapeutics in Retinal Diseases

Stef Bosch, Tatu Lajunen, Arto Merivaara, Kirsi Rilla, and Arto Urtti

Abstract Rapidly expanding research activity on extracellular vesicles (EV) holds promise for new biomedical applications in many tissues of our body. EVs are also emerging players in the retinal physiology, thereby increasing understanding on the mechanisms regulating pathophysiology of the eye and offering new possibilities for retinal therapeutics. This chapter summarizes the current knowledge of EVs in the posterior segment of the eye, participation of EVs in the pathogenesis of most common ocular diseases, and discusses the advancements, challenges, and future promises for EV-mediated ocular therapies. The origin, kinetics, ability to cross barriers, uptake, and transport of EVs through ocular cells will be discussed in light of the current knowledge.

The vitreous chamber of the eye is a highly protected biological compartment with limited communication with the rest of the body due to the blood–retinal barrier that restricts the exchange of EVs between the blood and the retina. These special features make the investigation of EVs in the homeostasis, pathophysiology, and therapy of the eye particularly interesting. Thus, deeper understanding of the role of EV in healthy and diseased ocular tissues and their kinetics, and interactions with the rest of the body, will significantly improve their potential utilization as drug delivery vehicles and therapeutic agents.

8.1 Introduction

The number of ophthalmic patients is increasing continuously in aging populations. Retinal diseases are a major cause of visual impairment and blindness worldwide affecting ≈ 400 million worldwide. Multifactorial diseases, such as age-related

S. Bosch · K. Rilla
Institute of Biomedicine, University of Eastern Finland, Kuopio, Finland

T. Lajunen · A. Merivaara · A. Urtti (✉)
School of Pharmacy, University of Eastern Finland, Kuopio, Finland
e-mail: arto.urtti@uef.fi

K. Rilla (ed.), *Extracellular Vesicles as Matrix Messengers*, Biology of Extracellular Matrix 15, https://doi.org/10.1007/978-3-031-68811-9_8

macular degeneration (AMD, 196 million), diabetic retinal diseases (>130 million), and glaucomatous retinopathy (80 million), are the most frequent causes of visual impairment. Number of retina patients will increase significantly. For example, the number of AMD patients is expected to reach 288 million by 2040 (Campochiaro 2021). Blindness may result from various mechanisms, including macular edema, vascular proliferation, and retinal cell death. In addition to the multifactorial diseases, retina is affected by multiple inherited genetic diseases, such as retinitis pigmentosa. These diseases are relatively rare, but unlike multifactorial common diseases, they may lead to visual impairment and even blindness already at young age.

Only a few retinal conditions can be treated with pharmaceuticals. Most notably, the neovascular AMD treatment with anti-VEGF (vascular endothelial growth factor) antibody (bevacizumab) and fab-fragment (ranibizumab) became available about 15 years ago leading to significant improvements in the treatment of AMD. Many patients with neovascular AMD are under-treated and many other retinal diseases are still without treatment (e.g., glaucomatous retinopathy, geographic atrophy, retinitis pigmentosa). Thus, new therapies are urgently needed.

Current therapeutics of the eye are mostly given locally. Topical eye drops are the most common way of ophthalmic drug administration, but they are effective only in the treatment of anterior eye segments (e.g. reduction of intraocular pressure, anterior tissue inflammation) (del Amo et al. 2017). The eye drop instillation results in low drug concentrations in the retina. Per oral or other systemic medications are rarely useful in the treatment of retina, because blood–retinal barrier limits access of entry from the blood stream into the retina (del Amo et al. 2017). Therefore, high doses should be used, which is not usually possible due to the adverse effects from other organs.

Intravitreal injections are the current method-of-choice for retinal delivery of therapeutics. Intravitreal injections result in almost direct drug delivery into the retina, but this invasive method of drug delivery imposes burden to the patients and healthcare, since ever increasing number of injections must be given to the patients with chronic retinal diseases (like AMD). For example, neovascular AMD is treated by anti-VEGF injections that are given at monthly or bimonthly intervals (del Amo et al. 2017). Important future challenges in retinal drug delivery include (1) prolongation of intravitreal injection interval; (2) targeted drug delivery to the retinal cells (especially in the case of biologics with intracellular sites of action); and (3) crossing of blood–retinal barrier to enable systemic or topical delivery.

Extracellular vesicles (EV) are an emerging class of therapeutics that is potentially useful also for the retinal treatment (Manai et al. 2024). The EVs have physiological functions in maintaining retinal health and thus they are possibly useful also as retinal therapeutics. EVs have similar size as nanomedicines, such as liposomes, polymeric nanoparticles, and viral gene vectors. Previously, medicinal nanoparticles and viral vectors have been investigated in retinal treatment, but their clinical use is still sparse. Only one gene therapy product (AAV-based sub-retinally injected Luxturna) (Gao et al. 2020) and photodynamic therapy (Visudyne) are in clinical use in retinal diseases (Khandhadia et al. 2023). Eventually, delivery of these nano-sized particles from vitreous into the retinal target cells is still poorly understood.

In this chapter, we are reviewing and discussing the current knowledge of EVs in the posterior eye segment, particularly in relation to their potential use as retinal therapeutics. Intravitreal injection is currently the most appealing option, because it is widely used in the retinal anti-VEGF therapies. However, converting EVs to intravitreal therapies is not an easy task, because pharmaceutical products must be effective and safe, but also stable, reproducible, and amenable to large-scale production. The following text will provide information about the exciting possibilities of EV-based retinal therapies as well as the unknown factors and limitations.

8.1.1 Retinal Cell Layers and EV Production in the Eye

The retina is a complex and delicate tissue located at the back of the eye and has an important role in vision. The retina has a thickness of about 250 μm and consists of 3 major cell types that are arranged in 10 distinct cell layers that are connected by synaptic layers, each contributing to the intricate processes of photoreception and signal transmission (Fig. 8.1). The 3 main types are glial cells, photoreceptor cells, and neuronal cells. The distinct layers, going from anterior to posterior are the inner limiting membrane, nerve fiber layer, ganglion cell layer, inner plexiform layer, inner nuclear layer, middle limiting membrane, outer plexiform layer, outer nuclear

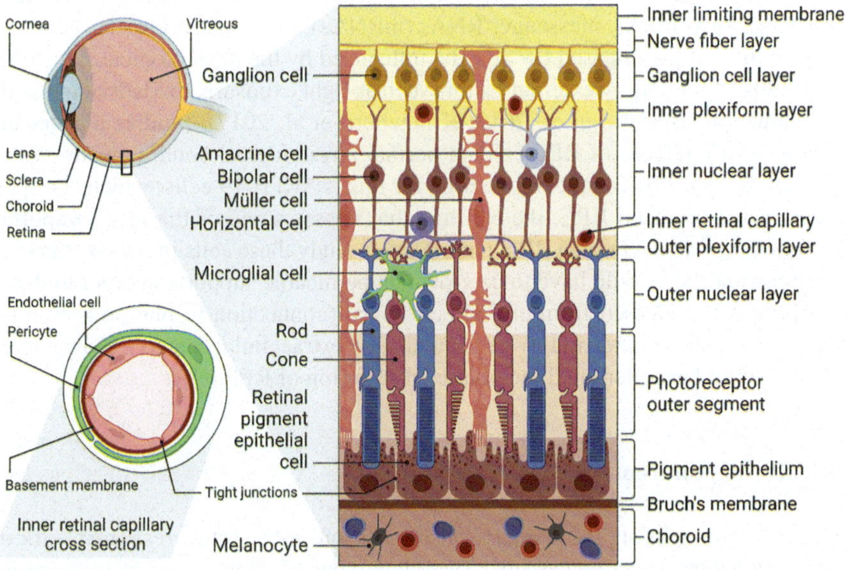

Fig. 8.1 Schematic presentation of the eye and a detailed structure of retinal cell layers and inner retinal capillaries forming the inner blood–retinal barrier. Created with BioRender.com

layer, external limiting membrane, and a layer of rods and cones. The inner limiting membrane borders the vitreous humor and acts as a diffusing membrane between the retina and the vitreous. The nerve fiber layer can be measurably smaller in patients with retinitis pigmentosa. The ganglion layer contains the retinal ganglion cells. The inner plexiform layer is the area that is made up of a dense network of fibrils that are formed by dendrites of ganglion cells and cells from the inner nuclear layer. The inner nuclear layer contains bipolar cells, horizontal cells, and amacrine cells. The outer plexiform layer contains synapses of rod and cone photoreceptor cells. The external limiting membrane contains the cell bodies of the photoreceptor cells and all these layers are supported by an outer layer of retinal pigment cells (RPE) (Hoon et al. 2014; Todorova et al. 2022; Mahabadi and Al Khalili 2023). Beyond their well-known functions, these cells also participate in intercellular communication through the secretion of EVs. Recent studies suggest that many different cell types of the retina contribute to EV production (Klingeborn et al. 2017; Country 2017; Yang et al. 2021).

8.1.1.1 Retinal Pigment Epithelial (RPE) Cells

Retinal pigment epithelial (RPE) cells are non-proliferating cells that form a monolayer between choriocapillaris/Bruch's membrane and photoreceptor cells (Röhlich 1970). The main function of RPE cells is to provide the photoreceptor cells with nutrients while at the same time get rid of waste products. The RPE layer also gives structural support to the retina (Strauss 2005). Recent studies have shown that RPE cells do indeed release EVs. These vesicles are known to contain various cargo, including microRNAs, messenger RNAs (mRNAs), proteins, and lipids. The cargo composition of the released EVs can be influenced by the physiological state of the RPE cells, as well as by external factors such as light exposure, oxidative stress, and inflammation (Biasutto et al. 2013; Klingeborn et al. 2017). Studies that try and check the EV release of RPE cells in normal physiological conditions are always difficult because the usual cell line used for this is ARPE-19 cells, which is a spontaneous immortalized RPE like cell line that misses some of the many important trademarks of actual human RPE cells. Also, to study these cells in a more "natural" environment they would have to be grown on permeable supports in order to distinguish the EVs released from either their apical (connection to photoreceptors) or basolateral side (connection to basal lamina of extracellular matrix). However, no studies so far have identified the polarized secretion of RPE EVs.

8.1.1.2 Photoreceptors

Photoreceptor cells are the primary sensory neurons in the retina responsible for the visual signal transduction cascade. Two main types of photoreceptors in the retina, rods and cones make up about 70% of all neural retinal cells. They consist of an outer segment, connecting cilium, inner segment, nucleus, and synaptic terminal. In many pathological states photoreceptors do release EVs that contain a variety of cellular materials depending on the underlying pathology. This is a possible mechanism to counteract the excess accumulation of harmful proteins or factors by expelling them within EVs (Lewis et al. 2022). It has been shown that photoreceptor cells also release EVs while in a healthy state in mice. Lewis and colleagues were able to visualize budding of EVs from the inner segment compartment of mouse photoreceptors into the interphotoreceptor space by using 3D-electron tomography (Lewis et al. 2022). Interestingly, they found that a fraction of EVs contain visual pigment rhodopsin, a protein that is necessary in rods to trigger visual transduction. These findings support the theory that EV release is a fundamental homeostatic process to get rid of damaged inner segment material.

8.1.1.3 Müller Glial Cells

Müller glial cells provide structure and metabolic support to the retina, by stretching over the entire retina between the inner and outer membranes. They also play an important role in retinal homeostasis. Müller glial cells from zebrafish are known to be able to regenerate after injury while mammalian counterparts have limited potential for this. In humans there is no evidence on the possible regenerative abilities of these cells. Müller cells are known to excrete EVs in both healthy and diseased state and the findings by Carapia and colleagues suggest that EVs derived from diseased Müller cells can negatively alter healthy Müller cells in mice (Carapia et al. 2023). Interestingly, Gao et al. showed that embryonic stem cell-derived EVs can trigger altered morphology and enhanced proliferation of healthy Müller glial cells in rats. This was associated with altered gene expression resembling retinal progenitor cells, thereby reactivating possible regenerative capacity of the cells (Gao et al. 2022).

EVs from Müller glial cells seem to originate mainly from two compartments: end feet facing the vitreous body and apical microvilli surrounding photoreceptor segments. Primary findings have shown that primary Müller glial cells do possess the ability to release both large and small EVs (Eastlake et al. 2021). Demais and colleagues showed that glial cells secrete a subset of EVs from their endfeet that face the vitreous as well as from their microvilli that surround photoreceptors in vivo in mice. The presence of CD63 in the endfeet of glial cells and in the apposed extracellular space suggests that EVs in this compartment originate from intraluminal spaces of the endfeet. EV-marker CD9 was also abundant on the apical microvilli of cells, which suggests CD9-positive EV shedding also from apical microvilli of Müller glial cells (Demais et al. 2022).

8.1.1.4 Ganglion Cells

Ganglion cells are one of the most important neural cells in the retina which transfer signals to the brain by their axons which make up the optic nerve. There is no direct evidence to show EV release by ganglion cells, but EVs derived from other cell types could have a neuroprotective effect on ganglion cells. Wang et al. have shown that EVs loaded by neuroprotective peptide PACAP38 promotes axonal regeneration in rats with traumatic optic neuropathy (Wang et al. 2021).

8.1.1.5 Amacrine and Bipolar Cells

Amacrine cells are intrinsic neuron cells that modify the output signals that are sent by postsynaptic ganglion cells. There is a large diverse group of amacrine cells but no specific type dominates in the retina (Masland 2012). Bipolar cells are an extra layer of neural cells that connect the primary sensory neurons (photoreceptor cells) to long projection neurons (ganglions). Both neuronal cells are known for their release of synaptic vesicles when transferring the signals but there is no literature to show the release of EVs by either of the cell types (Euler et al. 2014; Marshak et al. 2015).

8.1.2 EVs in the Vitreous Humor

8.1.2.1 Presence of Endogenous EVs in Vitreous

The vitreous humor is a transparent gel-like extracellular matrix network rich in water (98–99.7%) that fills the vitreous body of the eye. It consists of humor (considered as a filtrate of serum) and a molecular network of collagens (mainly types II and IX) which provides the vitreous with its gel-like consistency, and hyaluronic acid which stabilizes and stiffens the collagen network (Le Goff and Bishop 2008). Additionally, vitreous humor contains a large variety of different molecules including proteoglycans, fibrillin, and opticin (Mishra et al. 2023). It is normally acellular material apart from a few macrophage-like cells called hyalocytes located in the peripheral area of the vitreous body (Le Goff and Bishop 2008).

Extracellular vesicles (EVs) are nano-sized phospholipid membrane-derived particles present in all body fluids and extracellular matrices of tissues. Recent data show that EVs are also natural components of the eye and its fluids. There is evidence indicating the production of EVs in most of the ocular tissues such as cornea, ciliary body, lacrimal gland, retina, and trabecular meshwork (Manukonda et al. 2022), and EVs have been detected from ocular fluids like aqueous humor (Perkumas et al. 2007; Dismuke et al. 2015; Hsiao et al. 2021; Verta et al. 2023) and tears (Han et al. 2021; Shi et al. 2022).

Even though only few studies have reported the presence of EVs in the vitreous humor and even a smaller number of studies have isolated or characterized the vitreal EVs, recent data suggest that EVs are an important component of the vitreous body (Ragusa et al. 2015; Zhao et al. 2018) with potential functional role as regulators of retinal homeostasis. Zhao et al. detected vitreal EVs at concentrations of 1.3, 35, and 9 billion/ml in vitrectomy patients, postmortem donors, and mice, respectively. Their results obtained with proteomic analysis, ultrastructural imaging, and NTA analysis suggest that EVs are abundantly present in the normal vitreous (Zhao et al. 2018).

Understanding the contents and functions of EVs in vitreous humor could provide valuable insights into the pathophysiology of many eye diseases and may lead to the development of new diagnostic and therapeutic strategies.

8.1.2.2 Interactions of EVs with the Molecular Network of the Vitreous

EVs have been isolated and enriched from vitreous humor for further analysis and characterization by serial ultracentrifugation (Ragusa et al. 2015; Zhao et al. 2018). However, these studies do not show the actual localization of EVs in an intact vitreous in situ to detect their interactions with the ECM molecular network. Interestingly, ultrastructural analyses of the vitreous performed decades ago by electron microscopy suggest the existence of EVs in the vitreous. SEM micrographs indicate the interactions of EVs with collagen fibers and TEM micrographs suggest the possible EV secretion by hyalocytes directly to the surrounding matrix of the vitreous (Sebag 1992; Qiao et al. 2005). Now that the existence of EVs is generally accepted, further and more detailed studies would be valuable to confirm their location and interactions with the molecular network of the vitreous by in situ image analyses.

8.1.2.3 Molecular Contents of Vitreal EVs

Vitreal EVs carry typical EV markers, such as ALIX, HSP70, GAPDH, Annexin V (Zhao et al. 2018), tetraspanins CD9, CD63, CD81, and TSG101 (You et al. 2023), and also their size distribution and morphology are consistent with EVs from other body fluids.

Recent advances in proteomics (Murthy et al. 2014; Weber et al. 2023) and other analytical techniques continue to provide more detailed insights into the protein content and the changes in the proteome in retinal disease states (Santos et al. 2023). Proteomic analysis of EVs isolated from the vitreous humor of healthy patients has been performed (Zhao et al. 2018) and proteome of vitreal EVs from patients with proliferative diabetic retinopathy (PDR) shows certain proteins unique to PDR (Wang et al. 2023). These findings are expected to expand the diagnostic and therapeutic value of EVs and their putative function in the pathogenesis of ocular diseases.

Interestingly, the human vitreous humor proteome (Skeie et al. 2015) showed 47% similarity in protein composition as compared to the proteome of human retina (Velez et al. 2018). This suggests a remarkable contribution of retinal cells on the vitreal proteome. Because EVs are abundant in the vitreous, it is plausible that EVs are one of the major contributors to the vitreous proteome.

Recent study by Zhou et al. reported that retinal progenitor cells are the major cell type contributing to EV release from hiPSC-derived 3D human retinal organoids (Zhou et al. 2021). Interestingly, Demais et al. showed the release of EVs from endfeet and microvilli of retinal Müller cells facing the vitreous body in adult mice in vivo (Demais et al. 2022). This is supported by the idea that EVs originate from specialized plasma membrane protrusions such as cilia, microvilli, and filopodia of various cell types (Rilla 2021).

Despite these fascinating findings, the identification of cellular source of vitreal EVs has still challenges. Because vitreous humor is considered as a filtrate of serum, it may contain also EVs originating from the systemic circulation. The vitreal chamber of the eye has highly restricted connectivity with the rest of the body through the natural passages, such as the blood–retinal barrier that regulates the exchange between the blood and the retina. However, there is evidence that EVs are capable of crossing the barriers like blood–brain barrier in zebrafish model (Yang et al. 2015) and blood–retinal barrier (Elliott and He 2021). In addition, aqueous humor is rich in EVs and it is thus possible that aqueous humor EVs could be diffused from the anterior chamber to the vitreous. Thus, we cannot exclude the possibility that a population of vitreal EVs is originating from systemic circulation. Based on these preliminary findings and hypotheses, it is obvious that more comprehensive studies would provide valuable insight into the cellular origin and molecular composition of vitreal EVs.

8.1.2.4 The Functional Role of Vitreal Chamber EVs

In addition to acting as a natural structural element of the vitreous and cellular waste management, EVs may have a regulatory role in the homeostasis of the eye and in the intraocular cellular communication, immune modulation, regeneration, and biomolecular transport. There is evidence that EVs present in the vitreous may act as communicators between vitreous and retinal cells (Zhou et al. 2021). Additionally, exogenously administrated natural, engineered, or drug-loaded EVs may have potential in ocular therapies. The therapeutic potential of EVs depends on their lifespan, the efficiency of their penetration through a highly viscous vitreous body and barriers formed by the cells and the surrounding ECM. The ocular kinetics of EVs and potential factors regulating it will be discussed in more detail below.

8.1.3 Role of EVs in Ocular Diseases

8.1.3.1 Diabetic Retinopathy (DR)

DR is one of the possible microvascular complications of diabetes mellitus. Hyperglycemia for longer periods of time can damage the blood supply of the retina and increased oxidative stress that can lead to vision impairment and even blindness because of apoptosis of retinal cells. (Elkjaer et al. 2020). Hyperglycemia for longer periods has shown to increase circular RNA cPWWP2A expression in pericytes.

This circular RNA acts as a miRNA sponge. In specific, it inhibits miR-579 activity. This is known to increase pericyte cell death when upregulated so the increase of cPWWP2A can be seen as a form of pericyte protection. At the same time, it was shown that there is an increased crosstalk between pericytes and endothelial cells in hyperglycemic conditions. EVs that originated from pericytes showed to be in the endothelial cells and deliver this overexpressed cPWWP2A (Liu et al. 2019). Other EV analyses of DR have shown that they carry mainly cytokines and angiogenic factors. C-c chemokine receptor 5(CCR5) is enriched in these EVs and augments retinal injury (Tokarz et al. 2015, 2019; Mazzeo et al. 2018). Furthermore, Huang et al. demonstrated an increase in complement factor activating IgG as a cargo of plasma-derived EVs of diabetic mice. Their results showed increased vascular damage by these IgG-laden EVs, suggesting that plasma-derived EVs can contribute to the pathogenesis of DR (Huang et al. 2018).

8.1.3.2 Age-Related Macular Degeneration (AMD)

AMD is a progressive eye disease that affects the macula. This can lead to central vision loss in elderly individuals over 60 (GEHRS et al. 2006). The gradual death of retinal pigment epithelium and photoreceptor cells is the result of accumulated deposition of extracellular material, known as drusen (Gu et al. 2021). It has been seen that in AMD patients there is an increase in EVs from RPE cells and choroidal endothelial cells. There is an increase both in number and in size of EVs. The cargo of the vesicles also contained more pro-inflammatory cytokines and complement parts. Thus there may be a link that these EVs contribute to chronic inflammation, oxidative stress, and drusen formation, which are textbook features of AMD (Ebrahimi et al. 2014). EVs from RPE cells that are under stress have been shown to promote autophagy and neovascularization because of the increased amount of VEGFR2 that have been detected in as cargo in these EVs. Interestingly, EVs from healthy RPE cells seem to not have any of these complement factors at all (Liu et al. 2020). More recent findings found that 3 microRNAs (miR-486-5p and miR-626 upregulated, miR-885-5p downregulated) differ inside EVs from patients with AMD compared to healthy patients (Elbay et al. 2019). Biasutto et al. found that some phosphoproteins in ARPE-19 cells that were exposed to oxidative stress were changed. This can suggest that these proteins have a function toward cell apoptosis and metabolism (Biasutto et al. 2013).

8.1.3.3 Glaucoma

Glaucoma is a group name for optic neuropathies that are characterized by elevated intraocular pressure (IOP) by malfunctioning of the trabecular meshwork (TM) and the degeneration of retinal ganglion cells and their axons due to this increased pressure. This degeneration leads to progressive visual field loss (Aires and Santiago 2021). Increasing amount of data shows the presence of modulators of IOP in the aqueous humor. Analysis of EVs from TM cells showed an upregulation of myocilin, which is associated with glaucoma. TM cells that were treated with dexamethasone showed an increase in EV myocilin levels which suggests that EVs do have a functional role in control of IOP (Stamer et al. 2011). Additionally, expression of annexin A2, a heparin sulfate binding protein, was altered (Dismuke et al. 2016). A combination of these plus other factors can cause faults in the clearing of the TM cells and hereby increasing IOP.

Since the main source of increased IOP is the increase of aqueous humor (AH), this biofluid is the main source that is being investigated for biomarkers. Dismuke et al. have characterized the miRNA content of human AH derived from cataract surgeries and found multiple miRNAs including miR-486-5p, miR-184, and miR-204 which have been indirectly linked to regulation of proper outflow facility of the TM (Dismuke et al. 2015).

8.1.3.4 Retinitis Pigmentosa (RP)

RP is an inherited disease but clinical symptoms, onset time, and rate of progression are rather heterogeneous. RP is characterized by degeneration of photoreceptor cells that decrease night vision and visual field (Yi et al. 2020). RP is set with a primary loss of rods that later on lead to a secondary loss of cones. This secondary loss leads then to complete blindness (Chizzolini et al. 2011). Vesicle production in RP increases from the injured photoreceptor cells that influence the transport of rhodopsin, which is necessary for functioning photoreceptor cells (Muraoka et al. 2012). Additionally, they contain more miRNAs and angiogenic substances. Inhibition of these EVs has been shown to effectively increase photoreceptor cell survival (Vidal-Gil et al. 2019).

8.1.4 Kinetics of EVs in the Posterior Segment of the Eye

Therapeutic EVs, with or without drug loading, should be isolated, characterized, and delivered exogenously into the eye, preferably as intravitreal injection. EVs should remain in the vitreous long enough to exert clinically significant duration of action and they should permeate to the retina. Scientific literature includes more pharmacokinetic information about intravitreal liposomes and other nanoparticles as compared to EVs. Herein, we shall compare those two classes of nano-sized structures.

Intravitreally injected liposomes, nanoparticles, and EVs should be distributed in the vitreous by diffusion or convection (Tavakoli et al. 2021). Diffusion is believed to be the dominant mechanism in small rodent eyes, whereas the role of convection is more pronounced in the larger eyes (e.g. human, rabbit). Diffusion studies with porcine vitreous with single particle tracking indicate that vitreous humor has mesh size of $\approx$550 nm and net negative charge due to hyaluronic acid (Xu et al. 2013).

Diffusion of neutral and anionic liposomes and other lipid-based nanoparticles was modestly dependent on size; smaller liposomes (<50 nm) diffusing faster than larger ones (>200 nm) (Tavakoli et al. 2021).

Overall, liposomes and polymeric nanoparticles are eliminated from the vitreous almost completely via anterior chamber. After injection, they are distributed in all directions, often slowly, also toward the retinal surface. However, most of the mass transfer takes place toward the anterior chamber where they are eliminated via chamber angle (Sadeghi et al. 2022). Half-life of 50 nm liposomes in the vitreous humor of rabbits was about 1 week, while retention of intravitreally liposomes and nanoparticles in rat eyes varied from 4 to 65 days (Sadeghi et al. 2022).

Compared to the relatively rich information about nanoparticle kinetics, the distribution and elimination of EVs in the vitreous has been reported in only one publication. Matthew et al. injected fluorescently labeled EVs to the rat vitreous in vivo. Interestingly, the EVs were eliminated rapidly, the mean residence time of EVs in the vitreous was 2.5 days that is comparable to many nanoparticles as nearly 90% of the EVs would be eliminated in 10 days (Mathew et al. 2021). Diameter of EVs is typically 100–200 nm, allowing their distribution in the vitreous. Relatively short residence time may limit usefulness of EVs as intravitreal therapeutics.

Retinal permeation of EVs to rat retina was investigated with flat-mount retinae after intravenous injections in vivo (Mathew et al. 2021). Basically, EVs were distributed from the vitreous only to the inner side of the retina, mostly to ganglion cells and microglia. The peak concentrations were observed in 7 and 14 days for microglia and ganglion cells, respectively, while at 28 days no EVs were visible anymore in the retina. Apparently, distribution of EVs into the retina is limited by inner limiting membrane that was earlier shown to seriously limit permeation of liposomes with mean diameter of $\approx$100 nm (Tavakoli et al. 2020). Therefore, Mathew et al. proposed surface modification of EVs to prolong vitreal residence that would also promote retinal distribution of the EVs, thereby improving efficacy and prolonging the injection intervals (Mathew et al. 2021).

8.1.4.1 Uptake/Transport of EVs Across the Retinal Cells in Cell Culture Models

The uptake and transport of EVs into the retinal cells are crucial for EVs to deliver their beneficial effect at the correct site of action. The regenerative and therapeutic effects of EVs will be discussed later in this chapter, and here the focus is solely on the cellular uptake of EVs.

Receptor-mediated cell uptake is an important mechanism for EV cell uptake in ARPE19 cells (Nicholson et al. 2020, 2021). In particular, integrin and proteoglycan-mediated cell-uptake mechanisms were shown to be crucial for EV uptake. When integrins were blocked with RGD peptide, proteoglycans removed with heparinase, or neuraminidase inhibited with oseltamivir, the cell uptake was blocked. Furthermore, the lipid-based interaction was shown to be an insignificant cell-uptake mechanism. However, Mathew et al. (2019) showed that no statistically significant reduction in the uptake of mesenchymal stem cell (MSC) derived EVs was observed when the integrins of retinal (R28 cells) cells were blocked (Mathew et al. 2019). These contraindicatory results can indicate that the uptake mechanisms are dependent on the receiving cells as well as on the origin of EVs.

Various retinal cells show uptake of EVs. In a study by Mathew et al. Astrocytes, microglia, and retinal neurons demonstrated dose-dependent saturable EV uptake kinetics (Mathew et al. 2021). Müller cells internalize EVs and the cellular uptake is similar to different sizes of EVs despite different effects on Müller cells (Peng et al. 2018).

EVs have an important role in the pathophysiology and they can alter the behavior of the biological barrier in the retina. Interestingly, the H_2O_2 treatment which increases the local oxygen concentration significantly increased the release of EVs on the apical side (Shah et al. 2018). Furthermore, EVs derived from ARPE19 cells under oxidative stress (induced with H_2O_2) showed increased uptake by recipient ARPE19 cells (Nicholson et al. 2020, 2021). The EVs released after H_2O_2 induction reduced the transepithelial resistance (TER) of RPE by 25–35% while the EVs released from the untreated cells showed only a 5–10% change in TER indicating higher uptake of H_2O_2 treated EVs (Shah et al. 2018; Nicholson et al. 2020, 2021). The result was confirmed to be caused by the uptake of EVs because when the phagocytosis of EVs was prevented, no reduction in TER was observed. The biological barrier function is altered during the disease state and the EVs themselves can cause the effect. Furthermore, the uptake of therapeutic EVs can be altered during the pathological state.

8.1.4.2 Crossing Retinal Barriers (BRB and ILM)

Biological barriers are essential for the organs as the barriers protect the organs from physical, chemical, and biological damage such as physical shocks, harmful chemical substances, or microorganisms. In addition, the biological barriers are important factors that facilitate the homeostasis and transfer of information between the organ and the surrounding environment. Different biological barriers with different properties protect the eye. For example, dynamic barrier as tear fluid, physical barrier of tight junctions of cornea epithelium and conjunctiva, hydrogel structure of the vitreous, the tight junctions and structure of retinal pigment epithelium and astrocytes and endothelium surrounding the choroidal capillaries (del Amo et al. 2017; Ramsay et al. 2023).

Despite the importance of the biological barriers, their normal function hinders and challenges drug delivery to the point of interest. Reaching the retina through the anterior chamber is currently challenging due to the extremely low bioavailability of topically administered drugs (del Amo et al. 2017). In addition, systemic drug administration is not an efficient way to treat the retina due to the presence of inner and outer blood–retinal barriers (BRB), i.e. the tight structure of the RPE (outer BRB) and the astrocytes and endothelial cells surrounding the choroidal capillaries (inner BRB). Small and lipophilic compounds permeate RPE readily, while paracellular transportation of hydrophilic compounds decreases with increasing size. This means that tight junctions of RPE prevent the paracellular transportation of larger compounds. Intravitreal injections, for example anti-VEGF drugs to treat age-related macular degeneration, are currently the golden standard of administration to treat retinal disorders. However, if larger particles than proteins, i.e. nanoparticles, were used as intravitreal injections, the inner limiting membrane (ILM) in the vitreoretinal interphase can hinder their retinal drug delivery. The small particles can pass ILM by diffusion, but larger ones are stuck at the membrane. Hence, new approaches to cross the BRBs and ILM to treat retinal disorders are needed.

EVs are a promising option for crossing the biological barriers of an eye. EVs are capable of crossing the blood–brain barrier (BBB), which promotes the idea that they can pass also the BRB (Morad et al. 2019). Exosomes are shown to contain, for example, transferrin, insulin, and LDL which may improve their uptake through the BBB (Elliott and He 2021). Similar surface structures can improve the uptake of exosomes through the BRB through the transporters of the RPE (Hellinen et al. 2019). However, the studies correlating the EV and ocular barrier structures are limited.

After intravitreal injections exosome-adeno-associated viral (Exo-AAVs) have been shown to successfully transduce genes in mice (Wassmer et al. 2017). The vectors efficiently transduced a high number of bipolar cells and some photoreceptors. This indicates that the Exo-AAVs can cross the ILM and penetrate into the retinal layers efficiently. Furthermore, the Exo-AAVs are efficiently internalized by the cells. Exosomes have been shown to target actively choroidal neovascularization. Pollalis et al. (2022) showed that exosomes extracted from mice retina that are engineered with an anchor, spacer, and RGD ligand can penetrate the inner plexiform layer (IPL), inner nuclear layer (INL), outer plexiform layer (OPN), and outer nuclear layer (ONL) already after 1 and 7 days of the intravitreal injection (Pollalis et al. 2022). The exosomes with the RGD ligand were mostly found in ONL and RPE indicating their targeting into the CNV site. Furthermore, some exosomes were observed in the retina after systemic administration which suggests that the exosomes can cross also the BRB. The penetration of RGD-modified exosomes suggests that the integrin-binding improved the exosome uptake (Pollalis et al. 2022). This supports the earlier discussed results about the role of the receptor-mediated EV uptake, especially integrin and proteoglycan-mediated, cell uptake was shown to be an important mechanism of EVs cell uptake by ARPE19 cells (Nicholson et al. 2020).

8.1.5 EVs as Potential Therapeutics for Ocular Applications

8.1.5.1 EVs as Potential Therapeutics for Retinal Regeneration

EVs offer a safer option compared to stem cell therapies as they do not differentiate, or proliferate, and they do not cause similar immune reactions as cells (Kou et al. 2022). The research on EVs for the treatment of retinal disorders has focused mainly on EVs derived from mesenchymal stem cells (MSC), embryonic stem cells, and retinal cells. The research has focused on the reduction of damage to the retina and regeneration of damaged retina (Table 8.1).

MSC-derived EVs are the most studied EVs to treat retinal disorders and their efficacy against retinal ischemic disorders, optic nerve crush, and retinal laser injury has been evaluated. For these studies, MSC-derived EVs have been obtained from mouse adipose MSCs (Yu et al. 2016), human umbilical cord MSCs (Yu et al. 2016; Pan et al. 2019), and human bone marrow MSCs (Mead and Tomarev 2017; Mathew et al. 2019, 2023).

In retinal ischemic disorders the human bone marrow MSC-derived EVs have been shown to reduce cell death of dissociated retinal neurons, suppress retinal microglial activation, and reduce damage caused by hypoxia (Mathew et al. 2019, 2023). Interestingly, the hypoxic preconditioning of MSC EVs enhanced their anti-inflammatory and neuroprotective properties and showed a higher reduction of death of dissociated retinal neurons when compared to EVs produced in normoxia (Mathew et al. 2023). It appears that the EVs guide the cells toward homeostasis. Furthermore, the ischemic retina has been shown to uptake MSC-derived EVs more efficiently than the non-ischemic retina (Mathew et al. 2019). The human bone marrow MSC EVs promote neuroprotection and regeneration of retinal ganglion cells after optic nerve crush (Mead and Tomarev 2017), while human umbilical cord MSC EVs promote RGC survival, but no optic nerve axon regeneration (Pan et al. 2019). Pan et al. (2019) speculated that the differences in the regenerative properties may result from different severity of the damage or different miRNA composition of the EVs. Both mouse adipose MSCs and human umbilical cord MSCs show beneficial effects, such as better ERG response, reduced apoptosis, and fewer lesioned areas, in the mouse retinal injury model (Yu et al. 2016). The results are comparable to the beneficial effects of MSC cell therapies.

MSC EVs are internalized by retinal microvascular endothelial cells, glial cells, retinal ganglion cells, and mixed retinal neurons (Mead and Tomarev 2017; Mathew et al. 2019, 2023). The peak in vivo uptake in RGCs and microglia cells was observed 14 and 7 days after intravitreal injection, respectively (Mathew et al.

Table 8.1 EVs are used for retinal treatment

EV source	Disease/trauma	References
Mesenchymal stem cells	Retinal ischemic disorders	Mathew et al. (2019, 2023)
Bone marrow MSCs	Optic nerve crush	Mead and Tomarev (2017)
Human embryonic stem cells	Retinal degeneration	Ke et al. 2021)
Mouse retinal astrocytes	Laser-induced choroidal neovascularization	Hajrasouliha et al. (2013)
Mouse adipose MSCs and human umbilical cord MSCs	Retinal laser injury	Yu et al. (2016)
Human retinal progenitor cells	RCS rat (inherited retinal degeneration)	Chen et al. (2021)
Human umbilical cord MSCs	Optic nerve crush	Pan et al. (2019)

2021). Ischemic retina has been shown to increase the uptake of MSC EVs (Mathew et al. 2019). Interestingly, in the study of Mathew et al. (2019) MSC EVs were not observed deeper than ganglion cells in the retina, but reduced damage of cells was observed in deeper layers (Mathew et al. 2019). This can indicate that the endocytosed EVs promote other retinal cells to secrete protective substances that reduce the apoptosis of photoreceptors.

Different protective and regenerative actions of MSC EVs in retinal disorders have been observed. The MSC EVs have been shown to suppress injury-induced inflammation by downregulating monocyte chemotactic protein (MCP-1), tumor necrosis factor-alpha (TNF-α), and interleukin-6, and decrease ischemia-induced apoptotic-related gene expression (caspases 3 and 7, bax, bcl2, and p53) (Yu et al. 2016; Mathew et al. 2019, 2023). With the optic nerve crush, the results suggest that MSC EVs promote the regeneration of retinal ganglion cell axons through the delivery of their miRNA rather than their protein content (Mead and Tomarev 2017).

One approach to utilize EVs as therapeutics to regenerate the retina is to establish a method to generate proliferating progenitor cells. Didiano et al. (2020) experimented with zebrafish eyes and screened EVs from various sources that could induce Müller glial cells to dedifferentiate and generate proliferating progenitor cells (Didiano et al. 2020). Interestingly, their results indicated that EVs from cancer cells were also capable of inducing the regenerative response, and even with a higher response than stem cell-derived EVs. The EVs that induced the dedifferentiation and generated proliferating progenitor cells included EVs derived from, for example rat brain glioma, rat primary cortical glia, and mesencephalic dopaminergic neurons. Another approach to dedifferentiate Müller cells into retinal progenitor cells was based on human embryonic stem cell-derived EVs. Ke et al. (2021) aimed to alleviate retinal degeneration with human embryonic stem cell (hESC) derived EVs by inducing dedifferentiation of Müller cells (Ke et al. 2021). The hESC-EVs were shown to promote retinal Müller cell dedifferentiation into retinal progenitor cells in Royal College of Surgeons (RCS) rats. RCS rats show inherited retinal degeneration due to the inability of RPE cells to phagocytose shed photoreceptor outer segments (Strauss et al. 1998). The dedifferentiation of Müller cells into retinal progenitor cells was confirmed in vitro by observing a strong signal of retinal progenitor cell marker CHX10.

Human retinal progenitor cell EVs (hRPC-EVs) have been studied also as such (Chen et al. 2021). hRPC-EVs showed a beneficial effect on retinal degeneration of RCS rats. The hRPC-EVs protected photoreceptors from apoptosis and ONL of the retina after 2 and 4 weeks of intravitreal injection. The hRPC-EVs were observed to suppress the apoptosis factors Bax, and caspase 3, and increase the antiapoptotic Bcl2. In addition, the ocular inflammation was reduced. The results support the idea that exosomes inhibit neuroinflammation by inhibiting the activation of microglia in neurodegenerative diseases.

Hajrasouliha et al. (2013) investigated the effect of retinal astroglial cell (RAC) derived EVs against laser-induced choroidal neovascularization (Hajrasouliha et al. 2013). The EVs were administered with a subtenon injection and observed already after 15 minutes of the injection in the retina. The concentration increased over the

60 minutes of the experiment period. Also, Yu et al. (2016) showed rapid diffusion of EVs into the retina after intravitreal injection (Yu et al. 2016). They observed the MSC EVs in the retina already after 30 min of the injection, and already in 60 min time-point the diffusion of EVs through the neural retina to the RPE was observed. After that gradual spreading of EVs to the outer layers of the retina was observed. The rapid diffusion into the retina may indicate also other mechanisms present, such as convection. Interestingly, the EVs have been observed in the vitreous even after 4 weeks of IVT injection suggesting that the vitreous can act as a reservoir for EVs (Mathew et al. 2019).

RAC derived EVs showed inhibition of vascular leakage after laser-induced CNV (Hajrasouliha et al. 2013). The EVs contained multiple antiangiogenic proteins with elevated levels and inhibited the development of choroidal neovascularization after laser induction. In comparison, RPE EVs did not contain any antiangiogenic factors. Both RPE and RAC EVs were internalized by macrophages and mouse retinal microvascular endothelial cells. The daily doses of RAC EVs decreased the leakage to 0, compared to 42% of PBS injections, and that administration of EVs every second day showed 25% leakage. The leakage area was statistically significantly smaller in the treated groups than in the untreated. Hajrasouliha et al. (2013) concluded that endostatin is responsible for the prevention of the leakage. Their results suggested that endostatin alone is responsible for the suppression of retinal vascular leakage but not a reduction of CNV area.

8.1.5.2 EVs as Ocular Drug Carriers

In addition to EVs as functional therapeutics themselves, they have been studied as ocular drug delivery systems. Although ocular use of EVs as a DDS is quite rare, this option is attractive due to the site-specific tissue-homing capability, as discussed earlier, and generally suitably small particle size. EV surface properties can be modified, similarly to liposomes, by attaching targeting moieties with covalent and non-covalent binding. Furthermore, the lipid bilayer of the EVs may protect the cargo from degradation while retaining the high biocompatibility, especially in case of patient derived EVs. Generally, one of the roadblocks for EV usage as drug carriers are the problems with drug loading. Nevertheless, active pharmaceutical compounds can be loaded into EVs by select methods: e.g. loading high doses of the drug into the EV secreting cells, and loading the "empty" EVs directly by co-culturing with the drug, sonication, freeze-thawing, or electroporation (Elsharkasy et al. 2020). For example, EVs loaded with synthetic drugs, peptides, proteins, or miRNA have been explored thus far (Table 8.2). For EVs as drug delivery vehicles regardless of target tissue, the most common cargos have been nucleic acids (46.5%) and small drug molecules (39.5%) (Escudé Martinez de Castilla et al. 2021).

Drug Loading There are several ways to incorporate drug molecules into EVs, and selection depends on the drug properties and required loading capacity. Reddy et al. (2023) studied bevacizumab-loaded EVs for diabetic retinopathy treatment (Reddy

Table 8.2 Examples of ocular EV therapies loaded with pharmaceutical cargoes

EV source	Cargo	Ant./Post.; in vitro/in vivo	Disease	Effect	References
Mesenchymal stem cell-derived exosome	Ascorbic acid	Ant.; in vitro and in vivo	Dry eye disease	Improved corneal epithelium recovery, reduced inflammation, and tear secretion restoration	Ma et al. (2022a)
Human umbilical cord mesenchymal stem cells	Rapamycin	Ant.; in vitro and in vivo	Corneal injury	Reduction in corneal epithelial defects and stromal opacity	Ma et al. (2022b)
Rabbit adipose derived mesenchymal stem cells	microRNA 24-3p	Ant.; in vitro and in vivo	Corneal injury	Acceleration of corneal epithelial defect healing	Sun et al. (2023)
Umbilical mesenchymal stem cell-derived	N-acetylgalactosamine-6-sulfatase (GALNS)	Ant.; in vitro	Mucopolysaccharidosis IVA	Cargo was taken up by GALNS deficient fibroblasts	Flanagan et al. (2021)
Mesenchymal stem cells	IL-10	Post.; in vitro and in vivo	Experimental autoimmune uveitis	Modulation of T-cells in vitro and in vivo	Li et al. (2022b)
Gingival mesenchymal stem cells	microRNA-21-5p antagomir	Post.; in vitro and in vivo	Retinal ischemia-reperfusion injury	Anti-inflammation and neuroprotection	Yu et al. (2022)
Mesenchymal stem cells	Rapamycin	Post.; in vitro and in vivo	Experimental autoimmune uveitis	Reduced inflammatory cell infiltration	Li et al. (2022a)
Human umbilical cord mesenchymal stem cell	microRNA-126	Post.; in vitro and in vivo	Retinal inflammation	Suppressed HMGB1 pathway and inflammation	Zhang et al. (2019)
Exosome-associated viral vector	Green fluorescent protein	Post.; in vivo	–	Penetration to inner nuclear and outer plexiform layers after IVT	Wassmer et al. (2017)
Bone marrow mesenchymal stromal cells	Bevacizumab	Post.; in vitro and in vivo	Diabetic retinopathy	Inhibition of angiogenesis and suppression of VEGF expression	Reddy et al. (2023)
Rat retinal ganglion cell	PACAP38	Post.; in vitro and in vivo	Traumatic optic neuropathy	Enhanced ganglion cell survival and nerve fiber layer thickness	Wang et al. (2021)
Breast cancer cell lines	Doxorubicin	Post.; in vitro	Retinoblastoma	Cytotoxicity in vitro	Farhat et al. (2022)

et al. 2023). Bevacizumab is a humanized monoclonal antibody with a molecular mass of about 150 kDa. The researchers studied four methods for loading the drug into the EVs: simple incubation in RT, incubation with 0.2% saponin in a shaker, three freeze-thaw cycles, and sonication. The methods resulted in loading capacities of 61–74 μg and encapsulation efficiencies of 30–37%. Reddy et al. concluded that in this case the differences were not significant enough to warrant choosing one method over the other, and thus they used the simplest incubation method in the rest of their studies (Reddy et al. 2023).

Encapsulation of oligonucleotides is commonly performed by transfection of the EV-producing cells. In this method the pharmaceutical compound is already found inside the produced EVs and no additional loading step is necessary. Transfection of the cells can be done with viruses (Li et al. 2022b) or transfection reagents (Zhang et al. 2019; Yu et al. 2022). Alternatively, oligonucleotides can be post-encapsulated to empty EVs with electroporation (Sun et al. 2023). With these methods, encapsulation efficiency is not calculated, but the loading can be confirmed with fluorescent labels (Sun et al. 2023).

The loading of small molecular drugs for ocular therapies has been done by incubation (Ma et al. 2022b) and sonication (Li et al. 2022a) for rapamycin (*M* 914 g/mol). Ascorbic acid (*M* 176 g/mol) has been loaded to EVs by first embedding lipid bound polyethylene glycol (DSPE-PEG-SH) on the EV surface. After that, the researchers covalently bound the ascorbic acid to the end of the PEG chain with a reduction reaction (Ma et al. 2022a). Interestingly, in some cases small molecules have been incorporated to the EVs by incubating the EV-producing cells with the drug before harvesting the EVs. This method was utilized with cancer drug doxorubicin (*M* 544 g/mol) (Farhat et al. 2022), likely because the common pH gradient methods, that are used in loading the drug to liposomes (Mayer et al. 1990), are difficult to use with EVs.

Pharmacokinetics The cellular processes of ocular pharmacokinetics of EVs are covered in detail above. Here we discuss the physical barriers briefly and examples in the context of drug delivery. Cornea and sclera form a strong barrier for the entry of nano-sized and larger particles to the inner tissues of the eye. With the additional elimination routes through vascularized tissues, e.g. the conjunctiva, the penetration of EVs to the retina via topical administration is not possible. Thus, intraocular injections are needed for retinal therapies (Fig. 8.2). After intravitreal injection, lipid structured nanoparticles, like EVs, of suitable size (<200 nm) and charge (non-cationic) can diffuse freely in the vitreous (Tavakoli et al. 2021). Before reaching the neural retina, the injected EVs need to pass through the inner limiting membrane (ILM). The bovine ILM has been shown to block the penetration of liposomes above 100 nm (Tavakoli et al. 2020). However, EVs with a mean size of 140 nm were shown to cross the ILM in mice (Pollalis et al. 2022). This can result from the rat's leakier and more permeable ILM than the bovine ILM.

For the anterior segment drug delivery, the EVs may improve the permeation of the drug to the superficial cells by increasing the residence time on the ocular

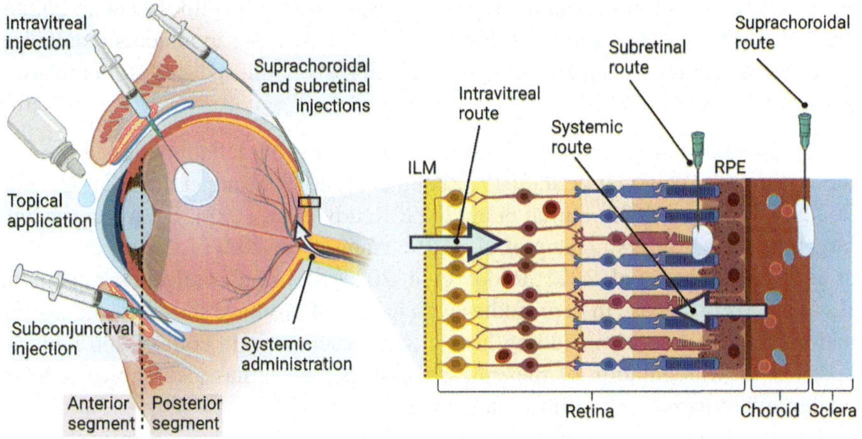

Fig. 8.2 Schematic presentation of ocular drug administration routes in relation to retinal delivery. Created with BioRender.com

surface. For example, the mucin layer in the tear film may be a significant barrier for the corneal absorption causing too high drug washout to the tear ducts. EVs with negative surface charge may improve mucin penetration and residence time on the eye. Nevertheless, topical drug delivery with EVs still usually requires multiple doses per day (Ma et al. 2022a, b). Extension of dosing frequency has been achieved by packing the EVs into hydrogels for better topical retention and subconjunctival injection (Sun et al. 2023). The actual ocular penetration processes of drug-loaded EVs have not been studied, and it is highly unlikely that intact EVs would penetrate the tight epithelial barriers of cornea and conjunctiva.

The most common drug delivery route to the posterior segment of the eye with EVs is intravitreal injection. Several in vivo studies with mice and rats have shown efficient drug delivery to retinal layers with EVs within 24 h post injection (Wassmer et al. 2017; Zhang et al. 2019; Wang et al. 2021; Li et al. 2022b; Reddy et al. 2023). As discussed earlier, species differences should be taken into account when considering the retinal penetration of the relatively large EVs, because of the higher leakiness of the retinal barriers in small rodents compared to the human retina. Likewise, drug delivery to the retina after subconjunctival injection of rapamycin EVs has been reported on mouse model (Li et al. 2022a), but translatability to clinical use remains to be shown.

Therapeutic Effects Ocular drug delivery with EVs has shown several benefits over treatment with plain drug in various diseases. Ma et al. (2022a) studied ascorbic acid surface decorated EVs for dry eye disease. They showed improved tear film breakup time and tear fluid production with ascorbic acid EVs compared to control treatments (saline, EVs without ascorbic acid, and free ascorbic acid). Furthermore, the reactive oxygen species scavenging of the modified EVs increased the corneal thickness and restored stromal layers in BAC model for dry eye disease. Corneal defects have also been treated with EVs encapsulating rapamycin (Ma et al. 2022b) and microRNA 24-3p (Sun et al. 2023). In both cases, the treatments resulted in accelerated corneal healing, less stromal fibrosis, and reduced inflammatory responses.

Several ocular drug-EV formulations aim at combatting inflammation in the posterior segment (Wassmer et al. 2017; Zhang et al. 2019; Wang et al. 2021; Li et al. 2022a, b; Yu et al. 2022; Farhat et al. 2022; Reddy et al. 2023). Examples include larger biological drugs (Zhang et al. 2019; Wang et al. 2021; Li et al. 2022b; Yu et al. 2022; Farhat et al. 2022; Reddy et al. 2023) and small molecular rapamycin (Li et al. 2022a). These in vivo studies with mice and rats have shown lower clinical and histological scores, including less retinal detachment, in comparison with the controls. Downregulation of the inflammatory pathways has also shown reduced ocular inflammatory cell infiltration (Li et al. 2022a, b).

Increased angiogenesis in the retina is a common complication in age-related macular degeneration and diabetic retinopathy, both serious diseases of the ocular posterior segment that can lead to vision loss. Drugs to reduce the amount of angiogenesis stimulating vascular endothelial growth factor (VEGF), like bevacizumab, are the standard treatment to prevent the uncontrolled formation of blood vessels in the retina. Reddy et al. (2023) showed that IVT treatment with bevacizumab encapsulated into EVs resulted in lower levels of retinal exudate and VEGF expression when compared to the free bevacizumab drug (Reddy et al. 2023). Importantly, this effect was seen with longer, 2 months vs. 1 month, than the free drug, indicating a possibility of less frequent injection frequency. This may lessen the burden on the medical system and adverse effect incidences.

Retinoblastoma is the most common ocular pediatric cancer. It can spread from the eye and eventually lead to death, if left untreated. Doxorubicin is a cancer drug that has been traditionally formulated into liposomes for longer circulation times and lower adverse effects. Farhat et al. (2022) studied the potency of doxorubicin in EVs for the treatment of retinoblastoma (Farhat et al. 2022). The EVs were efficiently taken up by Y79 retinoblastoma cell in vitro and anti-cancer effect was found to be significantly higher than free doxorubicin.

Overall, several studies have indicated clear benefits of EVs as ocular drug delivery systems. The field is still relatively new and shows potential, but many critical points of clinical translatability remain to be elucidated.

8.1.5.3 Engineered EVs for Therapies

Various properties of EVs have been modified to improve their characteristics in therapy. Generally, EVs surface is negatively charged and thus is electrostatically repulsed by negatively charged cell surfaces reducing the interaction with the target cell. Midekessa et al. (2020) reported that incorporation of cationic lipids or amphiphilic peptides increases the cellular uptake of EVs (Midekessa et al. 2020).

As EVs and liposomes share many properties starting from the basic lipid bilayer structure, incorporation of select properties from liposomes is an attractive option.

These, so-called hybrid EVs, have been developed and produced by various means including simple co-incubation, sonication, and breaking the nanoparticles to the bare compound level (Antimisiaris et al. 2018). Furthermore, engineering of the EV-producing cells can be used to alter the EV yields (Phan et al. 2018), increase the therapeutic potential (Segura et al. 2005; Lopatina et al. 2014), or improve targeting to specific cell types (Pollalis et al. 2022).

8.1.5.4 Storage Conditions and Freeze-Drying for EV Therapeutics

In order for EVs to become high-volume therapeutics for widespread application, suitable stability in storage conditions is paramount. Generally, deep freeze storage conditions (−80 °C) are recommended for EV storage (Jeyaram and Jay 2017), but the effects of storage conditions vary with the EV species. Higher storage temperatures than −80 °C tend to cause more rapid changes in the EV characteristics, but even at −80 °C unchanged storage duration varies from few days to several years depending on the EV source (Jeyaram and Jay 2017). The consensus is that EVs should be stored at 4 °C for maximum of only 2 days and aliquots should be made to minimize the need for freeze-thaw cycles.

Freeze-drying with cryoprotectants, e.g. trehalose and sucrose, is commonly used to decrease the osmotic damage during freezing and eventual thawing of the EV sample (Elsharkasy et al. 2020). In addition to preserving the physical characteristics of the EVs, freeze-drying also improves the stability of the EV bound proteins. Other factors to improve the stability of freeze-dried EVs are buffers with neutral pH (Cheng et al. 2019) and addition of poloxamer 188 copolymer (Trenkenschuh et al. 2022).

In case of drug-loaded EVs, the loading method may affect the properties of the EVs and furthermore the storage stability. For example, Johnsen et al. (2016) showed that loading EVs with RNA by electroporation caused aggregation (Johnsen et al. 2016). This may potentially have a detrimental effect on the long-term storage stability, as larger aggregates may function as attracting clusters for further aggregation.

8.2 Conclusion and Future Perspectives

EVs are important physiological messengers with specific biological functions that are being revealed in the retina. Thus, the EVs are potentially valuable therapeutics. However, several important considerations must be addressed prior to their use in the clinical setting. Firstly, production of EVs should be scaled up, show high reproducibility, and the final products should be well characterized in physical, chemical, and biological sense. Secondly, the EVs must be stable during shelf-life, convenient to use in the clinics, and show reproducible activity. Thirdly, we must generate better understanding of the delivery properties. Therefore, quantitative methods are important to determine the disposition of EVs to the retinal targets after delivery intravitreally or systemically. Fourthly, it is necessary to understand the pathways that mediate EV delivery across retinal barriers and into the retinal target cells. Finally, surface tailoring of the EVs toward optimized retinal distribution and optimized cargo loading methods are needed to achieve the full potential of EVs as retinal drug delivery systems and therapeutics.

Acknowledgments We acknowledge the financial support by the Academy of Finland to Research Flagship "GeneCellNano."

References

Aires ID, Santiago AR (2021) Microglial exosomes in retinal neuroinflammation: focus in glaucoma. Neural Regen Res 16:1801–1802. https://doi.org/10.4103/1673-5374.306084

Antimisiaris SG, Mourtas S, Marazioti A (2018) Exosomes and exosome-inspired vesicles for targeted drug delivery. Pharmaceutics 10:218. https://doi.org/10.3390/pharmaceutics10040218

Biasutto L, Chiechi A, Couch R et al (2013) Retinal pigment epithelium (RPE) exosomes contain signaling phosphoproteins affected by oxidative stress. Exp Cell Res 319:2113–2123. https://doi.org/10.1016/j.yexcr.2013.05.005

Campochiaro PA (2021) Retinal and choroidal vascular diseases: past, present, and future: the 2021 proctor lecture. Invest Ophthalmol Vis Sci 62:26. https://doi.org/10.1167/iovs.62.14.26

Carapia AK, Martinez-Colin EJ, Segura-Villalobos D et al (2023) Müller glia to Müller glia extracellular vesicle-dependent signaling induces multipotency genes nestin and lin28 expression in response to N-methyl-D-aspartate (NMDA) exposure. ASN Neuro 15:17590914231183272. https://doi.org/10.1177/17590914231183272

Chen M, Ren C, Ren B et al (2021) Human retinal progenitor cells derived small extracellular vesicles delay retinal degeneration: a paradigm for cell-free therapy. Front Pharmacol 12:1–14. https://doi.org/10.3389/fphar.2021.748956

Cheng Y, Zeng Q, Han Q, Xia W (2019) Effect of pH, temperature and freezing-thawing on quantity changes and cellular uptake of exosomes. Protein Cell 10:295–299. https://doi.org/10.1007/s13238-018-0529-4

Chizzolini M, Galan A, Milan E et al (2011) Good epidemiologic practice in retinitis pigmentosa: from phenotyping to biobanking. Curr Genomics 12:260–266. https://doi.org/10.2174/138920211795860071

Country MW (2017) Retinal metabolism: a comparative look at energetics in the retina. Brain Res 1672:50–57. https://doi.org/10.1016/j.brainres.2017.07.025

del Amo EM, Rimpelä AK, Heikkinen E et al (2017) Pharmacokinetic aspects of retinal drug delivery. Prog Retin Eye Res 57:134–185. https://doi.org/10.1016/j.preteyeres.2016.12.001

Demais V, Pohl A, Wunderlich KA et al (2022) Release of VAMP5-positive extracellular vesicles by retinal Müller glia in vivo. J Extracell Vesicles 11:e12254. https://doi.org/10.1002/jev2.12254

Didiano D, Abner JJ, Hinger SA et al (2020) Induction of a proliferative response in the zebrafish retina by injection of extracellular vesicles. Exp Eye Res 200:108254. https://doi.org/10.1016/j.exer.2020.108254

Dismuke WM, Challa P, Navarro I et al (2015) Human aqueous humor exosomes. Exp Eye Res 132:73–77. https://doi.org/10.1016/j.exer.2015.01.019

Dismuke WM, Klingeborn M, Stamer WD (2016) Mechanism of fibronectin binding to human trabecular meshwork exosomes and its modulation by dexamethasone. PLoS One 11:e0165326. https://doi.org/10.1371/journal.pone.0165326

Eastlake K, Lamb WDB, Luis J et al (2021) Prospects for the application of Müller glia and their derivatives in retinal regenerative therapies. Prog Retin Eye Res 85:100970. https://doi.org/10.1016/j.preteyeres.2021.100970

Ebrahimi KB, Fijalkowski N, Cano M, Handa JT (2014) Oxidized low-density-lipoprotein-induced injury in retinal pigment epithelium alters expression of the membrane complement regulatory factors CD46 and CD59 through exosomal and apoptotic bleb release. In: Ash JD, Grimm C, Hollyfield JG et al (eds) Retinal degenerative diseases. Springer, New York, pp 259–265

Elbay A, Ercan Ç, Akbaş F et al (2019) Three new circulating microRNAs may be associated with wet age-related macular degeneration. Scand J Clin Lab Invest 79:388–394. https://doi.org/10.1080/00365513.2019.1637931

Elkjaer AS, Lynge SK, Grauslund J (2020) Evidence and indications for systemic treatment in diabetic retinopathy: a systematic review. Acta Ophthalmol 98:329–336. https://doi.org/10.1111/aos.14377

Elliott RO, He M (2021) Unlocking the power of exosomes for crossing biological barriers in drug delivery. Pharmaceutics 13:1–20. https://doi.org/10.3390/pharmaceutics13010122

Elsharkasy OM, Nordin JZ, Hagey DW et al (2020) Extracellular vesicles as drug delivery systems: why and how? Adv Drug Deliv Rev 159:332–343. https://doi.org/10.1016/j.addr.2020.04.004

Escudé Martinez, de Castilla P, Tong L, Huang C et al (2021) Extracellular vesicles as a drug delivery system: a systematic review of preclinical studies. Adv Drug Deliv Rev 175:113801. https://doi.org/10.1016/j.addr.2021.05.011

Euler T, Haverkamp S, Schubert T, Baden T (2014) Retinal bipolar cells: elementary building blocks of vision. Nat Rev Neurosci 15:507–519. https://doi.org/10.1038/nrn3783

Farhat W, Yeung V, Kahale F et al (2022) Doxorubicin-loaded extracellular vesicles enhance tumor cell death in retinoblastoma. Bioengineering 9:671. https://doi.org/10.3390/bioengineering9110671

Flanagan M, Pathak I, Gan Q et al (2021) Umbilical mesenchymal stem cell-derived extracellular vesicles as enzyme delivery vehicle to treat Morquio A fibroblasts. Stem Cell Res Ther 12:276. https://doi.org/10.1186/s13287-021-02355-0

Gao J, Hussain RM, Weng CY (2020) Voretigene neparvovec in retinal diseases: a review of the current clinical evidence. Clin Ophthalmol 14:3855–3869. https://doi.org/10.2147/OPTH.S231804

Gao Y, Li H, Qin C et al (2022) Embryonic stem cells-derived exosomes enhance retrodifferentiation of retinal Müller cells by delivering BDNF protein to activate Wnt pathway. Immunobiology 227:152211. https://doi.org/10.1016/j.imbio.2022.152211

Gehrs KM, Anderson DH, Johnson LV, Hageman GS (2006) Age-related macular degeneration—emerging pathogenetic and therapeutic concepts. Ann Med 38:450–471. https://doi.org/10.1080/07853890600946724

Gu BJ, Huang X, Avula PK et al (2021) Deficits in monocyte function in age related macular degeneration: a novel systemic change associated with the disease. Front Med (Lausanne) 8:634177. https://doi.org/10.3389/fmed.2021.634177

Hajrasouliha AR, Jiang G, Lu Q et al (2013) Exosomes from retinal astrocytes contain antiangiogenic components that inhibit laser-induced choroidal neovascularization. J Biol Chem 288:28058–28067. https://doi.org/10.1074/jbc.M113.470765

Han J-S, Kim SE, Jin J-Q et al (2021) Tear-derived exosome proteins are increased in patients with thyroid eye disease. Int J Mol Sci 22:1115. https://doi.org/10.3390/ijms22031115

Hellinen L, Sato K, Reinisalo M et al (2019) Quantitative protein expression in the human retinal pigment epithelium: comparison between apical and basolateral plasma membranes with emphasis on transporters. Investig Ophthalmol Vis Sci 60:5022–5034. https://doi.org/10.1167/iovs.19-27328

Hoon M, Okawa H, Della Santina L, Wong ROL (2014) Functional architecture of the retina: development and disease. Prog Retin Eye Res 42:44–84. https://doi.org/10.1016/j.preteyeres.2014.06.003

Hsiao Y-P, Chen C, Lee C-M et al (2021) Differences in the quantity and composition of extracellular vesicles in the aqueous humor of patients with retinal neovascular diseases. Diagnostics 11:1276. https://doi.org/10.3390/diagnostics11071276

Huang C, Fisher KP, Hammer SS et al (2018) Plasma exosomes contribute to microvascular damage in diabetic retinopathy by activating the classical complement pathway. Diabetes 67:1639–1649. https://doi.org/10.2337/db17-1587

Jeyaram A, Jay SM (2017) Preservation and storage stability of extracellular vesicles for therapeutic applications. AAPS J 20:1. https://doi.org/10.1208/s12248-017-0160-y

Johnsen KB, Gudbergsson JM, Skov MN et al (2016) Evaluation of electroporation-induced adverse effects on adipose-derived stem cell exosomes. Cytotechnology 68:2125–2138. https://doi.org/10.1007/s10616-016-9952-7

Ke Y, Fan X, Hao R et al (2021) Human embryonic stem cell-derived extracellular vesicles alleviate retinal degeneration by upregulating Oct4 to promote retinal Müller cell retrodifferentiation via HSP90. Stem Cell Res Ther 12(1):21. https://doi.org/10.1186/s13287-020-02034-6

Khandhadia S, Thulasidharan S, Hoang NTV et al (2023) Real world outcomes of photodynamic therapy for chronic central serous chorioretinopathy. Eye 37:2548–2553. https://doi.org/10.1038/s41433-022-02370-2

Klingeborn M, Dismuke WM, Rickman CB, Stamer WD (2017) Roles of exosomes in the normal and diseased eye. Prog Retin Eye Res 59:158–177. https://doi.org/10.1016/j.preteyeres.2017.04.004

Kou M, Huang L, Yang J et al (2022) Mesenchymal stem cell-derived extracellular vesicles for immunomodulation and regeneration: a next generation therapeutic tool? Cell Death Dis 13:1–16. https://doi.org/10.1038/s41419-022-05034-x

Le Goff MM, Bishop PN (2008) Adult vitreous structure and postnatal changes. Eye (Lond) 22:1214–1222. https://doi.org/10.1038/eye.2008.21

Lewis TR, Phan S, Kim K-Y et al (2022) Microvesicle release from inner segments of healthy photoreceptors is a conserved phenomenon in mammalian species. Dis Model Mech 15:dmm049871. https://doi.org/10.1242/dmm.049871

Li H, Zhang Z, Li Y et al (2022a) Therapeutic effect of Rapamycin-loaded small extracellular vesicles derived from mesenchymal stem cells on experimental autoimmune uveitis. Front Immunol 13

Li Y, Ren X, Zhang Z et al (2022b) Effect of small extracellular vesicles derived from IL-10-overexpressing mesenchymal stem cells on experimental autoimmune uveitis. Stem Cell Res Ther 13:100. https://doi.org/10.1186/s13287-022-02780-9

Liu C, Ge H-M, Liu B-H et al (2019) Targeting pericyte–endothelial cell crosstalk by circular RNA-cPWWP2A inhibition aggravates diabetes-induced microvascular dysfunction. Proc Natl Acad Sci USA 116:7455–7464. https://doi.org/10.1073/pnas.1814874116

Liu J, Jiang F, Jiang Y et al (2020) Roles of exosomes in ocular diseases. Int J Nanomed 15:10519–10538. https://doi.org/10.2147/IJN.S277190

Lopatina T, Bruno S, Tetta C et al (2014) Platelet-derived growth factor regulates the secretion of extracellular vesicles by adipose mesenchymal stem cells and enhances their angiogenic potential. Cell Commun Signal 12:26. https://doi.org/10.1186/1478-811X-12-26

Ma F, Feng J, Liu X, et al (2022a) Ascorbic acid-coupled mesenchymal stem cell-derived exosomes ameliorate dry eye disease

Ma S, Yin J, Hao L et al (2022b) Exosomes from human umbilical cord mesenchymal stem cells treat corneal injury via autophagy activation. Front Bioeng Biotechnol 10

Mahabadi N, Al Khalili Y (2023) Neuroanatomy, retina. In: StatPearls. StatPearls Publishing, Treasure Island, FL

Manai F, Smedowski A, Kaarniranta K et al (2024) Extracellular vesicles in degenerative retinal diseases: a new therapeutic paradigm. J Control Release 365:448–468. https://doi.org/10.1016/j.jconrel.2023.11.035

Manukonda R, Attem J, Yenuganti VR et al (2022) Exosomes in the visual system: new avenues in ocular diseases. Tumour Biol 44:129–152. https://doi.org/10.3233/TUB-211543

Marshak DW, Chuang AZ, Dolino DM et al (2015) Synaptic connections of amacrine cells containing vesicular glutamate transporter 3 in baboon retinas. Vis Neurosci 32:E006. https://doi.org/10.1017/S0952523815000036

Masland RH (2012) The tasks of amacrine cells. Vis Neurosci 29:3–9. https://doi.org/10.1017/S0952523811000344

Mathew B, Ravindran S, Liu X et al (2019) Mesenchymal stem cell-derived extracellular vesicles and retinal ischemia-reperfusion. Biomaterials 197:146–160. https://doi.org/10.1016/j.biomaterials.2019.01.016

Mathew B, Torres LA, Gamboa Acha L et al (2021) Uptake and distribution of administered bone marrow mesenchymal stem cell extracellular vesicles in retina. Cells 10

Mathew B, Acha LG, Torres LA et al (2023) MicroRNA-based engineering of mesenchymal stem cell extracellular vesicles for treatment of retinal ischemic disorders: engineered extracellular vesicles and retinal ischemia. Acta Biomater 158:782–797. https://doi.org/10.1016/j.actbio.2023.01.014

Mayer LD, Tai LCL, Bally MB et al (1990) Characterization of liposomal systems containing doxorubicin entrapped in response to pH gradients. Biochim Biophys Acta 1025:143–151. https://doi.org/10.1016/0005-2736(90)90091-2

Mazzeo A, Beltramo E, Lopatina T et al (2018) Molecular and functional characterization of circulating extracellular vesicles from diabetic patients with and without retinopathy and healthy subjects. Exp Eye Res 176:69–77. https://doi.org/10.1016/j.exer.2018.07.003

Mead B, Tomarev S (2017) Bone marrow-derived mesenchymal stem cells-derived exosomes promote survival of retinal ganglion cells through miRNA-dependent mechanisms. Stem Cells Transl Med 6:1273–1285

Midekessa G, Godakumara K, Ord J et al (2020) Zeta potential of extracellular vesicles: toward understanding the attributes that determine colloidal stability. ACS Omega 5:16701–16710. https://doi.org/10.1021/acsomega.0c01582

Mishra D, Gade S, Glover K et al (2023) Vitreous humor: composition, characteristics and implication on intravitreal drug delivery. Curr Eye Res 48:208–218. https://doi.org/10.1080/02713683.2022.2119254

Morad G, Carman CV, Hagedorn EJ et al (2019) Tumor-derived extracellular vesicles breach the intact blood–brain barrier via transcytosis. ACS Nano 13:13853–13865. https://doi.org/10.1021/acsnano.9b04397

Muraoka Y, Ikeda HO, Nakano N et al (2012) Real-time imaging of rabbit retina with retinal degeneration by using spectral-domain optical coherence tomography. PLoS One 7:e36135. https://doi.org/10.1371/journal.pone.0036135

Murthy KR, Goel R, Subbannayya Y et al (2014) Proteomic analysis of human vitreous humor. Clin Proteomics 11:29. https://doi.org/10.1186/1559-0275-11-29

Nicholson C, Shah N, Ishii M et al (2020) Mechanisms of extracellular vesicle uptake in stressed retinal pigment epithelial cell monolayers. Biochim Biophys Acta Mol Basis Dis 1866:165608. https://doi.org/10.1016/j.bbadis.2019.165608

Nicholson C, Ishii M, Annamalai B et al (2021) J or H mtDNA haplogroups in retinal pigment epithelial cells: effects on cell physiology, cargo in extracellular vesicles, and differential uptake of such vesicles by naïve recipient cells. Biochim Biophys Acta Gen Subj 1865. https://doi.org/10.1016/j.bbagen.2020.129798

Pan D, Chang X, Xu M et al (2019) UMSC-derived exosomes promote retinal ganglion cells survival in a rat model of optic nerve crush. J Chem Neuroanat 96:134–139. https://doi.org/10.1016/j.jchemneu.2019.01.006

Peng Y, Baulier E, Ke Y et al (2018) Human embryonic stem cells extracellular vesicles and their effects on immortalized human retinal Müller cells. PLoS One 13:1–22. https://doi.org/10.1371/journal.pone.0194004

Perkumas KM, Hoffman EA, McKay BS et al (2007) Myocilin-associated exosomes in human ocular samples. Exp Eye Res 84:209–212. https://doi.org/10.1016/j.exer.2006.09.020

Phan J, Kumar P, Hao D et al (2018) Engineering mesenchymal stem cells to improve their exosome efficacy and yield for cell-free therapy. J Extracell Vesicles 7:1522236. https://doi.org/10.1080/20013078.2018.1522236

Pollalis D, Kim D, Nair GK et al (2022) Intraocular RGD-engineered exosomes and active targeting of choroidal neovascularization (CNV). Cells 11

Qiao H, Hisatomi T, Sonoda K-H et al (2005) The characterisation of hyalocytes: the origin, phenotype, and turnover. Br J Ophthalmol 89:513–517. https://doi.org/10.1136/bjo.2004.050658

Ragusa M, Barbagallo C, Statello L et al (2015) miRNA profiling in vitreous humor, vitreal exosomes and serum from uveal melanoma patients: pathological and diagnostic implications. Cancer Biol Ther 16:1387–1396. https://doi.org/10.1080/15384047.2015.1046021

Ramsay E, Lajunen T, Bhattacharya M et al (2023) Selective drug delivery to the retinal cells: biological barriers and avenues. J Control Release 361:1–19. https://doi.org/10.1016/j.jconrel.2023.07.028

Reddy SK, Ballal AR, Shailaja S et al (2023) Small extracellular vesicle-loaded bevacizumab reduces the frequency of intravitreal injection required for diabetic retinopathy. Theranostics 13:2241–2255. https://doi.org/10.7150/thno.78426

Rilla K (2021) Diverse plasma membrane protrusions act as platforms for extracellular vesicle shedding. J Extracell Vesicles 10:e12148. https://doi.org/10.1002/jev2.12148

Röhlich P (1970) The interphotoreceptor matrix: electron microscopic and histochemical observations on the vertebrate retina. Exp Eye Res 10:80–86. https://doi.org/10.1016/S0014-4835(70)80013-6

Sadeghi A, Ruponen M, Puranen J et al (2022) Imaging, quantitation and kinetic modelling of intravitreal nanomaterials. Int J Pharm 621:121800. https://doi.org/10.1016/j.ijpharm.2022.121800

Santos FM, Ciordia S, Mesquita J et al (2023) Proteomics profiling of vitreous humor reveals complement and coagulation components, adhesion factors, and neurodegeneration markers as discriminatory biomarkers of vitreoretinal eye diseases. Front Immunol 14:1107295. https://doi.org/10.3389/fimmu.2023.1107295

Sebag J (1992) Anatomy and pathology of the vitreo-retinal interface. Eye (Lond) 6(Pt 6):541–552. https://doi.org/10.1038/eye.1992.119

Segura E, Nicco C, Lombard B et al (2005) ICAM-1 on exosomes from mature dendritic cells is critical for efficient naive T-cell priming. Blood 106:216–223. https://doi.org/10.1182/blood-2005-01-0220

Shah N, Ishii M, Brandon C et al (2018) Extracellular vesicle-mediated long-range communication in stressed retinal pigment epithelial cell monolayers. Biochim Biophys Acta Mol Basis Dis 1864:2610–2622. https://doi.org/10.1016/j.bbadis.2018.04.016

Shi T-T, Zhao R-X, Xin Z et al (2022) Tear-derived exosomal biomarkers of Graves' ophthalmopathy. Front Immunol 13:1088606. https://doi.org/10.3389/fimmu.2022.1088606

Skeie JM, Roybal CN, Mahajan VB (2015) Proteomic insight into the molecular function of the vitreous. PLoS One 10:e0127567. https://doi.org/10.1371/journal.pone.0127567

Stamer WD, Hoffman EA, Luther JM et al (2011) Protein profile of exosomes from trabecular meshwork cells. J Proteome 74:796–804. https://doi.org/10.1016/j.jprot.2011.02.024

Strauss O (2005) The retinal pigment epithelium in visual function. Physiol Rev 85:845–881. https://doi.org/10.1152/physrev.00021.2004

Strauss O, Stumpff F, Mergler S et al (1998) The Royal College of surgeons rat: an animal model for inherited retinal degeneration with a still unknown genetic defect. Acta Anat 162:101–111. https://doi.org/10.1159/000046474

Sun X, Song W, Teng L et al (2023) MiRNA 24-3p-rich exosomes functionalized DEGMA-modified hyaluronic acid hydrogels for corneal epithelial healing. Bioactive Mater 25:640–656. https://doi.org/10.1016/j.bioactmat.2022.07.011

Tavakoli S, Peynshaert K, Lajunen T et al (2020) Ocular barriers to retinal delivery of intravitreal liposomes: impact of vitreoretinal interface. J Control Release 328:952–961. https://doi.org/10.1016/j.jconrel.2020.10.028

Tavakoli S, Kari OK, Turunen T et al (2021) Diffusion and protein corona formation of lipid-based nanoparticles in the vitreous humor: profiling and pharmacokinetic considerations. Mol Pharm 18:699–713. https://doi.org/10.1021/acs.molpharmaceut.0c00411

Todorova V, Merolla L, Karademir D et al (2022) Retinal layer separation (ReLayS) method enables the molecular analysis of photoreceptor segments and cell bodies, as well as the inner retina. Sci Rep 12:20195. https://doi.org/10.1038/s41598-022-24586-8

Tokarz A, Szuścik I, Kuśnierz-Cabala B et al (2015) Extracellular vesicles participate in the transport of cytokines and angiogenic factors in diabetic patients with ocular complications. Folia Med Cracov 4:35–48

Tokarz A, Żurakowski A, Stępień EŁ et al (2019) Retinopathy severity correlates with RANTES concentrations and CCR5-positive microvesicles in diabetes. Folia Med Crac 59(3):95–112

Trenkenschuh E, Richter M, Heinrich E et al (2022) Enhancing the stabilization potential of lyophilization for extracellular vesicles. Adv Healthc Mater 11:2100538. https://doi.org/10.1002/adhm.202100538

Velez G, Machlab DA, Tang PH et al (2018) Proteomic analysis of the human retina reveals region-specific susceptibilities to metabolic- and oxidative stress-related diseases. PLoS One 13:e0193250. https://doi.org/10.1371/journal.pone.0193250

Verta R, Saccu G, Tanzi A et al (2023) Phenotypic and functional characterization of aqueous humor derived extracellular vesicles. Exp Eye Res 228:109393. https://doi.org/10.1016/j.exer.2023.109393

Vidal-Gil L, Sancho-Pelluz J, Zrenner E et al (2019) Poly ADP ribosylation and extracellular vesicle activity in rod photoreceptor degeneration. Sci Rep 9:3758. https://doi.org/10.1038/s41598-019-40215-3

Wang T, Li Y, Guo M et al (2021) Exosome-mediated delivery of the neuroprotective peptide PACAP38 promotes retinal ganglion cell survival and axon regeneration in rats with traumatic optic neuropathy. Front Cell Dev Biol 9:659783. https://doi.org/10.3389/fcell.2021.659783

Wang J, Wang Z, Zhang Y, Li J (2023) Proteomic analysis of vitreal exosomes in patients with proliferative diabetic retinopathy. Eye (Lond) 37:2061–2068. https://doi.org/10.1038/s41433-022-02286-x

Wassmer SJ, Carvalho LS, György B et al (2017) Exosome-associated AAV2 vector mediates robust gene delivery into the murine retina upon intravitreal injection. Sci Rep 7:45329. https://doi.org/10.1038/srep45329

Weber S, Carruthers N, Gates C et al (2023) Mass spectrometry-based vitreous proteomics: validated methods and analysis pipeline. Methods Mol Biol 2678:157–167. https://doi.org/10.1007/978-1-0716-3255-0_11

Xu Q, Boylan NJ, Suk JS et al (2013) Nanoparticle diffusion in, and microrheology of, the bovine vitreous ex vivo. J Control Release 167:76–84. https://doi.org/10.1016/j.jconrel.2013.01.018

Yang T, Martin P, Fogarty B et al (2015) Exosome delivered anticancer drugs across the blood-brain barrier for brain cancer therapy in Danio Rerio. Pharm Res 32:2003–2014. https://doi.org/10.1007/s11095-014-1593-y

Yang S, Zhou J, Li D (2021) Functions and diseases of the retinal pigment epithelium. Front Pharmacol 12

Yi Z, Ouyang J, Sun W et al (2020) Comparative exome sequencing reveals novel candidate genes for retinitis pigmentosa. EBioMedicine 56:102792. https://doi.org/10.1016/j.ebiom.2020.102792

You Y, Tian Y, Yang Z et al (2023) Intradermally delivered mRNA-encapsulating extracellular vesicles for collagen-replacement therapy. Nat Biomed Eng 7:887–900. https://doi.org/10.1038/s41551-022-00989-w

Yu B, Shao H, Su C et al (2016) Exosomes derived from MSCs ameliorate retinal laser injury partially by inhibition of MCP-1. Sci Rep 6:1–12. https://doi.org/10.1038/srep34562

Yu Z, Wen Y, Jiang N et al (2022) TNF-α stimulation enhances the neuroprotective effects of gingival MSCs derived exosomes in retinal ischemia-reperfusion injury via the MEG3/miR-21a-5p axis. Biomaterials 284:121484. https://doi.org/10.1016/j.biomaterials.2022.121484

Zhang W, Wang Y, Kong Y (2019) Exosomes derived from mesenchymal stem cells modulate miR-126 to ameliorate hyperglycemia-induced retinal inflammation via targeting HMGB1. Invest Ophthalmol Vis Sci 60:294–303. https://doi.org/10.1167/iovs.18-25617

Zhao Y, Weber SR, Lease J et al (2018) Liquid biopsy of vitreous reveals an abundant vesicle population consistent with the size and morphology of exosomes. Transl Vis Sci Technol 7:6. https://doi.org/10.1167/tvst.7.3.6

Zhou J, Flores-Bellver M, Pan J et al (2021) Human retinal organoids release extracellular vesicles that regulate gene expression in target human retinal progenitor cells. Sci Rep 11:21128. https://doi.org/10.1038/s41598-021-00542-w

Chapter 9
AFM-Based Mechanobiology of Extracellular Vesicles

Debashish Paul, Saroj Saroj, Tanya Agrawal, Safoora Zaffar, and Tatini Rakshit

Abstract Extracellular vesicles (EVs) are lipid-bound nano-sized vesicles secreted by all cells. There are emerging pieces of evidence suggesting mechanical properties of single EVs offer important insights into their structure, biomarker roles, and potential therapeutic applications. Herein, we discussed the mechanical aspects of single EV particles using atomic force microscopy (AFM)-based approach. We discussed that AFM is not only an imaging tool of EVs that delivers high-resolution images but the label-free AFM force spectroscopy could also be valuable to study the mechano-responses of single EVs. We propose that the multi-parametric AFM structure-mechanical property analysis has important implications for EV biology research and clinical applications.

9.1 An Introduction to Extracellular Vesicles (EVs)

All cells secrete lipid bilayer-enclosed EVs during their normal or diseased state for cellular communication. EVs can be broadly divided into two categories, microvesicles and exosomes. Microvesicles are lipid bilayer-enclosed large pouches (50–1000 nm) formed via outward budding of the plasma membrane. Exosomes (40–160 nm) are endosome-derived vesicles (Fig. 9.1) (Verweij et al. 2022). This mystifying exosome biogenesis contributes to the diversity of its constituents. This exosome secretion mechanism could be a means of discarding excess cellular proteins to maintain cellular homeostasis, but recent findings suggest that exosomes carry mostly parent cell-specific nucleic acids, lipids, metabolites, cytosolic and membrane proteins, and glycans to either a neighboring or a distant cell which advocates for exosomes' role in cellular crosstalk (Rakshit and Pal 2024). The

D. Paul · S. Saroj · T. Agrawal · S. Zaffar · T. Rakshit (✉)
Department of Chemistry, Shiv Nadar Institution of Eminence, Delhi-NCR, India
e-mail: tatini.rakshit@snu.edu.in

K. Rilla (ed.), *Extracellular Vesicles as Matrix Messengers*, Biology of Extracellular Matrix 15, https://doi.org/10.1007/978-3-031-68811-9_9

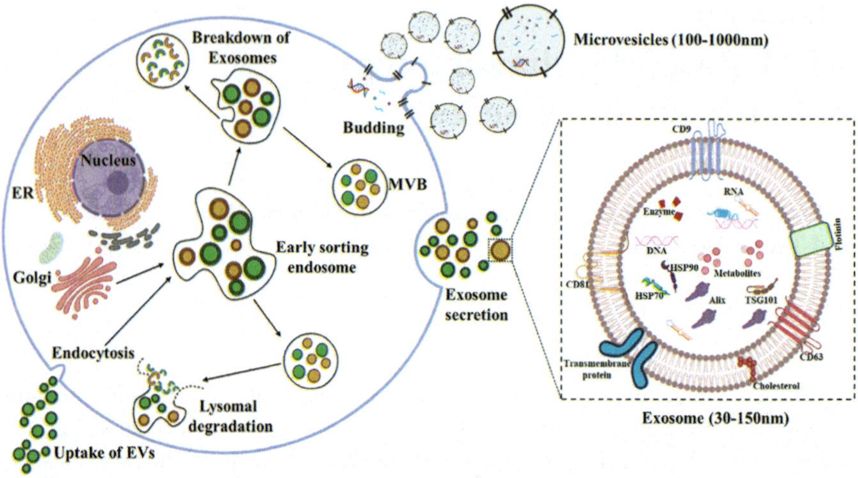

Fig. 9.1 Exosomes are formed via inward budding of the multivesicular body (MVB) membrane and released by means of the collapse of MVBs with the plasma membrane. Microvesicles are generated by outward budding followed by fission of the plasma membrane. A large variety of biomolecules are present in the interior and surface of EVs, for example nucleic acids, proteins, carbohydrates, metabolites, and lipids

packing of cargo in exosomes depends on the parent cell, the cell's metabolic state, and the cell microenvironment (Kalluri and LeBleu 2020; Datta et al. 2021, 2022). Exosomes are abundant in most of our body fluids, including plasma, urine, amniotic fluid, synovial fluid, feces, sweat, breast milk, and saliva. The tetraspanins, syntenin, Alix, etc. are enriched in both exosomes and microvesicles. In this chapter, we will describe both exosomes and microvesicles as "Extracellular Vesicle" or EVs. Interestingly, diseased cells secrete more EVs than normal cells, the physiological urge of diseased cells to shed more EVs is unknown (Akoto and Saini 2021).

9.2 Mechanobiology of EVs in Cancer Microenvironment

In this chapter, we mainly focus on small EVs such as exosomes. The sub-200 nm diameter EV size impedes the use of fluorescence microscopy to image single EVs and demands the use of nanoscale electron microscopy (EM) and atomic force microscopy (AFM). So far, only a handful of studies focused on biophysical dissections of single EVs apart from size, shape, counts, and bio-molecular-composition evaluations. In specific, the analysis of the mechanical properties (stiffness, elasticity, adhesion, etc.) of single EVs that have not been exploited much. This provides label-free critical understandings into the biophysical distinctions among EV subpopulations, and eventually their biological role and biomarker utilities. Within the biological settings, the mechanical aspects of nanoparticles (synthetic or EV) impact their uptake mechanism by the cells, vesicle circulation time, tissue-specific recognition, and the release of molecules. A biological nano-vesicle, such as an EV, in its lifetime, can be exposed similarly to multiple types of mechanical forces, for example, during their transport within the cell and in the extracellular matrix, uptake and secretion by cells, and adhesion of EVs to cell surfaces. These various stresses will distort the EVs, and the extent of distortion will depend on the EV's mechanical characteristics.

9.3 Atomic Force Microscopy (AFM): A Direct Method to Study the Mechanical Properties of EVs

AFM was first introduced in 1986. The elementary components of an AFM comprise a piezoelectric type of scanner, a cantilever containing a sharp probe, a laser, a photodiode sensor, and feedback electronics. AFM is based on a basic principle whereby the movements of a flexible cantilever with an atomically sharp probe are examined. The engagements of the flexible cantilever can be checked by changes in laser deflection from the reflective back-surface of the cantilever. A photodiode detector senses the variations in deflection of the laser. In many commercial AFM systems, the sample gets attached on the piezoelectric scanner, which moves in all 3D-directions by applying voltage to the piezoelectric scanner, while the cantilever stays in a fixed position. In other systems, the piezoelectric scanner is fixed to the AFM cantilever holder where it directly controls the movement of the cantilever while the sample remains fixed at the position. This simple AFM setup offers for both high-resolution imaging and measuring of the molecular interactions of biological samples (Fig. 9.2). AFM is a useful tool for measuring the nano-mechanical properties of EVs, since it offers pico-newton level force sensitivity and subnanometer resolved displacement accuracy. The primary benefits of using AFM are the simultaneous capability to determine topography of an EV sample at nanoscale resolution and measuring its effective stiffness/elasticity/adhesion forces at a precise location. Moreover, AFM's ability to investigate biological systems in real time under physiological buffer solutions and in vitro are major asset of its operation. AFM samples of EVs are prepared generally by the droplet casting on muscovite mica substrate since it offers an atomically flat surface. For different types of applications, gold on mica, ITO, silicon wafer substrates are occasionally used. The sample should be properly attached to the substrate beneath either by electrostatic attractive forces or by covalent bond formation in order to resist the raster scanning by the AFM flexible cantilever. Usually, silicon nitride soft cantilevers (spring constant 0.01 N/m–0.5 N/m) are used for AFM in fluid EV imaging and force spectroscopy experiments in fluid in order to prevent damage to the fragile EV sample. AFM force spectroscopy encompasses the direct measurement of forces between the cantilever tip and the EV sample surface which is a function of the distance between the tip and sample. The x-axis of a force curve denotes the relative vertical distance

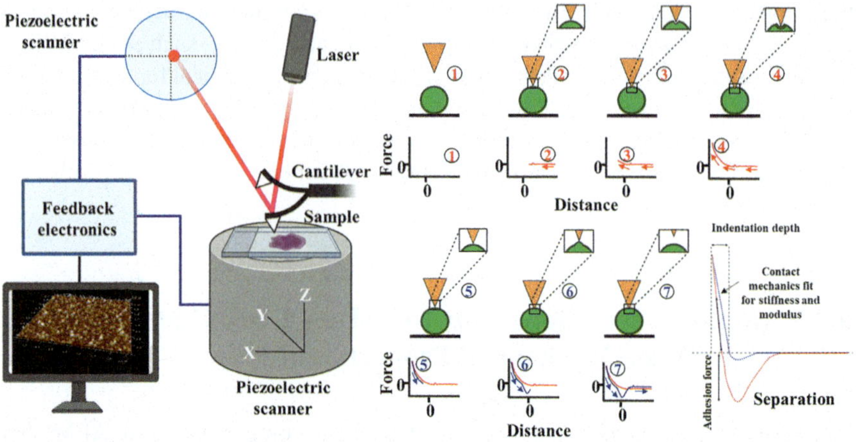

Fig. 9.2 (**Left**) Setup of the AFM components. (**Right**) Schematic representation of an approach (pushing, (2)–(4)) and retract (pulling, (5)–(7)) curve. (1) The tip is approaching toward the EV surface, (2) the tip makes initial contact, (3) the tip begins to push into the EV wall, there is a nonlinear change in force, which indicates the elasticity of the cell surface, (4) the tip pushes further the EV surface and generates a linear change in force, from which the EV stiffness is calculated. After that, (5) the tip begins to retract from the EV surface, (6) adhesive forces allow the tip and the EV surface to remain connected, (7) eventually, the attractive forces between the tip and the sample ceases, the tip and surface get separated which creates a snap-off event in which the force rapidly returns to zero

movement between the sample and the AFM tip, and the vertical axis shows the deflection of the cantilever as the tip is approached toward the sample surface makes physical contact with EV surface and pushes the EV against the hard surface and then goes away from the surface (Fig. 9.2). Various types of mechanical properties of the EV sample surface can be measured from the AFM force-separation curves (approach and retract). The stiffness of the EV sample can be determined from the slope of the force-separation curve considering the contact region. In order to convert stiffness (an extrinsic property) to quantitative Young's modulus (an intrinsic material property), the geometry of the tip and the tip-sample contact should be considered. One of the contact mechanics models (e.g., Hertz, DMT, JKR, Oliver-Pharr models, etc.) can be applied (Butt et al. 2005; LeClaire et al. 2021a). In addition to stiffness and Young's modulus, the adhesion force between an EV and any EV-recognition moiety attached with the AFM cantilever can be measured as the maximum negative force considering the AFM retract curve. The shape of the force-separation curve strongly relies on tip and sample characteristics and tip–sample interaction. Force-Distance (FD)-based AFM can record hundreds/thousands of force curves per topography image of a single EV in the form of multi-parametric (force) maps including adhesion, Young's modulus, etc. which calls for automated procedures for AFM data analysis, mostly available with the commercial AFM software (Rakshit et al. 2020).

9.4 Morphological and Mechanical Characteristics of EVs

EVs are smaller than 200 nm, and therefore, the structural and molecular characterization of these vesicles at the individual level is challenging. In 2010, the AFM investigation by Sharma et al. on human saliva-derived EVs reported high-resolution AFM topography images of single human saliva exosomes with tri-lobed structures (Fig. 9.3a). Using force spectroscopy with antiCD63 IgG functionalized AFM tips, specific detection of antigen CD63 on EV surface was shown by measuring unbinding forces between CD63 positive saliva EVs and CD63 antibody functionalized AFM tip (Fig. 9.3b), they also claimed that at the single-vesicle level, cancer EVs exhibit significantly increased CD63 surface densities compared to normal EVs (Sharma et al. 2010). In a recent report, the same group worked on mechanical dissections of breast cancer and normal breast epithelial cell-derived EV mechanical properties (LeClaire et al. 2021b) using AFM nanoindentation. The EVs derived from MCF10A (normal), MCF-7 (non-invasive), and MDA-MB-231 (invasive) cells were imaged on (3-aminopropyl)-triethoxysilane-functionalized (AP-functionalized) to locate single EVs prior to mechanical analysis (Fig. 9.3c–e). Subsequently, force-indentation curves were used to derive Young's modulus of single EVs. The results revealed that under identical forces of indentation, MDA-MB-231 EVs were ~40% more mechanically compliant than MCF-7 EVs (0.61 ± 0.03 MPa and 0.85 ± 0.07 MPa, respectively). MCF10A EVs had the highest elastic modulus (0.97 ± 0.07 MPa). The Young's modulus (E, MPa) was calculated by converting the force curves into force-indentation curves and fitting them with the Hertz–Sneddon model using the full length of the force-indentation curve from the point of contact to the maximum force set point. From the multi-parametric AFM force maps, they also reported higher adhesion forces for invasive EVs than non-invasive breast cancer EVs. In 2015, Whitehead et al. reported that the malignant (metastatic (FL3) and non-metastatic (T24)) bladder cancer cell lines derived EVs displayed a markedly reduced stiffness (95 MPa and 280 MPa) and adhesion but an increased complement activation compared to non-malignant cell line (HCV-29) EVs (1527 MPa). They performed quantitative nano-mechanical mapping atomic force microscopy (QNM AFM) to determine size and nano-mechanical properties of EVs at the single particle level (Fig. 9.4) (Whitehead et al. 2015).

9.5 Mechanical Behavior of EVs Through AFM Force Curves

AFM-based EV mechanical properties (stiffness, Young's modulus, etc.) are evaluated from the approach force curves. In a recent report, Yang et al. depicted beautifully how AFM-nano-indentation revealed the individual nano-elastic behavior of the EVs (Ye et al. 2021). They performed sequential nano-indentations at the

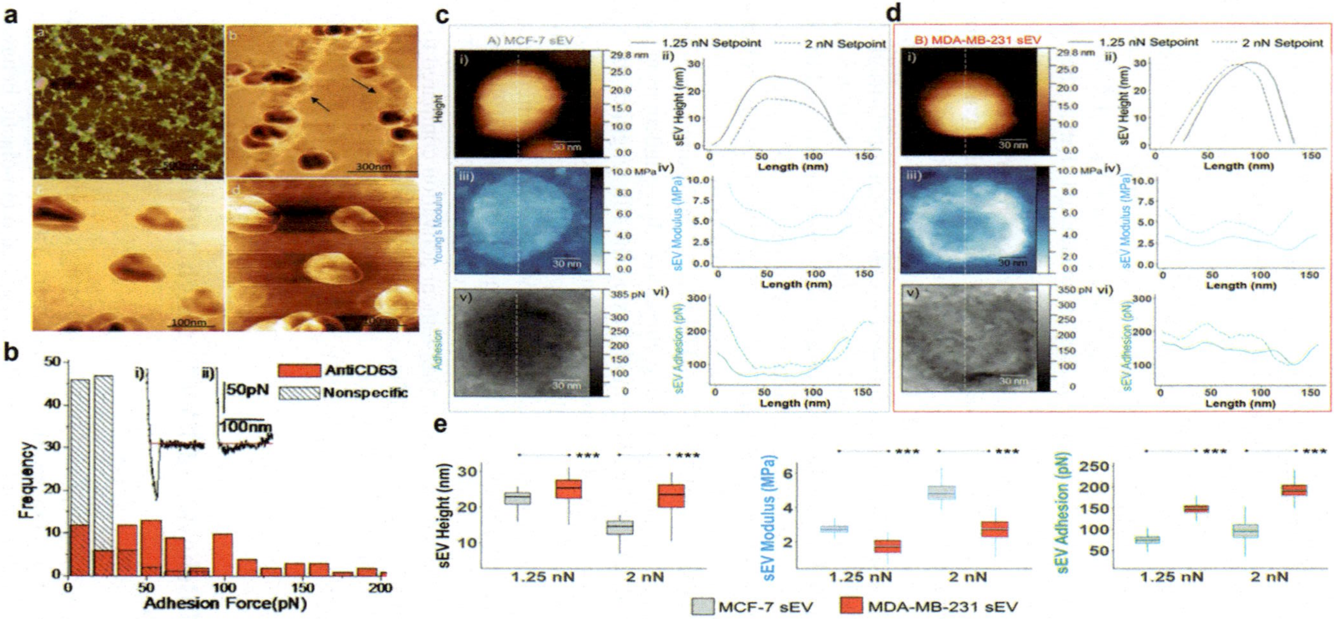

Fig. 9.3 (**a**) Tapping mode AFM showing the ultrastructure of individual saliva EVs. (**b**) Strong adhesive event between CD63 antibody functioned AFM tip and EV modified surface was measured, but no adhesion event for nonspecific antibody functionalized tip and EV was observed. The figure with caption is reprinted (adapted) with permission (Sharma et al. 2010). Copyright © 2010 American Chemical Society. (**c**, **d**) AFM nano-mechanical mapping of single EVs. Representative maps of individual EVs derived from two breast cancer cell lines and corresponding cross-sectional profiles, marked height, Young's modulus and surface adhesion force are shown. (**e**) Comparison of nano-mechanical properties for breast cancer cell EVs, box plots show differences in median ± std values for height, Young's modulus, and surface adhesion forces at 1.25 and 2 nN applied forces ($p < 0.001$, represented by ***). The figure with caption is reprinted (adapted) with permission (LeClaire et al. 2021b). Copyright © 2021 American Chemical Society

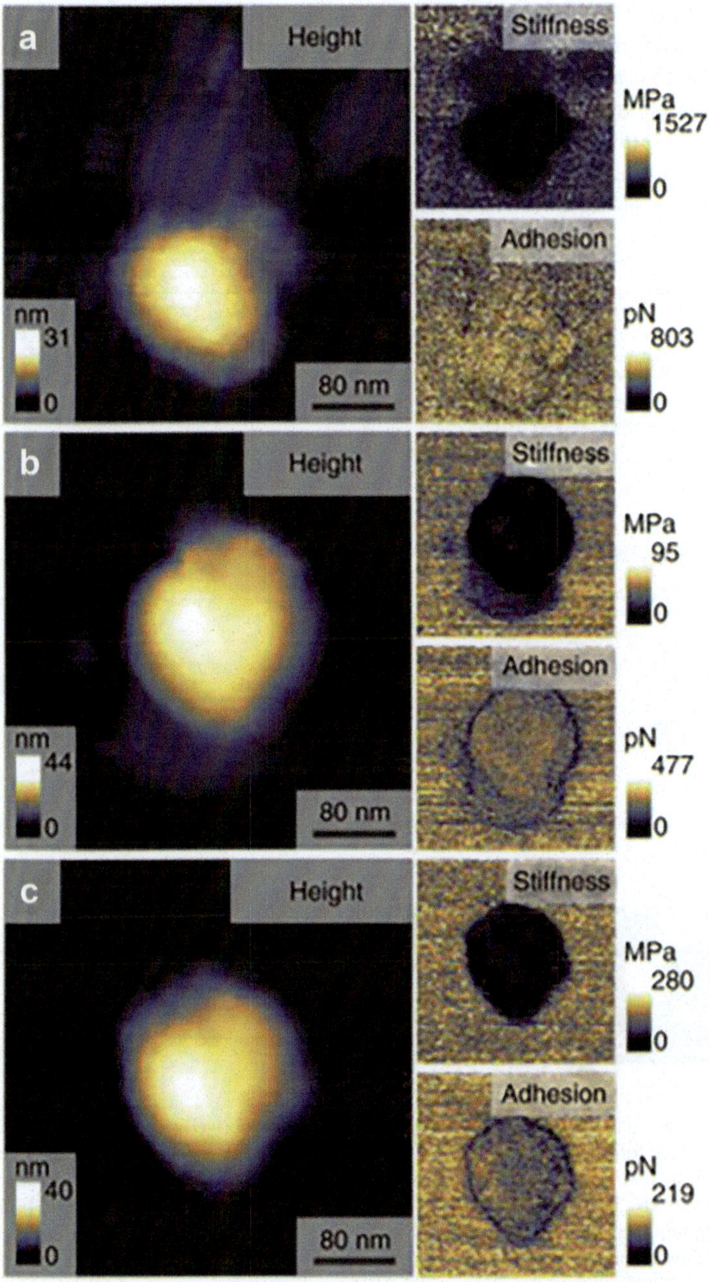

Fig. 9.4 (**a**) Atomic force microscopy height images and nano-mechanical maps showing stiffness and adhesion profile of EVs derived from non-malignant HCV-29, (**b**) malignant non-metastatic T24, and (**c**) malignant metastatic FL3 cell-derived EVs. Analysis was performed on $n = 50$ vesicles per cell line. The figure with caption is reprinted (adapted) from Whitehead et al. (2015) © 2015 Bradley Whitehead et al.

center point of the EVs with the force increased from 0.5 to 3 nN at 0.5 nN intervals at a loading rate of 250 nm s^{-1}. When the applying force was low (0.5 and 1.0 nN in the first and second indentations), they observed complete overlap between the approach and retract curves, indicating that the initial mechanical response of the EV was fully elastic. As they increased the force (1.5, 2.0, and 2.5 nN in the third, fourth, and fifth indentations, Fig. 9.5), after a linear stage indicating the deformation of the EV, they observed a plateau in the approach curve, indicating

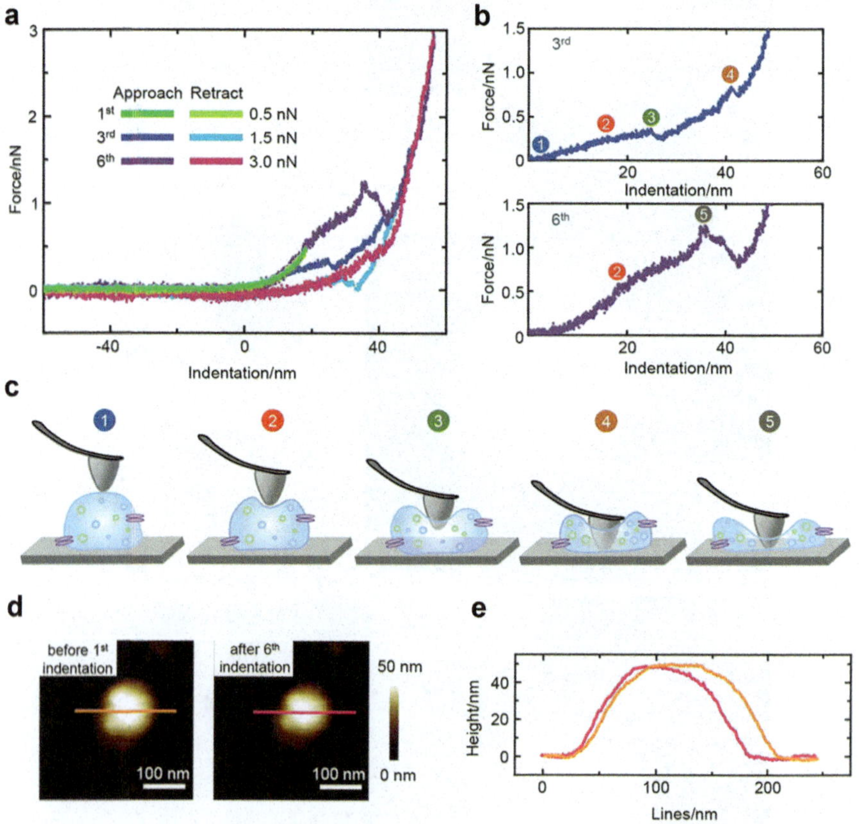

Fig. 9.5 The nano-mechanical behavior of the EVs derived from MCF-10A cells. (**a**) Representative force- indentation curves showing the sequential indentations on the same EV with increasing forces. The dark colors represent the approach curve and the light colors represent the retract curve. (**b**) The enlarged approach curves of the third (1.5 nN) and sixth (3.0 nN) indentations. The numberings indicate the nano-mechanical behavior of the EV at different stages. (**c**) Scheme illustration of the nano-mechanical response of the EV against tip indentation from steps 1–5, 1, nano-indentation starts; 2, inward tether force forms; 3, tip penetrates the upper membrane of the EV; 4, tip penetrates the lower membrane of the EV; 5, the upper and lower membranes of the EV are pushed together. (**d**) AFM topographic images and (**e**) the corresponding height profiles of the same EV before the first indentation (orange) and after the sixth indentation (pink). The figure with caption is reprinted (adapted) from Ye et al. (2021) *Adv. Sci.* 2021, 8, 2100825. © 2021 Advanced Science published by Wiley-VCH GmbH

decreased force response. Two breaks were observed after the plateau in the approach curve, suggesting that the tip penetrated the upper and lower lipid bilayer membranes of the adsorbed EV. When the force further increased (3.0 nN in the sixth indentation), only one break was observed in the approach curve, which suggested the upper and lower membranes were pushed together and penetrated by the tip. However, after sequential indentations, they checked the height and shape of the EVs, the values remained almost unchanged (Fig. 9.5). This suggested that after transient AFM tip penetration on the EV, the lipid membranes of the EV were self-repaired. Therefore, the integrity and elasticity of the EV were unaltered. This is due to the fluidity and dynamic organization of the lipids in the vesicle membrane (Ye et al. 2021).

9.6 Mechanobiology of Hyaluronan-Coated EVs

In our group we performed single-molecule level investigations on colon cancer cell-derived EVs using AFM to identify a critical colon cancer biomarker, hyaluronan (HA), a carbohydrate polymer (Paul et al. 2020, 2022). Important findings from Rilla group suggested previously that HA-coated EVs communicate crucial signals for malicious tumor development and metastasis in different types of cancers (Rilla et al. 2014; Mustonen et al. 2021).

We visualized the EVs using tapping mode AFM in the buffer. AFM topography and amplitude images revealed distinct multivesicular bodies (MVBs) with EVs trapped inside (Fig. 9.6a, b). EV surface is adorned with many tetraspanin protein receptor molecules. Using AFM unbinding force spectroscopy measurements, we probed the interactions between AFM tips functionalized with the CD9 (tetraspanin) antibody. To improve the reproducibility, we performed the unbinding experiments on the EV film prepared on mica using spin coating (Fig. 9.6c). As a control, BSA-coated mica surface was used in place of EV film and the same experiment was repeated with a bare tip. Specific interactions between antigen and antibody were found to be in the range of 60–200 pN for both the normal and cancer EVs, which was consistent with the previous reports. Retraction AFM force traces for both EV samples showed clear unbinding events (Fig. 9.6d). The rupture/unbinding force histogram with the relative frequency of its measurement was plotted and then fitted with a Gaussian function. This was done to determine the highest rupture force that is the measure of the strength of the bond between the CD9 protein (on the EV surface) and the CD9 antibody (on the tip). The unbinding forces for both cancer and normal EV samples were measured as 128 ± 12 and 109 ± 11 pN. The frequencies of unbinding events for both EVs samples are 50 and 35%, respectively, for 3 independent experiments, indicating that the observed unbinding forces are more frequently detected for cancer EVs. It also suggested that the cancer EV surface is having a substantial density of CD9 molecules compared to normal EVs. We also observed multiple rupture events in the unbinding force traces frequently for the cancer EV sample, which suggests the likelihood of having multiple tetraspanin

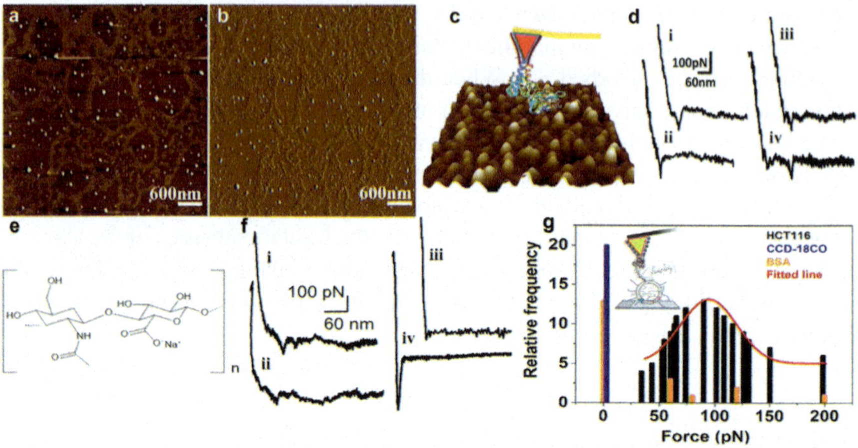

Fig. 9.6 (**a**, **b**) Release of EVs from a cluster of multivesicular bodies (MVBs) from cancer cells, (**c**) Scheme showing AFM tip functionalized with CD9 antibody approaching towards EV film on mica. Typical force (pN) curves as a function of separation (nm) for a single pulling experiment with an adhesive event between cancer (**d** (i)) and normal EVs (**d** (ii)) probed with an anti-CD9-functionalized tip (**d** (iii, iv)). (**e**) Structure of hyaluronan (HA). Cancer EVs probed with an HABP-functionalized tip (**f** (i, ii)) showing force (pN) as a function of separation (nm) for a single pulling experiment with clear adhesion events. No adhesion peaks were found for normal EVs for ~90% of the interactions (**f** (iii)), and nonspecific interactions were observed for BSA-HABP experiments a couple of times with an adhesion force of >400 pN (**f** (iv)). (**g**) Schematic representation of the experimental setup for unbinding experiments (inset). Histogram of rupture events showing that HABP-induced forces were distributed in the range of 30−150 pN for cancer EVs. The histogram is fitted with a Gaussian function. The figure with caption is reprinted (adapted) from with permission (Paul et al. 2020). Copyright © 2020 American Chemical Society

CD9 proteins present in close vicinity. For the control experiment with BSA protein film underneath, the rupture forces were nonspecific, and therefore the range of forces was either <50 pN or >300 pN. These experiments showcased AFM's capability in identifying tetraspanin protein of single EV particles validating its extracellular origin (Paul et al. 2020).

Next, we wanted to identify HA molecules (a negatively charged carbohydrate molecule, a nonsulfated glycosaminoglycan) (Fig. 9.6e) on the cancer EV outer surface using similar strategy with AFM force spectroscopy. We chemically modified the AFM tip (covalent bonds) with HA binding protein (HABP) molecules involving APTES-glutaraldehyde chemistry. We investigated both the cancer and normal (as control) EVs by preparing EV spin-coated films on the underlying mica surface. BSA film on mica was used as a control. Typical, unbinding events between HA on EV surface and HABP on AFM tip for the cancer EV sample are depicted in Fig. 9.6f (i, ii). Specific adhesion peaks in the range of 40–200 pN for the cancer EVs were measured and the mean rupture force was calculated from the histogram fitting with Gaussian function (82 ± 10 pN). This force indicated the bond strength between HA on EV surface and HABP molecules hanging from the AFM tip. This value was a little on the higher side of the values of unbinding forces reported previously. We supposed, during unbinding, that a few HA and HABP molecules were participating in a single event captured by AFM force trace. For normal EVs most of the adhesion forces were <30 pN (Fig. 9.6f iii) and 40–200 pN range was mostly not observed. For control BSA film, the adhesion forces were observed in the range of either >300 pN (Fig. 9.6f iv) or <30 pN. In case of cancer EV film, multiple interaction peaks were observed frequently with the HABP-functionalized tip (Fig. 9.6f). This clearly suggested that HA density on cancer EV surface is substantially high and they are in close proximity. The frequency of unbinding events for cancer EV samples was >75%, which was quite high and clearly pointed out that observed unbinding forces are specific for HA-HABP interactions ($p < 0.001$), whereas in the case of control BSA and normal EV samples the rupture events in this force regime were rare (Paul et al. 2020).

Next, we devised a scheme to measure the contour lengths of a single HA molecule on a cancer EV surface by force spectroscopy pulling strategy exploiting the interactions (non-covalent) between HA binding protein (HABP) and HA. Here, we proposed that HA chains on the EV surface were pulled by the HABP molecules attached on the tip (Fig. 9.7) which could be identified from the force-separation profiles, those curves were fitted with worm-like chain (WLC) model where

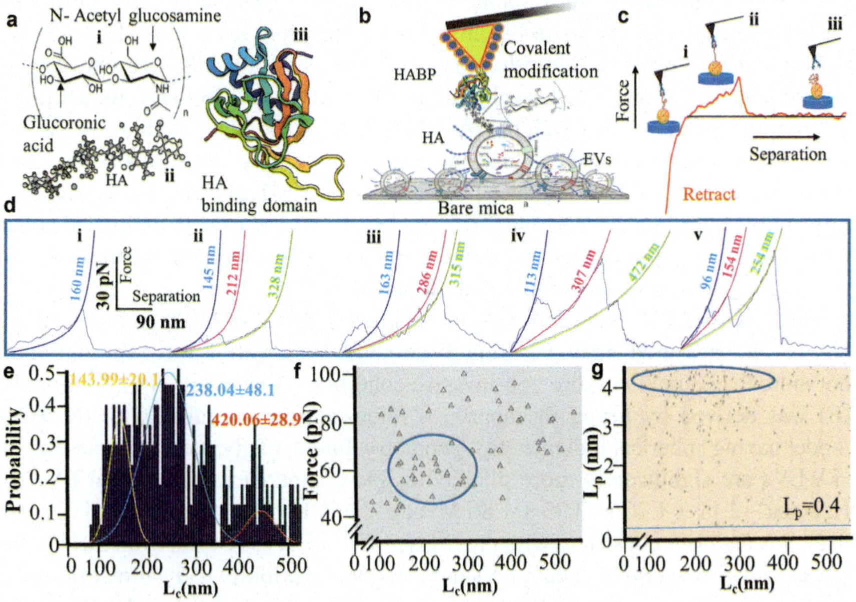

Fig. 9.7 (**a**) Structure of HA, PDB ID: 1HYA, crystal structure of HA binding domain (HABP), PDB ID: 5N86. (**b**) Model AFM pulling experiment, (**c**) A typical retract curve showing the HA-HABP interaction, (**d**) Sample retract curves for single (i) and multiple (ii)–(v) rupture events, WLC fitting. (**e**) Multi-Gaussian fit (blue, red, and yellow lines) predicts the mean ± SD. L_c values of HA. (**f**) Unbinding force vs L_c values. (**g**) L_p values of individual L_c measured. The figure with caption is reprinted (adapted) with permission (Paul et al. 2022). Copyright © 2022 American Chemical Society

persistence length (L_p) and the contour length (L_c) of HA, both were adjustable parameters using Fodis software. For most of the force curves, the L_p agreed well with the values previously reported for HA (4.1–4.4 nm) (Bano et al. 2016, 2018) (Fig. 9.7). In these force curve data one or a few HA molecule(s) were under stretching conditions by the HABP molecules on the tip, followed by a rupture of force between HA and HABP. This data actually provided the combined length of binding partners HA and HABP between the tip and substrate, that we termed as "contour length" (L_c) (Fig. 9.7c, d). The frequency of unbinding events was calculated to be quite high (~70%), which pointed out that the interactions we measured were specific between HA and HABP. Most of the unbinding events showed multiple interactions (Fig. 9.7d (ii)–(v)); and a few (only 5%) represented a single unbinding event (Fig. 9.7d (i)), the nonspecific interaction curves were minimal and excluded. We assumed that each tethering end (HA and HABP) was placed face to face with each other during unbinding (Fig. 9.7b). In practice, the interpretation of contour length (L_c) is complex. It would depend on mainly three factors. (1) Abundance of HA on the HA-EV surface and HABP molecules on AFM tip surface, (2) HABP relative location on the AFM tip when interacting with the EVs on mica, and (3) HABP binding site on HA homopolymer chain. All these factors would result in apparently different L_c values. The most probable L_c for HA was measured to be 238.0 ± 48.1 nm (Fig. 9.7e); L_c is the stretched polymer length not considering the bond angles, contrasting chain length, which is therefore < L_c. Now, L_c = 238 nm corresponded to ~200–225 nm long HA polymer with molecular weight range of 85–90 kDa, (L_c of one HA disaccharide is 1 nm, MW is~0.38 kDa). As per previous reports, low molecular weight HA (LMW-HA) (<200 kDa) facilitates cancer to grow and metastasize, whereas high molecular weight HA (200–2000 kDa) exists in normal conditions. HA degrades by hyaluronidases (Hyals) and forms small HA fragments (Petrey and de la Motte 2014; Paul et al. 2022) during pathological conditions. Therefore, this contour length measurement assay could be useful in sensing HA on EV surfaces in cancer microenvironments.

Next, with AFM nano-indentation (applied force 200–300pN) on a single HA-EV we measured single EV Young's modulus, indentation was kept minimum, not more than 3 nm to ensure non-invasive conditions (Fig. 9.8a). Young's modulus (E) was derived by fitting the approach force-indentation curve with the Hertz model having spherical indenter (tip) geometry. Interestingly, we found out that the HA-EVs are significantly more pliable/flexible compared to the normal EVs (no HA coat) (2.11 ± 1.70 vs 3.33 ± 1.80 MPa, $p < 0.001$) (Paul et al. 2022) (Fig. 9.8b, d, e), this data was consistent with previous reports on EVs (Ouasti et al. 2011). This intrinsic elastic nature of cancer EVs indicated that flexibility had a direct influence on EV's biological performance and biomarker functions in disease. We also investigated the contributing factors that could be facilitating this inherent resilient nature of HA-EVs in cancer. We found out that Hyals (HA degrading enzyme) showed an appreciable increase in elasticity of EVs from 2.11 ± 1.70 MPa (before Hyals treatment) to 4.24 ± 1.90 MPa ($p < 0.001$) (after Hyals treatment) (Fig. 9.8c, d, f) (Paul et al. 2022). These observations proved that accumulation of LMW-HA had direct correlation with the intrinsic flexibility of HA-EVs. It is known that HA is

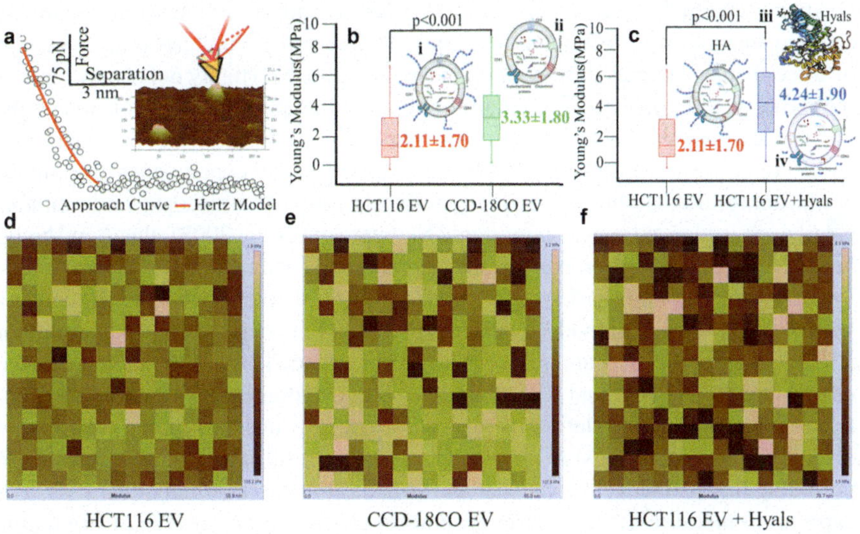

Fig. 9.8 (**a**) AFM nano-indentation on a single EV particle fitted with Hertz model with spherical indenter geometry. (**b**) Measured Young's modulus (**e**) values for cancer (HCT 116) and normal (CCD-18CO) EVs. (**c**) E values before and after Hyals treatment on cancer EVs (iv). (**d–f**) The elasticity maps on single EV particles for HCT116, CCD-18CO, and Hyals treated HCT116 EVs. The figure with caption is reprinted (adapted) with permission (Paul et al. 2022). Copyright © 2022 American Chemical Society

hygroscopic (Ouasti et al. 2011), and HA presence within tissues causes swelling (Nagy et al. 2018). We proposed that the hydroscopic small chain HA employed and ordered the water molecules on EV surface and surroundings, thus HA with water network acted as a cushion on the EV's outer surface. This could be strengthened with shorter and labile HA fragments (LMW-HA) due to less viscosity. The HA receptor proteins and proteoglycans could also be involved (Paul et al. 2022).

9.7 Conclusions and Future Perspectives

AFM mechanobiology has great potential in elucidating EV biology and dynamics at the single particle level. However, we should consider few points carefully while dealing with mechanobiology experiments with AFM. Given the small size of the EVs (less than 200 nm), substrate effects may complicate the measurements of the mechanical properties of EVs. Therefore, while performing AFM nano-indentation, the indentation depth more than 10% of the vesicle height should be avoided, because it can lead to overestimation of Young's modulus values due to the stiff underlying substrate contribution (Schillers et al. 2017). Furthermore, the choice of imaging substrates has shown to affect EV deformation on the surface (Vorselen et al. 2020). EVs usually carry a negative surface charge and get deformed more on positively-charged surfaces such as poly-l-lysine and deform less on negatively charged surfaces such as unmodified mica. This feature complicates the collection of force-distance curves during imaging (Ito et al. 2018). Highly deformed vesicles will be very soft and therefore, determining the initial point of contact between the EV surface and the AFM tip becomes challenging (Ito et al. 2018). Therefore, choice of substrate is critical in AFM force spectroscopy experiments. In order to get meaningful data with AFM single-molecule force spectroscopy, calibration of cantilevers is also extremely important (Neuman and Nagy 2008), this step should be carefully performed on the chosen bare substrate prior to the EV experiments. We also need efforts in order to standardize the AFM force spectroscopy methods with detailed written and video protocols globally to increase the reproducibility of EV mechanobiology research. In the future, better knowledge of the biomechanical characteristics of individual EVs should be able to address many outstanding questions regarding the correlation of physical characteristics and biological roles of diverse classes of EVs. AFM has shown to be a useful nanoscale instrument for unfolding the distinctive mechanical and structural features of individual EVs. The capacity to measure the label-free mechanical characteristics of EVs in circulation and analyze them quantitatively in bio fluids including blood, urine, and saliva could have remarkable impact for the discovery of orthogonal disease biomarkers (Woo et al. 2016). For targeted administration in clinical applications, a detailed characterization of EV mechanical properties could enhance drug delivery applications since it is now well-reported in the literature that nanoparticle elasticity has profound effect on its cellular uptake efficiency (Hui et al. 2020). Additionally, in order to realize the potential AFM mechanobiology in EV diagnostics and therapeutics, we should aim to combine machine learning techniques with AFM-EV investigations to generate multi-parametric (height-adhesion-Young's modulus) fingerprints of diverse classes of EVs that will guide future studies and reveal critical patterns and unique characteristics in EV structures.

Acknowledgments DP thanks SERB-CRG for JRF. SS, TA, and SD thank Shiv Nadar Institution of Eminence for the fellowship. TR thanks SERB (CRG/2019/007013), DST/INSPIRE/04/2017/002375, and Shiv Nadar Institution of Eminence, Delhi-NCR for generous funding and research facilities.

References

Akoto T, Saini S (2021) Role of exosomes in prostate cancer metastasis. Int J Mol Sci 22. https://doi.org/10.3390/ijms22073528

Bano F, Banerji S, Howarth M et al (2016) A single molecule assay to probe monovalent and multivalent bonds between hyaluronan and its key leukocyte receptor CD44 under force. Sci Rep 6:1–14. https://doi.org/10.1038/srep34176

Bano F, Tammi MI, Kang DW et al (2018) Single-molecule unbinding forces between the polysaccharide hyaluronan and its binding proteins. Biophys J 114:2910–2922. https://doi.org/10.1016/j.bpj.2018.05.014

Butt HJ, Cappella B, Kappl M (2005) Force measurements with the atomic force microscope: technique, interpretation and applications. Surf Sci Rep 59:1–152. https://doi.org/10.1016/j.surfrep.2005.08.003

Datta B, Paul D, Pal U, Rakshit T (2021) Intriguing biomedical applications of synthetic and natural cell-derived vesicles: a comparative overview. ACS Appl Bio Mater. https://doi.org/10.1021/acsabm.0c01480

Datta B, Paul D, Dey T et al (2022) Importance of extracellular vesicle derived RNAs as critical colorectal cancer biomarkers. ACS Bio Med Chem Au 2:222–235. https://doi.org/10.1021/acsbiomedchemau.1c00043

Hui Y, Yi X, Wibowo D et al (2020) Nanoparticle elasticity regulates phagocytosis and cancer cell uptake. Sci Adv 6:1–10. https://doi.org/10.1126/sciadv.aaz4316

Ito K, Ogawa Y, Yokota K et al (2018) Host cell prediction of exosomes using morphological features on solid surfaces analyzed by machine learning. J Phys Chem B 122:6224–6235. https://doi.org/10.1021/acs.jpcb.8b01646

Kalluri R, LeBleu VS (2020) The biology, function, and biomedical applications of exosomes. Science 367. https://doi.org/10.1126/science.aau6977

LeClaire M, Gimzewski J, Sharma S (2021a) A review of the biomechanical properties of single extracellular vesicles. Nano Sel 2:1–15. https://doi.org/10.1002/nano.202000129

LeClaire M, Wohlschlegel JA, Chang H et al (2021b) Nanoscale extracellular vesicles carry the mechanobiology signatures of breast cancer cells. ACS Appl Nano Mater 4:9876–9885. https://doi.org/10.1021/acsanm.1c02299

Mustonen AM, Capra J, Rilla K et al (2021) Characterization of hyaluronan-coated extracellular vesicles in synovial fluid of patients with osteoarthritis and rheumatoid arthritis. BMC Musculoskelet Disord 22:1–12. https://doi.org/10.1186/s12891-021-04115-w

Nagy N, De La Zerda A, Kaber G et al (2018) Hyaluronan content governs tissue stiffness in pancreatic islet inflammation. J Biol Chem 293:567–578. https://doi.org/10.1074/jbc.RA117.000148

Neuman KC, Nagy A (2008) Single-molecule force spectroscopy: optical tweezers, magnetic tweezers and atomic force microscopy. Nat Methods 5:491–505. https://doi.org/10.1038/nmeth.1218

Ouasti S, Donno R, Cellesi F et al (2011) Network connectivity, mechanical properties and cell adhesion for hyaluronic acid/PEG hydrogels. Biomaterials 32:6456–6470. https://doi.org/10.1016/j.biomaterials.2011.05.044

Paul D, Roy A, Nandy A et al (2020) Identification of biomarker hyaluronan on colon cancer extracellular vesicles using correlative AFM and spectroscopy. J Phys Chem Lett 11:5569–5576. https://doi.org/10.1021/acs.jpclett.0c01018

Paul D, Paul A, Mukherjee D et al (2022) A mechanoelastic glimpse on hyaluronan-coated extracellular vesicles. J Phys Chem Lett 13(36):8564–8572. https://doi.org/10.1021/acs.jpclett.2c01629

Petrey AC, de la Motte CA (2014) Hyaluronan, a crucial regulator of inflammation. Front Immunol 5:1–13. https://doi.org/10.3389/fimmu.2014.00101

Rakshit T, Pal S (2024) Extracellular vesicles for drug delivery and theranostics in vivo. JACS Au 4(2):318–327. https://doi.org/10.1021/jacsau.3c00611

Rakshit T, Melters DP, Dimitriadis EK, Dalal Y (2020) Mechanical properties of nucleoprotein complexes determined by nanoindentation spectroscopy. Nucleus 11:264–282. https://doi.org/10.1080/19491034.2020.1816053

Rilla K, Siiskonen H, Tammi M, Tammi R (2014) Hyaluronan-coated extracellular vesicles- a novel link between hyaluronan and cancer. Adv Cancer Res 123:121–148. https://doi.org/10.1016/B978-0-12-800092-2.00005-8

Schillers H, Rianna C, Schäpe J et al (2017) Standardized nanomechanical atomic force microscopy procedure (SNAP) for measuring soft and biological samples. Sci Rep 7:1–9. https://doi.org/10.1038/s41598-017-05383-0

Sharma S, Rasool HI et al (2010) Structural-mechanical characterization of nanoparticle exosomes in human saliva, using correlative AFM, FESEM, and force spectroscopy. ACS Nano 4:1921–1926. https://doi.org/10.1021/nn901824n

Verweij FJ, Bebelman MP, George AE et al (2022) ER membrane contact sites support endosomal small GTPase conversion for exosome secretion. J Cell Biol 221. https://doi.org/10.1083/jcb.202112032
Vorselen D, Piontek MC, Roos WH, Wuite GJL (2020) Mechanical characterization of liposomes and extracellular vesicles, a protocol. Front Mol Biosci 7:1–14. https://doi.org/10.3389/fmolb.2020.00139
Whitehead B, Wu LP, Hvam ML et al (2015) Tumour exosomes display differential mechanical and complement activation properties dependent on malignant state: implications in endothelial leakiness. J Extracell Vesicles 4. https://doi.org/10.3402/jev.v4.29685
Woo JR, Sharma S, Gimzewski J (2016) The role of isolation methods on a nanoscale surface structure and its effect on the size of exosomes. J Circ Biomark 5:1–9. https://doi.org/10.5772/64148
Ye S, Li W, Wang H et al (2021) Quantitative nanomechanical analysis of small extracellular vesicles for tumor malignancy indication. Adv Sci 8:1–10. https://doi.org/10.1002/advs.202100825

Chapter 10
Bioaffinity Recognition of Extracellular Vesicle Glycosylations

Khirul Islam and Janne Leivo

Abstract Extracellular vesicles (EVs) are rapidly gaining attention for their functional role in intercellular communication and their potential value as noninvasive biomarkers of various pathologies. EV glycosylations have been identified as potent functional molecules that govern the vesicles' biological roles and dictate their pathophysiological relevance. In this chapter, we review the glycosylation landscape of EVs and their biological function, along with the analytical tools used to study them. In addition, the most widely used carbohydrate-specific natural and synthetic affinity scaffolds currently used to study the glycosylations on EVs are described. The development of specific bioaffinity tools, such as glycan-specific antibodies, remains challenging due to the inherent low immunogenicity of glycans. Alternative strategies, including the use of recombinant immunoglobulins, aptamers, and molecular imprinted polymer derivatives, are actively explored to overcome these hurdles and harness the full potential of EV glycosylations in both basic glycobiology research and clinical applications.

Abbreviations

AAL	*Aleuria aurantia lectin*
CDR	Complementarity-determining region
EGF	Epidermal growth factor
ER	Endoplasmic reticulum
Eu	Europium
EV	Extracellular vesicle
GAG	Glycosaminoglycan
GPI	Glycosylphosphatidylinositol

K. Islam · J. Leivo (✉)
University of Turku and in FLAMES Flagship Research Centre, Turku, Finland

FICAN West Cancer Research Centre, Turku, Finland
e-mail: jpleiv@utu.fi

K. Rilla (ed.), *Extracellular Vesicles as Matrix Messengers*, Biology of Extracellular Matrix 15, https://doi.org/10.1007/978-3-031-68811-9_10

Ig	Immunoglobulin
JNP	Janus nanoparticle
LRR	Leucine-rich repeats
mAb	Monoclonal antibody
MIP	Molecularly imprinted polymer
MS	Mass spectrometry
NP	Nanoparticle
scFv	Single-chain antibody
sdAb	Single-domain antibody
SNA	*Sambucus nigra lectin*
UEA	*Ulex europaeus agglutinin*
VLR	Variable lymphocyte receptors
WGA	*Wheat germ agglutinin*

10.1 Introduction

Extracellular Vesicles (EVs) play a vital role in intercellular communication by encapsulating a complex composition that reflects their cellular origin. EVs and the surrounding biocorona are covered with diverse biomolecular content, including glycosylations, either as complexed with surface proteins or enzymatically synthesized to glycoproteins, glycolipids, and glycosaminoglycans. The remarkable structural diversity of glycosylations resulting from non-templated enzymatic processes proposes fundamental challenges in their analysis, particularly with reagents based on affinity recognition. The diversity is a result of the various monosaccharides involved, the branching within the carbohydrate chains, and the configuration and position of glycosidic linkages. Modifications like sulfation, phosphorylation, and acetylation add layers of complexity, resulting in a multifaceted "glycan-code" that is challenging to interpret.

EV glycosylations are not only structural embellishments; they are functional molecules that govern the vesicles' biological roles and dictate their pathophysiological relevance. Their intrinsic reciprocal link to cellular metabolism and the resulting changes in glycan expression during disease progression could also serve as early indicators and monitoring markers of diseases. As a result, increasing attention has been invested in using glycan structures on EVs as biomarkers for diagnosis and disease management, and they stand at the forefront of therapeutic innovations, including vaccine development and innovative treatment approaches for diseases that present unique glycan functionality.

Recognizing glycan structures with other molecules is challenging, and the selective discernment of glycan variations, including subtle isomeric differences, is even more challenging. The development of specific bioaffinity tools, such as glycan-specific antibodies, remains challenging due to the inherent low immunogenicity of glycans. Alternative strategies, including the use of lectins, aptamers, and molecular imprinted polymer derivatives, are actively explored to overcome these hurdles and harness the full potential of EV glycosylations in both basic glycobiology research and clinical applications. In this chapter, we introduce the basic concepts of carbohydrate recognition with the most used bioaffinity molecules and review the recent literature related to their use to study EV glycosylations.

10.2 Classification of Glycosylations and Their Biological Function

Carbohydrates are abundantly found on the surface of eukaryotic cell membranes as structural, physical, and functional constituents of the extracellular space. Glycans play a crucial role in a wide range of cellular functions, from cell signaling and recognition to adhesion. These complex molecules, made up of carbohydrates, undergo a series of enzymatic reactions in the endoplasmic reticulum (ER) and Golgi apparatus by glycosyltransferases and glycosidases. Glycosylations, which enzymatically attach glycans to proteins, can be classified into five main types (Fig. 10.1): N-linked and O-linked, C-linked, phosphoglycans, and glypiation.

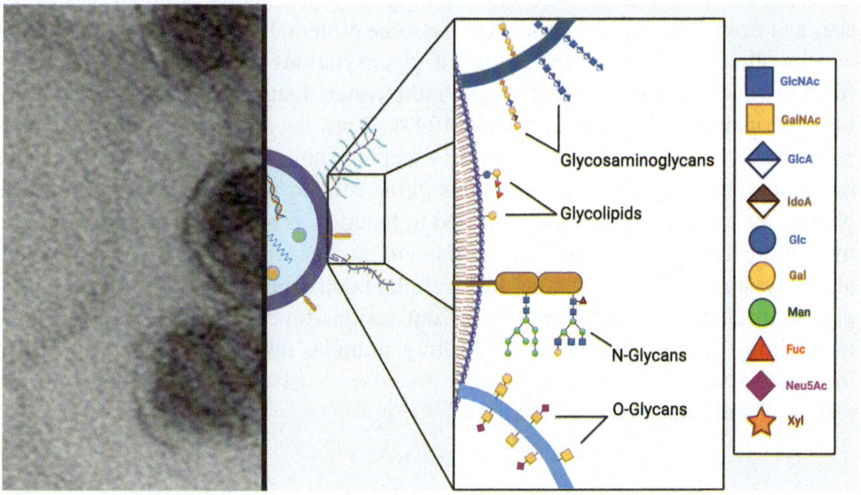

Fig. 10.1 Major glycan types and classification of glycosylations found on EVs. The ten monosaccharide types found in humans form the glycans through covalent glycosidic bonds in proteins and lipids. The main structural classification of glycosylations can be done according to the linkage to protein, lipid, or glycan, which they are enzymatically attached to. N-linked glycans are the most abundant glycosylation found on proteins through the nitrogen (N) of the asparagine side chains. O-linked glycans are the main constituents of Mucins and are attached to the oxygen (O) atom of Serine or Threonine residues. Glycolipids are classified according to the hydrocarbon backbone: glycerolipids contain glycerol and sphingolipids, sphingosine. Glycosaminoglycans (GAG) are a class of long and linear chains attached to proteoglycans. There are four main classes of GAGs based on the repeating core units (heparan sulfate, chondroitin sulfate, keratan sulfate I/II and hyaluronic acid)

N-linked glycans are attached to the nitrogen atom of asparagine residues in proteins, and O-linked glycans are attached to the oxygen atom of serine or threonine residues. Phosphate linkages can be found on proteins where the glycan is attached through the phosphate of a phosphoserine. Glypiation, which consists of the addition of glycosylphosphatidylinositols (GPIs) membrane anchor is used to attach proteins to the outer leaflet of the cell membrane. These GPIs are composed of two fatty acid tails that are inserted into the membrane and linked to a head group, which in turn is attached to a series of four saccharides. At the carboxyl end of the protein, a phosphoethanolamine residue is linked to the GPI, forming a stable covalent bond between the protein and the membrane. Glycolipids are attached to specific lipid moieties, with ceramide and glycerol moieties being the most common. Ceramide, which is composed of a fatty acid linked to a sphingosine backbone, is the predominant lipid moiety in glycolipids. When one or more sugar residues are attached, it forms the basis of glycosphingolipids. The category of glycosphingolipids includes cerebrosides, which have a single sugar molecule, and gangliosides, which are characterized by complex oligosaccharides that often contain sialic acid. Another less common lipid base for glycolipids is glycerol. In this case, the carbohydrate is attached to a glycerol backbone that is also esterified with fatty acids. Glycans are highly diverse, and their structures can vary depending on the cell type and tissue. Diversity can be found in all levels of biological functionality, from species to tissues and from cells to glycovariants of the same protein. From an evolutionary point of view, the highly conserved nature of glycosylations has resulted in numerous functionalities ranging from structural and physical features to intra- and intercellular recognition (Gagneux and Varki 1999). It has been postulated that glycosylations have biological functionality in escaping the ever-changing biomolecular recognition of pathogens: the variations in the cellular metabolism result in subtle phenotypic changes, which are proposed to function as a mask from infection. The hydrophilic nature of glycans enables them to interact with water molecules, which also plays a crucial role in maintaining cellular homeostasis. The diversity and complexity of glycans allow them to carry out various functions in the body, such as maintaining cell structure and modulating immune responses. Glycans are also involved in various specific biological processes, such as inflammation, immune response, and cancer progression.

10.3 Glycocalyx of Extracellular Vesicles

The perivesicular glycocalyx of EVs is formed from both the enzymatically synthesized and the biocorona surrounding the vesicle. Seminal studies to study the EV glycocalyx were carried out with lectin microarrays (Batista et al. 2011) and mass spectrometry which highlighted the biomolecular complexity of EVs also on the metabolic level. It soon became evident that the preanalytical variables, such as the selection of collection or separation methodology, led to method-dependent glycosylation patterns challenging the data interpretation (Freitas et al. 2019). Similar to the glycocalyx of the host cell, EV carbohydrates come in diverse forms of glycans, which can vary from linear to complex or highly branched structures. For a detailed review on the effect of EV glycosylations and the clinical implication, see review by Williams et al. (2018). All major classes of glycans have been found from EV preparations: N-glycans, O-glycans, glycolipids, glycosylphosphatidylinositol (GPI)-anchors, and glycosaminoglycans summarized in Table 10.1 (Costa 2017; Huang et al. 2021; Macedo-da-Silva et al. 2021). From these, N-and O-glycosylation are the most common forms of protein glycosylation and about 80 percent of all human proteins are glycosylated. The difference between N- and O-linked glycosylation is where the sugar molecule is attached. In N-linked glycosylation, glycan is attached to an asparagine residue via the side chain nitrogen atom. The N-glycans can be broadly classified into several classes: (a) complex oligosaccharides that contain multiple sugar types such as fucose, galactose, and N-acetylneuraminic acid; (b) high-mannose oligosaccharides that contain mannose and GlcNAc residues; as well as (c) hybrid branches of both complex oligosaccharides and high mannose. Among these three N-glycosylation subtypes, hybrid and complex type classes may contain several branches, such as antennary. Several studies demonstrated that N-glycosylation subtypes are heavily expressed on EVs (Staubach et al. 2012; Escrevente et al. 2013; Saraswat et al. 2015). Of particular note, mucins, which are multidomain proteins produced by epithelial cells, are the most common proteins with O-linked glycosylation and play an important role in various biological processes including infections, immune response, gastrointestinal protection, cell signaling and adhesion, tumor progression and metastasis (Corfield 2015; Reynolds et al. 2019; Malaker et al. 2022). In addition, other heavily N- and O-linked glycosylated proteins have been identified on EVs, such as carcinoembryonic antigen (CEA), a disintegrin and metalloproteinase domain-containing protein 10 (ADAM10), cell adhesion molecules (EpCAM, CD24), integrins (ITGA3, ITGAV, ITGB1, etc.), mucins (MUC1, MUC3, and MUC16), and carbohydrate antigen (CA19-9) (Escrevente et al. 2008; Lugini et al. 2016; Yamamoto et al. 2019).

Apart from above-mentioned major glycoprotein and glycolipid, other forms of glycosylation are also observed in mammalian cells that consist of GPI and glycosaminoglycans (GAGs). GPI-linked proteins enable signal transduction in malignancies. As an example, GPI-anchored glycoprotein CD109 is found on EVs and enhances EGF signaling and downregulates TGF-β1 signaling in glioblastoma cells (Zhang et al. 2015; Sakakura et al. 2016). Glycosaminoglycans are heavily glycosylated and covalently attached to proteoglycans. Each GAG consists of repeated two-sugar units (disaccharides) and has different degrees of sulfation. Based on their core disaccharide units, GAGs are primarily classified into four groups, including heparin/heparan sulfate, chondroitin sulfate/dermatan sulfate, keratan sulfate, and hyaluronic acid (Prydz 2015). The sulfation patterns aligned with GAG chains on proteoglycans have been found to function in EV biogenesis and overexpress in cancer (Baietti et al. 2012; Rilla et al. 2014; Melo et al. 2015).

The vesicular lipid bilayer of EVs anchors a series of glycolipids distinct from the glycans of the intra- and extravesicular proteins. Glycosphingolipid and glycoglycerolipid are two prominent glycolipids, typically conjugated to ceramide and

Table 10.1 Glycans carried by EVs

Classification	Glycan moieties	EV source	Use of EV	References
N-glycan	High mannose	Cell lines, human serum and urine	Biomarker, detection, capture, uptake, therapy	Batista et al. (2011), Gerlach et al. (2013, 2017), Escrevente et al. (2013), Gomes et al. (2015), Zou et al. (2017), Surman et al. (2018, 2019), Horrevorts et al. (2019)
	Branched	Cell lines	Biomarker, detection	Escrevente et al. (2013), Surman et al. (2018, 2019), Harada et al. (2019), Freitas et al. (2019), Matsuda et al. (2020)
	Bisected	Cell lines, human urine	Biomarker, detection	Gerlach et al. (2013), Gomes et al. (2015), Surman et al. (2018, 2019), Freitas et al. (2019)
	Complex	Cell lines, human serum & urine	Detection, biomarker	Batista et al. (2011), Gerlach et al. (2013), Zou et al. (2017), Surman et al. (2018), Lv et al. (2019), Harada et al. (2019)
Sialylation	a2,3 or a2,6-linked sialic acid	Cell lines, human serum & urine	Biomarker, detection, capture, uptake, biodistribution	Batista et al. (2011), Escrevente et al. (2011, 2013), Gerlach et al. (2013, 2017), Gomes et al. (2015), Zou et al. (2017), Surman et al. (2018, 2019), Feng et al. (2018), Lv et al. (2019), Royo et al. (2019), Islam et al. (2019), Williams et al. (2019), Freitas et al. (2019), Guo et al. (2020), Matsuda et al. (2020)
O-glycan	O-GlcNAc	Cell lines	Biomarker discovery	Royo et al. (2019)
Polylactosamine	GlcNAc/GalNAc	Cell lines	Detection	Batista et al. (2011)
Glycoproteins	ITGA3, CD133, ADAM10, CD24, EpCAM, LGALS3BP, EMMPRIN MUC1	Cell lines, urine	Biomarker, detection, capture	Runz et al. (2007), Escrevente et al. (2013), Im et al. (2014), Niu et al. (2019), Pan et al. (2019), Sakaue et al. (2019), Islam et al. (2022)

Table 10.1 (continued)

Classification	Glycan moieties	EV source	Use of EV	References
Glycolipid	Sphingomyelin, glycerolipid, glycerophospholipid, cholesterol, phospholipid	Cell lines	EV biomarker	Llorente et al. (2013), Harada et al. (2020)
GAG		Cell lines, human plasma	Capture	Escrevente et al. (2013), Rilla et al. (2014), Melo et al. (2015)
Fucosylation (core and/or terminal)	Fucα1−6GlcNAc Fucα1−3GlcNAc	Cell lines, human serum, urine	Biomarker discovery, detection	Gerlach et al. (2013), Costa et al. (2018), Surman et al. (2018), Harada et al. (2019), Freitas et al. (2019), Islam et al. (2022)
T-antigen	Gal-(1,3)-GalNAc-Thr/Ser	Cell lines	Biomarker discovery, detection	Gomes et al. (2015), Feng et al. (2018)
Tn-antigen	GalNAc-Thr/Ser	Cell lines	Biomarker discovery	Feng et al. (2018)
sTn-antigen	Neu5Ac (2,6)-GalNAc-Thr/Ser	Cell lines	Biomarker discovery	Feng et al. (2018), Freitas et al.(2019)

diacylglycerol, respectively. Glycolipids are found in membranes and play significant roles in biological processes, such as cell membrane structure, cell–cell recognition, and their adhesion (Hanafusa et al. 2020; Jala et al. 2022). Moreover, glycolipids on EVs surface play an important role in mediating interactions between EVs and recipient cells (Williams et al. 2019; He et al. 2022).

10.4 Lectins

The most widely used bioaffinity molecules fall into the very broad definition of lectins. Lectins can be found in all kingdoms of life and can be classified based on their structural properties or biological functions, including cell adhesion, signaling, and pathogen recognition. Lectins are a heterogeneous group of proteins with a long history of more than 135 years in research. Peter Stillmark first discovered lectin in plants in 1888 in his doctoral thesis at the University of Tartu, Estonia (reviewed in Sharon 2004). While working with the extracts from castor bean seeds, he obtained a protein preparation that was capable of agglutinating animal red blood cells. Over time, seeds of many plants were found to contain such protein preparation; researchers called it "agglutinin," later renamed as lectin. The terminology "lectin" came from the Latin term "legere," which means to select or to choose (Boyd and

Shapleigh 1954). Lectins are carbohydrate-binding proteins that are found abundantly in nature. Lectins are supposed to be present in all living organisms that synthesize glycans. Initially, lectins were thought to have huge potential to define only a group of plant agglutinins. Lately, several studies discovered that lectins are not only limited to plants but also originate from animals and microorganisms, including bacteria, algae, and fungi (Kobayashi and Kawagishi 2014; Nies et al. 2018). It has been well established that lectins have distinct sugar-binding abilities and roles in biological recognition processes at the molecular and cellular levels (Rutishauser and Sachs 1975). Particularly, lectins have been widely used as the basic tools for glycobiology in the detection, isolation, and structural analysis of glycoproteins (Sharon 2004). Based on their binding to glycan structures, lectins have been primarily classified as either glucose, galactose, mannose, GlcNAc, GalNAc (N-acetylgalactosamine), L-fucose, sialic acid-specific or complex glycan-binding lectins (Lis and Sharon 1986; Hashim et al. 2017). Different lectin groups show distinct binding specificities toward glycans based on their major binding sites. Due to their broad selectivity, lectins have also been used to study the glycosylations on EVs. Moreover, the lectins on the surface of EVs have a functional role during tumorigenesis and pathophysiological conditions (Barrès et al. 2010).

Lectins can be classified in different ways based on their source, function, specificity, or structure. They have been isolated from various species of plants, animals, or microorganisms. Lectins can also be categorized based on their binding specificity toward single or multiple types of carbohydrates, such as mannose, sialic acid, GalNAc, GlcNAc, fucose, galactose, T-antigen, and novel sialyl Lewis antigen (Mishra et al. 2019). However, some lectins, such as Ulex europaeus agglutinin 1 (UEA-1), are grouped based on their specificity for blood group antigens. In addition, some proteins contain lectin-like domains along with other functional domains. For example, the extracellular domains of cell surface receptors like epidermal growth factor (EGF) and fibronectin type III domain can have lectin-like properties. Structurally, lectins can be classified into C-, I-, S-, R-, L-, and P-type families based on their amino acid sequence and tertiary structure (Fig. 10.2). These lectin families often have distinct carbohydrate recognition domains (CRDs), which are responsible for their binding specificity to glycan. Additionally, the structure of a lectin can be complex and may involve multiple subunits, which can be monomers, dimers, tetramers, or large oligomers (Fig. 10.2). Lectins and lectin-based assays have been used in various applications related to the detection and analysis of EV glycosylations. This section addresses the most widely used lectin-based methods in EV analysis, including lectin microarray (Kuno et al. 2005; Krishnamoorthy et al. 2009; Batista et al. 2011; Gerlach et al. 2013). lectin blotting (Escrevente et al. 2011), lectin affinity chromatography (Echevarria et al. 2014; Gerlach et al. 2017; Harada et al. 2019; Yamamoto et al. 2019), and lectin-based immunoassay (Islam et al. 2019).

Lectin microarray involves immobilizing a selection of lectins on a solid surface, such as a glass slide or a microchip, allowing a high-throughput analysis of the glycosylation pattern of the analyte. Lectin microarray has distinct advantages over other technologies used in glycan analysis: (i) It enables the analysis of multiple

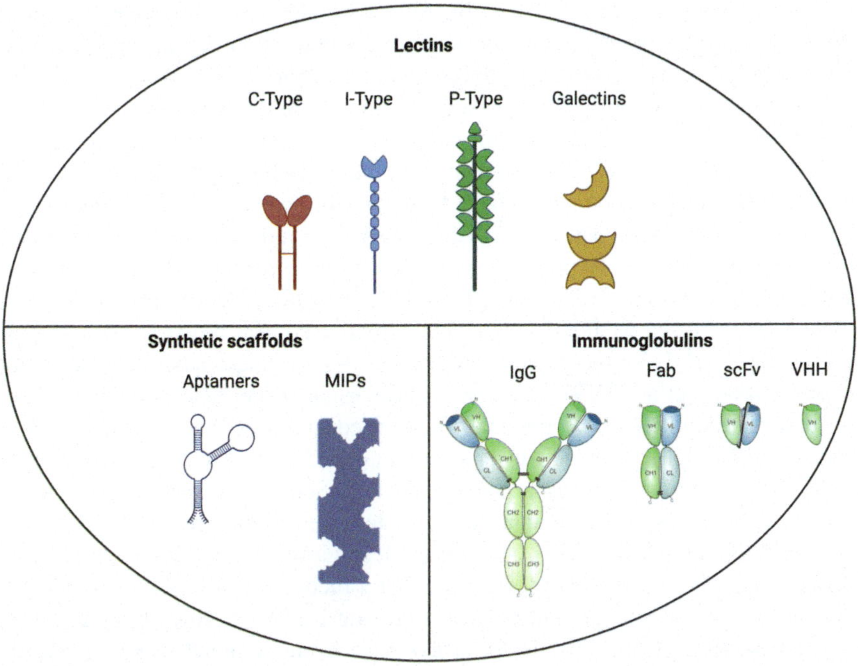

Fig. 10.2 Structural classification and schematic structure of the three main bioaffinity recognition elements used to detect carbohydrates. (Top) The four most abundant mammalian lectins display their corresponding CRDs on their surface. (Left) Synthetic binding scaffolds can be based on nucleotides such as DNA or RNA aptamers or molecularly imprinted polymers (MIPs). (Right) Structure of Immunoglobulin G (IgG) and antibody fragments. Immunoglobulins consist of individual domains and the functional regions of the antibody where the heavy chains are designated as VH, CH1-3 (Green), and light chains VL, CL (Blue)

samples for their glycosylation profiling simultaneously, making it a medium to high-throughput method. (ii) It allows sensitive and quick glycan profiling in both pure and crude glycoproteins. (iii) It provides a comprehensive view of glycosylation patterns overcoming analytical challenges due to the complex and diverse structure of glycans, including long branches that are difficult to analyze using conventional methods. (iv) Unlike some conventional methods like mass spectrometry (MS) and liquid chromatography (LC), which require glycan release for analysis, the lectin array can analyze glycan patterns directly on glycoproteins, reducing sample preparation steps and saving time. The lectin microarray technology introduced in 2005 by Kuno et al. used a panel of immobilized lectins that allow lectins to bind to glycans from different sample matrices (Kuno et al. 2005). In the EV field, the first report was in 2009, where the Mahal group used lectin microarray to analyze surface glycans of EVs from T cells (Krishnamoorthy et al. 2009). A related study from the same group reported that EVs contain conserved glycan signatures on their surface, abundantly consisting of complex N-glycans, high mannose, α-2,6-sialic acids, and polylactosamine. Progressively, several groups have used lectin microarray technology to study the glycosylation on EVs, typically applying one of two widely used approaches: (1) Surface-immobilized lectins are applied to capture EVs selectively based on their glycosylation patterns. Then captured EVs are detected using labeling techniques such as fluorescent dyes, antibodies, or other molecular probes. (2) EVs are first captured using antibodies targeting specific EV-surface proteins. Once EVs are captured, the lectin microarray is used to detect specific glycosylations on EV surface. Many groups have used lectin microarray to study EV-glycosylation patterns from complex body fluids. In a study, Gerlach et al. used a panel of lectins to profile urine-EVs aiming to detect autosomal dominant polycystic kidney disease. There are several other studies of lectin arrays being used to detect other diseases. For example, lectin microarray was used to identify highly glycosylated CD133 glycoprotein on ascites-EVs to detect advanced pancreatic cancer (Sakaue et al. 2019). In another study, using serum of pancreatic patients, lectin array identified six lectins that can be used to detect pancreatic cancer even when commercial pancreatic cancer serum marker CA19-9 was negative (Yokose et al. 2020). Similarly, using a panel of lectins, Saito et al. have found specific EV-glycan patterns in human induced pluripotent stem cells compared to EVs from control cells (Saito et al. 2018). Moreover, a lectin array was used to identify therapeutic EVs, where Hayashi et al. showed that a fucose-binding lectin could differentiate MSC-EVs compared to control source-derived EVs (Hayashi et al. 2022). Apart from glycan profiling, lectin arrays have been extended to investigate the process of EV biogenesis and uptake. Using a panel of lectins Shimoda et al. found that sialic acid binding lectins abundantly bound to EVs from human adipose mesenchymal stem cells (ADSC-EVs) compared to corresponding cell membrane, highlighting the possibility of sialic acid involvement in biogenesis and cellular uptake (Shimoda et al. 2017). Similarly, another study showed that EV-glycans detected by lectin array have the power to define EV heterogeneity which can control biodistribution and cellular uptake efficiency (Shimoda et al. 2022). Several additional studies showed that changes in glycosylation patterns during biogenesis can influence the properties of EVs such as their cargo loading, stability, and targeting (Williams et al. 2019; Clos-Sansalvador et al. 2022).

Lectin blotting involves transferring separated EVs onto a membrane aiming to create a snapshot of the EV proteins in an easier way to analyze, followed by probing with lectins to detect glycan structures. It provides insight into the glycosylation status of EV proteins, aiding in the characterization of EVs for a better understanding of their biogenesis, function, and potential therapeutic application. Lectin blotting is often used in combination with other techniques to confirm specific glycosylation patterns on EVs. The studies have confirmed that lectin blotting can be used to distinguish disease-specific glycan patterns from patient samples (Tan et al. 2020). In a study using lectin blotting, Nishida-Aoki et al. showed that the removal of a particular glycan from the EV surface has an inhibitory effect on EV uptake (Nishida-Aoki et al. 2020). Similarly, changes in glycan structure can influence the recruitment of specific glycoproteins into EVs (Liang et al. 2014).

EV separation methods remain at the center of debate for high-quality and reproducible EV studies. The traditional methods for EV separation are time-consuming, tedious, and expensive. Moreover, some methods may result in EV damage or aggregation (Wang and Sun 2014; Boriachek et al. 2018). Using lectin-based affinity capture has become a valuable technique to separate EVs, overcoming the limitations of traditional methods (Echevarria et al. 2014). Echevarria et al. found lectins such as STL, WGA, and LEL, especially STL, to be highly effective in binding urinary EVs. By combining biotin-streptavidin and magnetic beads, this method ensures efficient and pure EV isolation, particularly making it beneficial for studying EV composition and discovering biomarkers (Benecke et al. 2022). Overall, lectin affinity capture is a versatile and efficient method for the separation of glycan-enriched subpopulations of EVs, providing valuable insights from a metabolic perspective.

Lectins have been found to be useful in bioaffinity detection, where they are used to capture or detect EVs based on their unique glycan profiles. Nanoparticles can increase the binding affinity of certain lectins to detect low-abundant biomarkers or disease-specific subsets of EVs based on lectin-glycan interactions. Choi et al. used lectins with specific affinity for sialic acid (such as lectin SNA) and fucose (such as AAL) to bind to bifunctional Janus nanoparticles (JNPs) that helped them bind to EVs. This lectin-JNPs approach is strongly bound to EVs from pancreatic cancer sources, comparable to those of commercial pancreatic marker CA19-9 (Choi et al. 2021). By using lectins in their nanoparticle-assisted assay, researchers have been able to detect EVs with high specificity and sensitivity. This technology has the potential to discover new glycan targets on EVs, providing a better understanding of health and disease and paving the way for improved diagnostics.

10.5 Glycan-Binding Immunoglobulins and Lymphocyte Receptors

Glycans are central to immune homeostasis, with key functionality in extracellular matrix organization, protection from immune recognition, and modulation of membrane receptor signaling (Johnson et al. 2013). Although glycan-binding proteins endogenously expressed on immune cells can be considered lectins due to their relevance to human health and disease, these proteins are reviewed separately in this section. Particularly, various lymphocyte antigen receptors, such as immunoglobulins on B cells and T-cell receptors on T cells, have been extensively studied and applied in biomedical disciplines. The receptors generated by each lymphocyte have a unique antigen specificity, which is determined by the structure of their antigen-binding site. Of particular interest are immunoglobulins, or antibodies, which are a group of vertebrate immune-system-related glycoproteins secreted by the plasma B cells. The primary biological function of antibodies as a part of the immune system is to bind and inactivate foreign, harmful pathogens such as infectious agents and microbial toxins. This is accomplished by high sequential differentiation on the amino acid level, which ensures the efficient recognition of various foreign pathogens. Antibodies also have a regulatory role in the effector cell and complement activation. The immunoglobulins are classified into five categories according to the biological functionality and the structure of the constant heavy regions. The five different classes are IgA (α), IgD (δ), IgE (ε), IgG (γ), and IgM (μ), which are further divided into a set of sub-classes. In addition, the constant light chains can be divided into two structural categories: lambda (λ) and kappa (κ). The tertiary protein structure of antibodies consists of four polypeptide chains, two heavy and two light chains, connected by disulfide bridges (Fig. 10.2). The four connected domains form a symmetrical and flexible Y-shaped tertiary structure, approximately 440 amino acids long and 150 kDa in weight, which is the basic unit of an antibody molecule. The two structural regions, the constant region (C-region) and the variable region (V-region), of the antibody have different functionalities in immune regulation (Fig. 10.1). The C-region interacts with cell surface receptors and is involved in the complement and B-cell activation process during the immune response. The V-region, on the other hand, which contains six polypeptide loops, is the complementarity-determining region (CDR) that is involved in the antigen binding. However, it has been proposed that the two regions have more complex interactions than the proposed basic model of two regions with independent functionalities (Woof and Burton 2004). These interactions have been reported to have an effect both on the affinity and specificity of the antibody. The six CDR loops are located in the peripheral surface of the heavy and light V-regions in the antibody structure, three in each loop (CDR-L1-3 and CDR-H1-3) intervened by conserved immunoglobulin framework β-sheets, which define the orientation of the CDR loops. The high amino acid sequence and loop length diversity of the CDR regions form the basis of the efficient recognition of a wide selection of structurally variable antigens. This is especially the case for the third loop of both chains (CDR-L3 and CDR-H3), which have been found to be mainly responsible for the antibody–antigen interaction. The six CDR loops are highly variable in length and amino acid composition and form the antigen-binding site, or paratope of the antibody, which is structurally very adaptable to bind different targets (for a comprehensive review, see Janeway et al. 2001). The isolation of the antibody gene made it possible to clone and recombinantly express antibody fragments in other organisms, enabling the generation of smaller antibody fragments that can be fused with other functional proteins. Single-chain antibodies (scFv) and single-domain antibodies (sdAb) are examples of smaller antibody fragments that have been produced. The fragmentation of the intact IgG structure into smaller binding units and the expression, either individually or in fusion with other proteins or peptides, has brought versatility to the downstream applications of recombinant antibodies.

Carbohydrate-specific or glycan-binding antibodies can be found in all immunoglobulin classes. These antibodies are often classified as natural antibodies, presumed to arise without exposure to exogenous antigens (Kappler and Hennet 2020). Exposure to intestinal microbiota affects both the selectivity of the antibodies and cross-reaction with self-antigens, which is believed to contribute to the development of autoimmune disorders. This is a result of T-cell independent maturation process where the B-cell response against the antigen devoid of affinity maturation results primarily in IgM and IgG3 class immunoglobulins (Stein 1992). Human glycosylation relies on a combination of ten monosaccharides, whereas bacterial glycosylation involves over one hundred distinct monosaccharides. This diversity in monosaccharide composition and glycosidic linkages yields a vast repertoire of carbohydrate antigens that can stimulate human antibody production. A considerable portion of the naturally occurring, glycan-binding immunoglobulins in human circulation are IgM and IgG class, recognizing various antigens manifesting substantial inter-individual variability (Huflejt et al. 2009; Bovin et al. 2012). While most circulating glycan-binding antibodies have low specificity to larger glycoconjugates, some monosaccharide antigens, such as α-rhamnose and N-glycolylneuraminic acid (NeuGc), are associated with high antibody titers. This variation in antibody specificity and titer is crucial for understanding the diverse roles of glycan-specific antibodies with a myriad of physiological and pathophysiological functionality (Alter et al. 2018). The interaction of glycan-recognizing antibodies is like lectin–carbohydrate binding: the interactions are mainly derived from hydrogen bonds, van der Waals and electrostatic interactions with an emphasis on the use of charged and aromatic amino acids (Weis and Drickamer 1996). The entities responsible for glycan binding must discern between a vast array of carbohydrate structures, including closely related isomers. Even minute differences, such as the stereochemistry of a single hydroxyl group, can complicate selective recognition. The strong solvation of glycans in water further exacerbates this challenge, as receptors must compete with the extensive hydrogen bond network surrounding the glycans.

The development of glycan-specific antibodies is challenging due to several factors. Carbohydrates are poorly immunogenic due to inherent T-cell-independent immune response, which makes it difficult to develop high-affinity antibodies with immunization-based methods (Polonskaya et al. 2019). The resulting monoclonal antibodies against carbohydrate targets with immunization result in unpredictable immune responses, often with specificity to the biomolecular content surrounding the glycan (Martínez-Sáez et al. 2015). Unlike proteins, which elicit a robust and predictable immune response, immunization with glycans often results in weak affinity for the antigen. This is partly because many glycan structures on pathogens resemble those found in the host, leading to immune tolerance. The immune system, trained to avoid attacking the body's own cells, may not respond aggressively to glycans that it perceives as self-antigens. Furthermore, the sheer structural diversity and complexity of glycans for the targeted antibody development for all specific glycan epitopes would require systematic synthesis of the glycans, which would require approaches unavailable with current technologies. Using recombinant DNA and antibody engineering technologies and the synthetic biology approaches related to them can bring significant progress to the field. Following the development of hybridoma technology to generate monoclonal antibodies, the display of recombinant fragments of antibodies on filamentous phages and the construction of single-pot naïve, semi-synthetic, and synthetic libraries may bring novel, high-throughput approaches for the development of glycan-specific recombinant antibodies and antibody libraries (Smith 1985; McCafferty et al. 1990). Efforts to categorize and validate glycan-binding antibodies have been carried out. For example, a database for anti-glycan reagents (DAGR) has been established, consisting of >1100 monoclonal antibody entries with glycan targets, host species, and isotypes (Sterner et al. 2016). Analyzing the database shows one of the key limitations of monoclonal antibodies: by immunizing animals, the specificity is limited to the particularly against sulfated glycans. To some extent, this spans from the very fundamental issues of glycan-recognizing biomolecules used in glycobiology in general: the definition, specificity, and quality of the antibodies are ambiguous, un-complementary, and, for a significant part, not available. The issue partially spans from the use of different clone names for the same monoclonal antibodies provided by commercial sources and/or research groups. For further reading, the availability of glycan-binding antibodies and the fundamentals of carbohydrate recognition of antibodies are reviewed in Haji-Ghassemi et al. (2015).

To address the challenges related to developing high-quality antibodies against glycans, researchers have turned to the unique immune system of the sea lamprey (*Petromyzon marinus*). Lampreys, being jawless vertebrates, express variable lymphocyte receptors (VLRs) composed of leucine-rich repeats (LRRs), which differentiate into plasma cells secreting circulating antigen-specific VLR proteins upon antigenic stimulation (Pancer et al. 2004). So far, three classes of VLRs have been identified, defined as VLRA, VLRB, and VLRC (Kasamatsu et al. 2010). These VLRs are generated by a gene conversion-like assembly mechanism that produces many distinct receptors (Gunn et al. 2018). By engineering the proteins and expressing them as VLRB-Ig chimeras, termed smart anti-glycan reagents (SAGRs), the study demonstrated specificities for distinct glycan structures, including various linkages, functional groups, and unique presentations of terminal glycan motifs (McKitrick et al. 2020).

Recently, there has been a growing interest in studying the glycosylations of EVs with antibodies and other non-lectin binders. However, the extent of research on EV glycosylation is still not as comprehensive as that on nucleic acids and proteins. As is the case in glycobiology, most of the studies on EV glycosylation using bioaffinity reagents rely on lectin recognition. Although glycan-binding antibodies have been used in EV research, there is a lack of systematic studies where antibody-based recognition of glycans has not been conducted. The majority of the glycan-targeting studies using antibodies aim to use the EV glycosylations as biomarkers of various diseases. This has shed light on the ill-characterized status of some circulatory biomarkers used in commercial immunoassays. For example, conventional biomarkers such as carbohydrate antigen 19-9 (CA19-9) are a glycosylated epitope recognized by monoclonal antibody (Borenstein-Katz et al. 2021). Although CA19-9 has been extensively studied and is generally considered a biomarker for the management of pancreatic ductal adenocarcinoma, the carrier proteins and presence on EVs are incomplete. Recent reports have indicated that CA19-9 is, at least to some extent, expressed on circulatory EVs (Javeed et al. 2015; Sancho-Albero et al. 2020). Similarly, CA125 and CA15-3, well-established biomarkers for ovarian and breast cancer, respectively, have been identified in various EV preparations (Mitić et al. 2018; Li et al. 2021; Saroj et al. 2023). Although finding mucins on the surface of cancer cell-secreted EVs is not surprising, the clinical significance and the extent of expression in the circulation of these mucinous biomarkers remain elusive.

10.6 Synthetic and Non-protein Scaffolds for Glycan Recognition

Over the years, synthetic and glycobiology scientists have channeled their efforts into designing artificial receptors that can either mimic or surpass binding the properties of antibodies and lectins as glycan-recognizing molecules. Strategies involving synthetic and non-protein scaffolds, such as aptamers and synthetic derivatives, are being actively explored, offering hope for breakthroughs in glycan recognition. This section reviews the most widely used artificial and non-protein binding scaffolds used to detect and analyze glycans.

Aptamers are single-stranded DNA or RNA structures that specifically recognize a range of target biomolecules, from small haptens to complex multimeric protein structures (Tombelli et al. 2005). Their versatility in selection processes, ease of chemical synthesis, and small size make them highly attractive also for glycan-targeting applications (Levy-Nissenbaum et al. 2008; Sun et al. 2010). Aptamers can be selected in vitro from oligonucleotide sequence libraries, offering a high degree of customization and high specificity to various molecular ligands or protein targets. The molecular structure of aptamers, based on the nucleic acid binding to the target carbohydrate, relies solely on hydrophobic and hydrogen bonds, often making the interaction weak for carbohydrate targets (Yang et al. 1998; Hasegawa et al. 2016). Compared to proteins, which rely on charged and aromatic amino acids in carbohydrate recognition, achieving adequate specificity and selectivity for glycan target aptamers often requires customization. This can be achieved through the functional groups of aptamers, which can be chemically modified with chemical groups to overcome limitations. For example, the addition of boronic acid groups to aptamers has been found to significantly enhance their binding properties for glycan targets (Li et al. 2008). The affinity of boronic acid toward carbohydrates itself is a well-known phenomenon in glycan recognition (James et al. 1996; Li et al. 2015). Boronic acid is often used as the main constituent of Molecularly Imprinted Polymers (MIPs), which are synthetic materials engineered to mimic natural receptors' molecular recognition abilities. These polymers are specifically tailored to recognize and bind to glycan structures, making them powerful tools for various carbohydrate targets (Xing et al. 2017). The technology behind glycan-recognizing MIPs involves creating a molecular imprint of the target glycan on the polymer matrix, serving as a molecular recognition site that is highly selective for the target glycan. Their robustness and ability to mimic biological receptors like antibodies make glycan-recognizing MIPs an exciting alternative for protein scaffolds in terms of selectivity, stability, and ease of preparation in various glycobiology applications (Bie et al. 2015; Luo et al. 2017).

The study of the glycocalyx of EVs is a relatively new field. As such, reports on the use of synthetic and non-protein scaffolds to study EVs are scarce and mostly focused on the detection and diagnosis of various pathologies (Jiang et al. 2024). Aptamers, however, have been used in several studies to separate and analyze the phenotype of EVs (Zhu et al. 2023; Deng et al. 2023). The versatility of the aptamer libraries and innovative selection technologies has led to the development of approaches where instead of identifying or detecting individual biomarkers with aptamers, the fingerprint or bioaffinity profile of the entire pool is analyzed. Nucleic acid methods such as the SELEX system (Hornung et al. 2020) can be used in the phenotypic profiling of EVs, without prior knowledge of the specific target (Domenyuk et al. 2017; Hornung et al. 2020).

10.7 Conclusion and Future Perspectives

The study of EV glycosylations is hindered by the unmet challenges that affect all disciplines of glycobiology: the need for well-characterized, high-quality affinity reagents to provide detailed information about carbohydrates in a reproducible manner. In order to fully elucidate the functional role of EV glycosylations with systematic studies, new bioaffinity reagents that provide sufficient specificity are needed. To overcome the problems derived from glycans' structural complexity and diversity, we likely need fundamental technological advancements to characterize carbohydrates.

The main limitation in the use of antibodies in glycan analysis stems from the fact that carbohydrates are non-immunogenic, and the generation of high-quality binding reagents with clearly defined specificity is exceptionally challenging. Glycans exhibit an extraordinary level of structural complexity and diversity arising from the 10 monosaccharide building blocks, the different ways these sugars can link together, and the branching patterns they can form. Different from proteins and nucleic acids (DNA or RNA), which are linear polymers with a defined sequence dictated by a template, glycans are assembled by enzymatic synthesis in a template-free manner. This process leads to a heterogeneous population of molecules, even for a single glycoprotein, known as glycoforms or glycovariants. Antibodies need to be highly specific to recognize subtle differences in glycan structures, such as different linkages (α or β), the position of glycosidic bonds (1–2, 1–3, 1–4, etc.), and branching patterns. This specificity is essential for distinguishing between closely related glycan structures, which often have distinct biological functions. The complexity of glycans presents a significant challenge for immunogen design, as it is difficult to ensure that the immune system will produce antibodies that target the desired glycan structure with high specificity and affinity (Gillmann et al. 2023).

Understanding the structure and function of carbohydrates and glycans in the extravesicular space of EVs is increasingly important for developing novel therapeutic strategies for disease management. Glycan engineering has emerged as one of the potent strategies for developing vaccines and targeted delivery of EVs (Royo et al. 2019). Equally, the glycosylations on the surface of EVs are increasingly recognized for their roles in various biological and pathological events, and thereby mediating intercellular communication. As the presence of diverse glycans can influence the stability of the vesicles, their biodistribution, uptake by recipient cells, and immune modulation, these studies have opened new pathways in understanding how EVs contribute to various biological processes (Macedo-da-Silva et al. 2021). The growing interest in the clinical application of EVs has led to extensive research in glycoengineering, where the glycosylations of EVs are endo-or exogenously modified for the desired effect (Shimoda et al. 2022).

In conclusion, interpreting the complex "glycan-code" of EVs and the surrounding biocorona will, in the future, help us understand intercellular communication and the full lifecycle of EVs in vivo. We need to conduct systematic studies to explore the full reciprocity between the glycosylation of host cells and the EVs they shed. Currently, the use of artificial receptors for EV-glycan recognition and analysis is relatively unexplored. It is preferable to use a combination of all the affinity reagents discussed in this chapter to cover a significant portion of the glycocalyx of the extravesicular space of EVs. Once we start unraveling this intricate code, we can also develop applications that make us aware of the potent functionality of glycans. Once the "glycan-code" of EVs starts to unravel, applications that make use of the potent functionality of glycans will follow, and the results will likely be transformative.

Acknowledgments The authors would like to acknowledge Research Council of Finland's Flagship InFLAMES (decision numbers: 337530 and 357910).

References

Alter G, Ottenhoff THM, Joosten SA (2018) Antibody glycosylation in inflammation, disease and vaccination. Semin Immunol 39:102–110. https://doi.org/10.1016/j.smim.2018.05.003

Baietti MF, Zhang Z, Mortier E et al (2012) Syndecan–syntenin–ALIX regulates the biogenesis of exosomes. Nat Cell Biol 14:677–685. https://doi.org/10.1038/ncb2502

Barrès C, Blanc L, Bette-Bobillo P et al (2010) Galectin-5 is bound onto the surface of rat reticulocyte exosomes and modulates vesicle uptake by macrophages. Blood 115:696–705. https://doi.org/10.1182/blood-2009-07-231449

Batista BS, Eng WS, Pilobello KT et al (2011) Identification of a conserved glycan signature for microvesicles. J Proteome Res 10:4624–4633. https://doi.org/10.1021/pr200434y

Benecke L, Chiang D, Ebnoether E et al (2022) Isolation and analysis of tumor-derived extracellular vesicles from head and neck squamous cell carcinoma plasma by galectin-based glycan recognition particles. Int J Oncol 61:133. https://doi.org/10.3892/ijo.2022.5423

Bie Z, Chen Y, Ye J et al (2015) Boronate-affinity glycan-oriented surface imprinting: a new strategy to mimic lectins for the recognition of an intact glycoprotein and its characteristic fragments. Angew Chem Int Ed 54:10211–10215. https://doi.org/10.1002/anie.201503066

Borenstein-Katz A, Warszawski S, Amon R et al (2021) Biomolecular recognition of the glycan neoantigen CA19-9 by distinct antibodies. J Mol Biol 433:167099. https://doi.org/10.1016/j.jmb.2021.167099

Boriachek K, Islam MN, Möller A et al (2018) Biological functions and current advances in isolation and detection strategies for exosome nanovesicles. Small 14:1702153. https://doi.org/10.1002/smll.201702153

Bovin N, Obukhova P, Shilova N et al (2012) Repertoire of human natural anti-glycan immunoglobulins. Do we have auto-antibodies? Biochim Biophys Acta 1820:1373–1382. https://doi.org/10.1016/j.bbagen.2012.02.005

Boyd WC, Shapleigh E (1954) Antigenic relations of blood group antigens as suggested by tests with lectins. J Immunol 73:226–231. https://doi.org/10.4049/jimmunol.73.4.226

Choi Y, Park U, Koo H-J et al (2021) Exosome-mediated diagnosis of pancreatic cancer using lectin-conjugated nanoparticles bound to selective glycans. Biosens Bioelectron 177:112980. https://doi.org/10.1016/j.bios.2021.112980

Clos-Sansalvador M, Garcia SG, Morón-Font M et al (2022) N-glycans in immortalized mesenchymal stromal cell-derived extracellular vesicles are critical for EV–cell interaction and functional activation of endothelial cells. Int J Mol Sci 23:9539. https://doi.org/10.3390/ijms23179539

Corfield AP (2015) Mucins: a biologically relevant glycan barrier in mucosal protection. Biochim Biophys Acta 1850:236–252. https://doi.org/10.1016/j.bbagen.2014.05.003

Costa J (2017) Glycoconjugates from extracellular vesicles: structures, functions and emerging potential as cancer biomarkers. Biochim Biophys Acta Rev Cancer 1868:157–166. https://doi.org/10.1016/j.bbcan.2017.03.007

Costa J, Gatermann M, Nimtz M et al (2018) N-glycosylation of extracellular vesicles from HEK-293 and glioma cell lines. Anal Chem 90:7871–7879. https://doi.org/10.1021/acs.analchem.7b05455

Deng J, Liu C, Sun J (2023) DNA-based nanomaterials for analysis of extracellular vesicles. Adv Mater 2303092. https://doi.org/10.1002/adma.202303092

Domenyuk V, Zhong Z, Stark A et al (2017) Plasma exosome profiling of cancer patients by a next generation systems biology approach. Sci Rep 7:42741. https://doi.org/10.1038/srep42741

Echevarria J, Royo F, Pazos R et al (2014) Microarray-based identification of lectins for the purification of human urinary extracellular vesicles directly from urine samples. ChemBioChem 15:1621–1626. https://doi.org/10.1002/cbic.201402058

Escrevente C, Morais VA, Keller S et al (2008) Functional role of N-glycosylation from ADAM10 in processing, localization and activity of the enzyme. Biochim Biophys Acta 1780:905–913. https://doi.org/10.1016/j.bbagen.2008.03.004

Escrevente C, Keller S, Altevogt P, Costa J (2011) Interaction and uptake of exosomes by ovarian cancer cells. BMC Cancer 11:108. https://doi.org/10.1186/1471-2407-11-108

Escrevente C, Grammel N, Kandzia S et al (2013) Sialoglycoproteins and N-glycans from secreted exosomes of ovarian carcinoma cells. PLoS One 8:e78631. https://doi.org/10.1371/journal.pone.0078631

Feng Y, Guo Y, Li Y et al (2018) Lectin-mediated in situ rolling circle amplification on exosomes for probing cancer-related glycan pattern. Anal Chim Acta 1039:108–115. https://doi.org/10.1016/j.aca.2018.07.040

Freitas D, Balmaña M, Poças J et al (2019) Different isolation approaches lead to diverse glycosylated extracellular vesicle populations. J Extracell Vesicles 8:1621131. https://doi.org/10.1080/20013078.2019.1621131

Gagneux P, Varki A (1999) Evolutionary considerations in relating oligosaccharide diversity to biological function. Glycobiology 9:747–755. https://doi.org/10.1093/glycob/9.8.747

Gerlach JQ, Krüger A, Gallogly S et al (2013) Surface glycosylation profiles of urine extracellular vesicles. PLoS One 8:e74801. https://doi.org/10.1371/journal.pone.0074801

Gerlach JQ, Maguire CM, Krüger A et al (2017) Urinary nanovesicles captured by lectins or antibodies demonstrate variations in size and surface glycosylation profile. Nanomedicine 12:1217–1229. https://doi.org/10.2217/nnm-2017-0016

Gillmann KM, Temme JS, Marglous S et al (2023) Anti-glycan monoclonal antibodies: basic research and clinical applications. Curr Opin Chem Biol 74:102281. https://doi.org/10.1016/j.cbpa.2023.102281

Gomes J, Gomes-Alves P, Carvalho S et al (2015) Extracellular vesicles from ovarian carcinoma cells display specific glycosignatures. Biomolecules 5:1741–1761. https://doi.org/10.3390/biom5031741

Gunn RJ, Herrin BR, Acharya S et al (2018) VLR recognition of TLR5 expands the molecular characterization of protein antigen binding by non-Ig-based antibodies. J Mol Biol 430:1350–1367. https://doi.org/10.1016/j.jmb.2018.03.016

Guo Y, Tao J, Li Y et al (2020) Quantitative localized analysis reveals distinct exosomal protein-specific glycosignatures: implications in cancer cell subtyping, exosome biogenesis, and function. J Am Chem Soc 142:7404–7412. https://doi.org/10.1021/jacs.9b12182

Haji-Ghassemi O, Blackler RJ, Martin Young N, Evans SV (2015) Antibody recognition of carbohydrate epitopes. Glycobiology 25:920–952. https://doi.org/10.1093/glycob/cwv037

Hanafusa K, Hotta T, Iwabuchi K (2020) Glycolipids: linchpins in the organization and function of membrane microdomains. Front Cell Dev Biol 8:589799. https://doi.org/10.3389/fcell.2020.589799

Harada Y, Kizuka Y, Tokoro Y et al (2019) N-glycome inheritance from cells to extracellular vesicles in B16 melanomas. FEBS Lett 593:942–951. https://doi.org/10.1002/1873-3468.13377

Harada Y, Nakajima K, Suzuki T et al (2020) Glycometabolic regulation of the biogenesis of small extracellular vesicles. Cell Rep 33:108261. https://doi.org/10.1016/j.celrep.2020.108261

Hasegawa H, Savory N, Abe K, Ikebukuro K (2016) Methods for improving aptamer binding affinity. Molecules 21:421. https://doi.org/10.3390/molecules21040421

Hashim OH, Jayapalan JJ, Lee C-S (2017) Lectins: an effective tool for screening of potential cancer biomarkers. PeerJ 5:e3784. https://doi.org/10.7717/peerj.3784

Hayashi Y, Yimiti D, Sanada Y et al (2022) The therapeutic capacity of bone marrow MSC-derived extracellular vesicles in Achilles tendon healing is passage-dependent and indicated by specific glycans. FEBS Lett 596:1047–1058. https://doi.org/10.1002/1873-3468.14333

He X, Guan F, Lei L (2022) Structure and function of glycosphingolipids on small extracellular vesicles. Glycoconj J 39:197–205. https://doi.org/10.1007/s10719-022-10052-0

Hornung T, O'Neill HA, Logie SC et al (2020) ADAPT identifies an ESCRT complex composition that discriminates VCaP from LNCaP prostate cancer cell exosomes. Nucleic Acids Res 48:4013–4027. https://doi.org/10.1093/nar/gkaa034

Horrevorts SK, Stolks DA et al (2019) Glycan-modified melanoma-derived apoptotic extracellular vesicles as antigen source for anti-tumor vaccination. Cancers 11:1266. https://doi.org/10.3390/cancers11091266

Huang D, Chen J, Hu D et al (2021) Advances in biological function and clinical application of small extracellular vesicle membrane proteins. Front Oncol 11:675940. https://doi.org/10.3389/fonc.2021.675940

Huflejt ME, Vuskovic M, Vasiliu D et al (2009) Anti-carbohydrate antibodies of normal sera: findings, surprises and challenges. Mol Immunol 46:3037–3049. https://doi.org/10.1016/j.molimm.2009.06.010

Im H, Shao H, Park YI et al (2014) Label-free detection and molecular profiling of exosomes with a nano-plasmonic sensor. Nat Biotechnol 32:490–495. https://doi.org/10.1038/nbt.2886

Islam MK, Syed P, Lehtinen L et al (2019) A nanoparticle-based approach for the detection of extracellular vesicles. Sci Rep 9:10038. https://doi.org/10.1038/s41598-019-46395-2

Islam MK, Dhondt B, Syed P et al (2022) Integrins are enriched on aberrantly fucosylated tumour-derived urinary extracellular vesicles. J Extracell Biol 1:e64. https://doi.org/10.1002/jex2.64

Jala RCR, Vudhgiri S, Kumar CG (2022) A comprehensive review on natural occurrence, synthesis and biological activities of glycolipids. Carbohydr Res 516:108556. https://doi.org/10.1016/j.carres.2022.108556

James TD, Sandanayake KRAS, Shinkai S (1996) Saccharide sensing with molecular receptors based on boronic acid. Angew Chem Int Ed Engl 35:1910–1922. https://doi.org/10.1002/anie.199619101

Janeway Jr, CA, Travers P, Walport M (2001) Immunobiology: the immune system in health and disease. New York: Garland Science

Javeed N, Sagar G, Dutta SK et al (2015) Pancreatic cancer–derived exosomes cause paraneoplastic β-cell dysfunction. Clin Cancer Res 21:1722–1733. https://doi.org/10.1158/1078-0432.CCR-14-2022

Jiang Z, Luo K, Yang G et al (2024) An electrochemiluminescent sensor based on glycosyl imprinting and aptamer for the detection of cancer-related extracellular vesicles. Anal Chem 96:2550–2558. https://doi.org/10.1021/acs.analchem.3c04991

Johnson JL, Jones MB, Ryan SO, Cobb BA (2013) The regulatory power of glycans and their binding partners in immunity. Trends Immunol 34:290–298. https://doi.org/10.1016/j.it.2013.01.006

Kappler K, Hennet T (2020) Emergence and significance of carbohydrate-specific antibodies. Genes Immun 21:224–239. https://doi.org/10.1038/s41435-020-0105-9

Kasamatsu J, Sutoh Y, Fugo K et al (2010) Identification of a third variable lymphocyte receptor in the lamprey. Proc Natl Acad Sci 107:14304–14308. https://doi.org/10.1073/pnas.1001910107

Kobayashi Y, Kawagishi H (2014) Fungal lectins: a growing family. In: Hirabayashi J (ed) Lectins. Springer, New York, pp 15–38

Krishnamoorthy L, Bess JW, Preston AB et al (2009) HIV-1 and microvesicles from T cells share a common glycome, arguing for a common origin. Nat Chem Biol 5:244–250. https://doi.org/10.1038/nchembio.151

Kuno A, Uchiyama N, Koseki-Kuno S et al (2005) Evanescent-field fluorescence-assisted lectin microarray: a new strategy for glycan profiling. Nat Methods 2:851–856. https://doi.org/10.1038/nmeth803

Levy-Nissenbaum E, Radovic-Moreno AF, Wang AZ et al (2008) Nanotechnology and aptamers: applications in drug delivery. Trends Biotechnol 26:442–449. https://doi.org/10.1016/j.tibtech.2008.04.006

Li M, Lin N, Huang Z et al (2008) Selecting aptamers for a glycoprotein through the incorporation of the boronic acid moiety. J Am Chem Soc 130:12636–12638. https://doi.org/10.1021/ja801510d

Li D, Chen Y, Liu Z (2015) Boronate affinity materials for separation and molecular recognition: structure, properties and applications. Chem Soc Rev 44:8097–8123. https://doi.org/10.1039/C5CS00013K

Li P, Shan B, Xu X et al (2021) Exploration of potential diagnostic value of protein content in serum small extracellular vesicles for early-stage epithelial ovarian carcinoma. Front Oncol 11. https://doi.org/10.3389/fonc.2021.707658

Liang Y, Eng WS, Colquhoun DR et al (2014) Complex N-linked glycans serve as a determinant for exosome/microvesicle cargo recruitment. J Biol Chem 289:32526–32537. https://doi.org/10.1074/jbc.M114.606269

Lis H, Sharon N (1986) Lectins as molecules and as tools. Annu Rev Biochem 55:35–67. https://doi.org/10.1146/annurev.bi.55.070186.000343

Llorente A, Skotland T, Sylvänne T et al (2013) Molecular lipidomics of exosomes released by PC-3 prostate cancer cells. Biochim Biophys Acta 1831:1302–1309. https://doi.org/10.1016/j.bbalip.2013.04.011

Lugini L, Valtieri M, Federici C et al (2016) Exosomes from human colorectal cancer induce a tumor-like behavior in colonic mesenchymal stromal cells. Oncotarget 7:50086–50098. https://doi.org/10.18632/oncotarget.10574

Luo J, Huang J, Cong J et al (2017) Double recognition and selective extraction of glycoprotein based on the molecular imprinted graphene oxide and boronate affinity. ACS Appl Mater Interfaces 9:7735–7744. https://doi.org/10.1021/acsami.6b14733

Lv J, Wang Z, Li F et al (2019) Reverse capture for selectively and sensitively revealing the N-glycome of serum exosomes. Chem Commun 55:14339–14342. https://doi.org/10.1039/C9CC06742F

Macedo-da-Silva J, Santiago VF, Rosa-Fernandes L et al (2021) Protein glycosylation in extracellular vesicles: structural characterization and biological functions. Mol Immunol 135:226–246. https://doi.org/10.1016/j.molimm.2021.04.017

Malaker SA, Riley NM, Shon DJ et al (2022) Revealing the human mucinome. Nat Commun 13:3542. https://doi.org/10.1038/s41467-022-31062-4

Martínez-Sáez N, Castro-López J, Valero-González J et al (2015) Deciphering the non-equivalence of serine and threonine *O*-glycosylation points: implications for molecular recognition of the Tn antigen by an anti-MUC1 antibody. Angew Chem Int Ed 54:9830–9834. https://doi.org/10.1002/anie.201502813

Matsuda A, Kuno A, Yoshida M et al (2020) Comparative glycomic analysis of exosome subpopulations derived from pancreatic cancer cell lines. J Proteome Res 19:2516–2524. https://doi.org/10.1021/acs.jproteome.0c00200

McCafferty J, Griffiths AD, Winter G, Chiswell DJ (1990) Phage antibodies: filamentous phage displaying antibody variable domains. Nature 348:552–554. https://doi.org/10.1038/348552a0

McKitrick TR, Eris D, Mondal N et al (2020) Antibodies from lampreys as smart anti-glycan reagents (SAGRs): perspectives on their specificity, structure, and glyco-genomics. Biochemistry 59:3111–3122. https://doi.org/10.1021/acs.biochem.9b01015

Melo SA, Luecke LB, Kahlert C et al (2015) Glypican-1 identifies cancer exosomes and detects early pancreatic cancer. Nature 523:177–182. https://doi.org/10.1038/nature14581

Mishra A, Behura A, Mawatwal S et al (2019) Structure-function and application of plant lectins in disease biology and immunity. Food Chem Toxicol 134:110827. https://doi.org/10.1016/j.fct.2019.110827

Mitić N, Kosanović M, Milutinović B et al (2018) Nano-sized CA125 antigen glycocamouflage: mucin - extracellular vesicles alliance to watch? Arch Biochem Biophys 653:113–120. https://doi.org/10.1016/j.abb.2018.06.017

Nies VJM, Struik D, Wolfs MGM et al (2018) TUB gene expression in hypothalamus and adipose tissue and its association with obesity in humans. Int J Obes 42:376–383. https://doi.org/10.1038/ijo.2017.214

Nishida-Aoki N, Tominaga N, Kosaka N, Ochiya T (2020) Altered biodistribution of deglycosylated extracellular vesicles through enhanced cellular uptake. J Extracell Vesicles 9:1713527. https://doi.org/10.1080/20013078.2020.1713527

Niu L, Song X, Wang N et al (2019) Tumor-derived exosomal proteins as diagnostic biomarkers in non-small cell lung cancer. Cancer Sci 110:433–442. https://doi.org/10.1111/cas.13862

Pan D, Chen J, Feng C et al (2019) Preferential localization of MUC1 glycoprotein in exosomes secreted by non-small cell lung carcinoma cells. Int J Mol Sci 20:323. https://doi.org/10.3390/ijms20020323

Pancer Z, Amemiya CT, Ehrhardt GRA et al (2004) Somatic diversification of variable lymphocyte receptors in the Agnathan sea lamprey. Nature 430:174–180. https://doi.org/10.1038/nature02740

Polonskaya Z, Savage PB, Finn MG, Teyton L (2019) High-affinity anti-glycan antibodies: challenges and strategies. Curr Opin Immunol 59:65–71. https://doi.org/10.1016/j.coi.2019.03.004

Prydz K (2015) Determinants of glycosaminoglycan (GAG) structure. Biomol Ther 5:2003–2022. https://doi.org/10.3390/biom5032003

Reynolds IS, Fichtner M, McNamara DA et al (2019) Mucin glycoproteins block apoptosis; promote invasion, proliferation, and migration; and cause chemoresistance through diverse pathways in epithelial cancers. Cancer Metastasis Rev 38:237–257. https://doi.org/10.1007/s10555-019-09781-w

Rilla K, Siiskonen H, Tammi M, Tammi R (2014) Hyaluronan-coated extracellular vesicles—a novel link between hyaluronan and cancer. In: Advances in cancer research. Elsevier, pp 121–148

Royo F, Cossío U, Ruiz De Angulo A et al (2019) Modification of the glycosylation of extracellular vesicles alters their biodistribution in mice. Nanoscale 11:1531–1537. https://doi.org/10.1039/C8NR03900C

Runz S, Keller S, Rupp C et al (2007) Malignant ascites-derived exosomes of ovarian carcinoma patients contain CD24 and EpCAM. Gynecol Oncol 107:563–571. https://doi.org/10.1016/j.ygyno.2007.08.064

Rutishauser U, Sachs L (1975) Cell-to-cell binding induced by different lectins. J Cell Biol 65:247–257. https://doi.org/10.1083/jcb.65.2.247

Saito S, Hiemori K, Kiyoi K, Tateno H (2018) Glycome analysis of extracellular vesicles derived from human induced pluripotent stem cells using lectin microarray. Sci Rep 8:3997. https://doi.org/10.1038/s41598-018-22450-2

Sakakura H, Mii S, Hagiwara S et al (2016) CD109 is a component of exosome secreted from cultured cells. Biochem Biophys Res Commun 469:816–822. https://doi.org/10.1016/j.bbrc.2015.12.063

Sakaue T, Koga H, Iwamoto H et al (2019) Glycosylation of ascites-derived exosomal CD133: a potential prognostic biomarker in patients with advanced pancreatic cancer. Med Mol Morphol 52:198–208. https://doi.org/10.1007/s00795-019-00218-5

Sancho-Albero M, Sebastián V, Sesé J et al (2020) Isolation of exosomes from whole blood by a new microfluidic device: proof of concept application in the diagnosis and monitoring of pancreatic cancer. J Nanobiotechnol 18:150. https://doi.org/10.1186/s12951-020-00701-7

Saraswat M, Joenväära S, Musante L et al (2015) N-linked (N-) glycoproteomics of urinary exosomes. Mol Cell Proteomics 14:2298. https://doi.org/10.1074/mcp.A114.040345ERR

Saroj S, Paul D, Ali A et al (2023) Probing aberrantly glycosylated Mucin 1 in breast cancer extracellular vesicles. ACS Appl Bio Mater 6:4944–4951. https://doi.org/10.1021/acsabm.3c00651

Sharon N (2004) History of lectins: from hemagglutinins to biological recognition molecules. Glycobiology 14:53R–62R. https://doi.org/10.1093/glycob/cwh122

Shimoda A, Tahara Y, Sawada S et al (2017) Glycan profiling analysis using evanescent-field fluorescence-assisted lectin array: importance of sugar recognition for cellular uptake of exosomes from mesenchymal stem cells. Biochem Biophys Res Commun 491:701–707. https://doi.org/10.1016/j.bbrc.2017.07.126

Shimoda A, Miura R, Tateno H et al (2022) Assessment of surface glycan diversity on extracellular vesicles by lectin microarray and glycoengineering strategies for drug delivery applications. Small Methods 6:2100785. https://doi.org/10.1002/smtd.202100785

Smith GP (1985) Filamentous fusion phage: novel expression vectors that display cloned antigens on the Virion surface. Science 228:1315–1317. https://doi.org/10.1126/science.4001944

Staubach S, Schadewaldt P, Wendel U et al (2012) Differential glycomics of epithelial membrane glycoproteins from urinary exovesicles reveals shifts toward complex-type N-glycosylation in classical galactosemia. J Proteome Res 11:906–916. https://doi.org/10.1021/pr200711w

Stein KE (1992) Thymus-independent and thymus-dependent responses to polysaccharide antigens. J Infect Dis 165:S49–S52. https://doi.org/10.1093/infdis/165-Supplement_1-S49

Sterner E, Flanagan N, Gildersleeve JC (2016) Perspectives on anti-glycan antibodies gleaned from development of a community resource database. ACS Chem Biol 11:1773–1783. https://doi.org/10.1021/acschembio.6b00244

Sun W, Du L, Li M (2010) Aptamer-based carbohydrate recognition. Curr Pharm Des 16:2269–2278. https://doi.org/10.2174/138161210791792877

Surman M, Hoja-Łukowicz D, Szwed S et al (2018) Human melanoma-derived ectosomes are enriched with specific glycan epitopes. Life Sci 207:395–411. https://doi.org/10.1016/j.lfs.2018.06.026

Surman M, Hoja-Łukowicza D, Szwed S et al (2019) An insight into the proteome of uveal melanoma-derived ectosomes reveals the presence of potentially useful biomarkers. Int J Mol Sci 20:3789. https://doi.org/10.3390/ijms20153789

Tan Z, Cao L, Wu Y et al (2020) Bisecting GlcNAc modification diminishes the pro-metastatic functions of small extracellular vesicles from breast cancer cells. J Extracell Vesicles 10:e12005. https://doi.org/10.1002/jev2.12005

Tombelli S, Minunni M, Mascini M (2005) Analytical applications of aptamers. Biosens Bioelectron 20:2424–2434. https://doi.org/10.1016/j.bios.2004.11.006

Wang D, Sun W (2014) Urinary extracellular microvesicles: isolation methods and prospects for urinary proteome. Proteomics 14:1922–1932. https://doi.org/10.1002/pmic.201300371

Weis WI, Drickamer K (1996) Structural basis of lectin-carbohydrate recognition. Annu Rev Biochem 65:441–473. https://doi.org/10.1146/annurev.bi.65.070196.002301

Williams C, Royo F, Aizpurua-Olaizola O et al (2018) Glycosylation of extracellular vesicles: current knowledge, tools and clinical perspectives. J Extracell Vesicles 7:1442985. https://doi.org/10.1080/20013078.2018.1442985

Williams C, Pazos R, Royo F et al (2019) Assessing the role of surface glycans of extracellular vesicles on cellular uptake. Sci Rep 9:11920. https://doi.org/10.1038/s41598-019-48499-1

Woof JM, Burton DR (2004) Human antibody–Fc receptor interactions illuminated by crystal structures. Nat Rev Immunol 4:89–99. https://doi.org/10.1038/nri1266

Xing R, Wang S, Bie Z et al (2017) Preparation of molecularly imprinted polymers specific to glycoproteins, glycans and monosaccharides via boronate affinity controllable–oriented surface imprinting. Nat Protoc 12:964–987. https://doi.org/10.1038/nprot.2017.015

Yamamoto M, Harada Y, Suzuki T et al (2019) Application of high-mannose-type glycan-specific lectin from Oscillatoria Agardhii for affinity isolation of tumor-derived extracellular vesicles. Anal Biochem 580:21–29. https://doi.org/10.1016/j.ab.2019.06.001

Yang Q, Goldstein IJ, Mei H-Y, Engelke DR (1998) DNA ligands that bind tightly and selectively to cellobiose. Proc Natl Acad Sci 95:5462–5467. https://doi.org/10.1073/pnas.95.10.5462

Yokose T, Kabe Y, Matsuda A et al (2020) O-glycan-altered extracellular vesicles: a specific serum marker elevated in pancreatic cancer. Cancers 12:2469. https://doi.org/10.3390/cancers12092469

Zhang J-M, Murakumo Y, Hagiwara S et al (2015) CD109 attenuates TGF-β1 signaling and enhances EGF signaling in SK-MG-1 human glioblastoma cells. Biochem Biophys Res Commun 459:252–258. https://doi.org/10.1016/j.bbrc.2015.02.093

Zhu L, Tian W, Yuan L et al (2023) Aptamer-based extracellular vesicle isolation, analysis and therapeutics. Interdiscip Med 1:e20220019. https://doi.org/10.1002/INMD.20220019

Zou G, Benktander JD, Gizaw ST et al (2017) Comprehensive analytical approach toward glycomic characterization and profiling in urinary exosomes. Anal Chem 89:5364–5372. https://doi.org/10.1021/acs.analchem.7b00062

Chapter 11
Imaging of EVs in the 3D Matrix Networks

Heikki Kyykallio and Kirsi Rilla

Abstract Extracellular vesicles (EVs) play a crucial role in intercellular and inter-tissue communication by transferring bioactive molecules, such as proteins, lipids, and nucleic acids, between cells. However, to understand their biological functions, and to evaluate their potential in diagnostic and therapeutic implications, it is crucial to visualize and analyse their morphology, kinetics, and biodistribution in 3D environment. This chapter describes the current advances, discusses the challenges of EV imaging in 3D environment, and describes examples for most commonly used labelling methods, imaging techniques, and in vitro and in vivo models for EV imaging as components of the extracellular matrix. The research activity in EV imaging techniques is increasing strongly at the moment, which is promising for unravelling their roles in various biological processes, developing diagnostic tools, exploring therapeutic applications, and advancing our understanding of cellular communication and disease mechanisms.

11.1 Introduction

Extracellular vesicles (EVs) are small, membrane-bound particles that are released by cells not just into the body fluids but also into extracellular environment of our tissues. The extracellular matrix (ECM) is a complex network of proteins and other molecules that surrounds cells and provides structural support for tissues. The cells embedded within the ECM that are responsible for production of ECM molecules, such as fibroblasts and mesenchymal stem cells, also actively produce EVs. These EVs may play a role in modulating the composition and organization of the ECM, as well as influencing the behaviour of other cells within the ECM. EVs have been found to interact with the ECM and its molecules in various ways, influencing the composition, organization, and function of ECM. For example, EVs can bind and

H. Kyykallio · K. Rilla (✉)
Institute of Biomedicine, University of Eastern Finland, Kuopio, Finland
e-mail: kirsi.rilla@uef.fi

K. Rilla (ed.), *Extracellular Vesicles as Matrix Messengers*, Biology of Extracellular Matrix 15, https://doi.org/10.1007/978-3-031-68811-9_11

interact with ECM components such as collagen, fibronectin, and glycosaminoglycans. These interactions can affect the distribution and organization of the ECM, influencing cell adhesion, migration, and signalling. On the other hand, EV interaction with extracellular environment may have an effect on the morphology, surface molecular composition, and motility of the cells (Rilla et al. 2019).

Overall, the interactions between EVs and the ECM are an important area of research in the field of EV biology. Understanding these interactions could provide insight into the mechanisms underlying various biological processes and could also have implications for the development of new therapies for a range of diseases. Advanced imaging techniques have revolutionized EV research by providing the means to visualize, characterize, and understand the functions of EVs in intercellular communication. However, most imaging experiments are performed in isolated EV preparations or in conventional two-dimensional (2D) cell cultures. While conventional 2D imaging techniques offer valuable information on EVs, 3D imaging techniques can reveal additional insights on their structure, spatial localization in their native microenvironment, distribution, heterogenicity, and functions within complex biological systems. This chapter summarizes the current methodology and models for 3D imaging of EVs and discusses the challenges and advances of EV visualization and analysis in 3D environment.

11.2 Key Interactions Between EVs and Matrix Molecules in Tissues

In tissues, EVs can directly bind or interact with the ECM components, such as collagen, fibronectin, laminin, and hyaluronic acid. This binding can be mediated by surface receptors on EVs and adhesion molecules such as integrins (Altei et al. 2020) to the ECM components. The interaction between EVs and ECM components can influence the distribution and organization of the matrix and may also regulate cellular processes such as adhesion, migration, tissue remodelling, and regeneration processes (Rilla et al. 2019).

Additionally, EVs can incorporate ECM-related molecules into their membranes or cargo during their biogenesis or release. For example, EVs derived from cancer cells can acquire matrix metalloproteinases (MMPs), enzymes that degrade ECM components and facilitate cancer cell invasion and metastasis (Sun et al. 2023). EVs secreted by certain cell types, such as mesenchymal stem cells (Lee et al. 2023) and fibroblasts (Dong et al. 2023), can facilitate ECM remodelling and tissue repair and participate in prevention of scarring (Li et al. 2022a). These EVs can enhance the synthesis of new ECM components and stimulate the migration and proliferation of recipient cells involved in tissue repair processes. Additionally, the mechanical properties of the ECM, including its stiffness, can be influenced by EVs. EVs released by cancer cells, for example, can promote ECM stiffening by inducing the deposition of excessive ECM components or triggering cellular responses that lead to increased ECM production. On the other hand, altered matrix stiffness can impact EV secretion affecting cellular behaviour and contribute to disease progression (Wu et al. 2023). The interactions between EVs and matrix molecules are complex and multifaceted, and further understanding of these interactions is required to shed light on the role of EVs in physiological and pathological processes and for therapeutic implications in various diseases.

11.3 Why Imaging of EVs in 3D Environment?

After the release of EVs, the surrounding microenvironment and biological barriers between organs affect EV distribution, diffusion, and influence their physiological functions in vivo. Thus, studying EVs in 3D environments provides a more comprehensive understanding of their functions, interactions, and therapeutic potential than traditional 2D cell culture systems, facilitating the translation of EV research into clinical applications. Despite recent advances in the isolation of EVs from 3D cell cultures (Kyykallio et al. 2022) and from human tissues (Jang et al. 2019; Crescitelli et al. 2021), isolation of EVs from complex matrices or tissue samples is still challenging. The complexity of tissues makes it difficult to isolate EVs specifically from the tissue of interest, and EVs derived from different cell types within the tissue may vary in size, surface markers, and cargo. Contamination from cellular debris, proteins, lipids, and other extracellular components can interfere with EV isolation and downstream analysis. Additionally, EVs are present at relatively low concentrations in tissues, which makes it challenging to isolate sufficient numbers of EVs for downstream analysis, especially when working with small volume tissue samples. Because EVs are delicate structures that can be sensitive to harsh processing conditions, such as mechanical disruption or high shear forces, the isolation process may cause structural damage or aggregation of EVs. Thus, it is crucial to investigate EVs in their natural, authentic environment in situ. However, this requires advanced imaging techniques and expertise, which will be discussed in this chapter.

Modern biological imaging techniques enable in situ characterization of EVs in a 3D environment which more closely mimics the physiological conditions in which EVs interact and function with their natural environment. Imaging provides a possibility to analyse EVs without complex, time-consuming, and tedious isolation protocols. Furthermore, it allows the investigation of EV movements, interactions, and cargo delivery within a more biologically relevant context, leading to a better understanding of their behaviour and role in normal physiological processes and disease states (Verweij et al. 2021). For example, studying EVs in 3D tumour models or tumour tissue allows for the investigation of their roles in tumour growth, metastasis, immune evasion, and drug resistance. Furthermore, it provides insights into the complex interplay of EVs between tumour cells, stromal cells, and the surrounding microenvironment (Loconte et al. 2023). Additionally, EVs have emerged as potential vehicles for targeted drug delivery and therapeutics. Imaging analyses of EVs in 3D environments can help assess their ability to penetrate tissues, overcome biological barriers, and deliver therapeutic cargo to specific tissues (Gupta et al. 2020). It also enables the evaluation of EV-based drug delivery systems within a more representative tissue context, aiding in the development of more effective, safe, and targeted therapies.

11.4 Challenges of EV Imaging in 3D Environment

Three-dimensional imaging techniques offer valuable insights into complex biological structures and processes. However, there are several challenges associated with imaging of EVs in 3D environment that need to be addressed. First, achieving adequate spatial resolution for detection of small-sized EVs in 3D environment can be challenging, especially when imaging large volumes or complex structures. Balancing resolution with imaging depth is a critical consideration in 3D imaging. Secondly, the challenge of EVs low contrast due to small size, little cargo, and lack of specificity in complex tissue context and limitations in labelling are restricting the use of techniques.

Additionally, when dealing with thick samples, light penetration into the sample may be a limiting factor. Thus, maintaining high-quality imaging throughout the depth of a sample is often a challenge. Light scattering and absorption by biological tissues can limit the penetration depth, leading to reduced image quality and signal attenuation. This issue becomes more pronounced in dense samples. Techniques such as multiphoton microscopy or adaptive optics can help overcome these challenges to some extent (Perrin et al. 2019). Additionally, optical imaging requires tissue transparency, and the opacity of the sample can be improved by tissue clearing techniques (Tian et al. 2021). Use of transparent organisms, such as zebrafish discussed later in this chapter, offers one solution for these light penetration issues.

An additional challenge is possible limited depth of penetration of the fluorochromes: especially larger molecules may diffuse more slowly through thick samples and have limited penetration depth into thick samples. This can lead to incomplete or uneven staining and may require longer incubation times for adequate penetration. Additionally, tissue autofluorescence, which is the intrinsic fluorescence emitted by endogenous molecules in biological samples, can interfere with the detection of fluorochrome signals, making it challenging to distinguish the specific signal of small particles from the background. This can reduce the signal-to-noise ratio and impact the quality of the images. Also, incomplete or inappropriate fixation and permeabilization steps can hinder the penetration of fluorochromes. Proper sample preparation, including appropriate fixation, permeabilization, clearing, and labelling methods need to be selected to preserve sample integrity for successful staining in thick samples while providing adequate contrast and specificity for imaging. For example, specific fixation methods, such as 1-ethyl-3-(3-dimethylaminopropyl) carbodiimide (EDC) fixation, have been shown to preserve EVs for visualization in normal and cancer tissues in situ more efficiently than traditional formalin fixation alone (Gupta et al. 2019).

Acquiring high-resolution 3D images for EV detection requires scanning multiple focal planes or capturing large image stacks. This process can be time-consuming, especially when imaging large volumes or dynamic processes in live samples. Rapid imaging techniques, such as light-sheet microscopy or two-photon microscopy, can help to mitigate this challenge to some extent. Additionally, 3D imaging generates large amounts of data, which may be challenging in terms of data storage, processing, and analysis. Proper data management strategies, including compression algorithms, efficient storage systems, and computational resources, are necessary to handle the significant data volumes generated during 3D imaging experiments. Advanced visualization techniques, such as volume rendering or virtual reality, can aid in the interpretation and presentation of 3D image data. Addressing all these challenges often requires a combination of technical advancements, method optimization, and expertise in specific imaging techniques, analysis, and interpretation of results.

11.5 Potential of Imaging Techniques for the Visualization and Analysis of EVs in 3D Environment

Despite the limitations described above, advanced high-resolution imaging techniques, used alone or in combination, provide insights into the size, morphology, composition, and cargo of EVs, advancing our understanding of their morphology and spatial distribution, intercellular communication and their roles in various physiological and pathological processes. The most suitable techniques for EV imaging are described in this section.

Fluorescence microscopy techniques, such as confocal microscopy and super-resolution microscopy, are widely used for imaging EVs. Fluorescent dyes or antibodies can be used to label specific EV markers or cargo molecules, allowing for the visualization and characterization of EVs in complex biological samples. Fluorescence microscopy techniques can provide information about EV localization, distribution, and cargo content.

Confocal microscopy is a widely used technique for imaging EVs in 3D matrices. It uses laser scanning to generate optical sections at different depths within the sample eliminating the off-focus signal. By acquiring a series of images at different focal planes, a 3D image stack can be reconstructed. Confocal microscopy provides high-resolution imaging and allows for the visualization of fluorescent EVs released from cells within the 3D matrices, such as basement membrane extract gels (Rilla et al. 2013) and type I collagen-based 3D gels (Deen et al. 2016). However, confocal microscopy is limited in its ability to penetrate deep into thick samples due to light scattering and absorption by the sample. As the depth increases, the image quality and resolution decrease, resulting in reduced image contrast and signal-to-noise ratio. Specific techniques, such as multiphoton and superresolution microscopy, help to overcome some of these limitations.

Multiphoton microscopy is an advanced imaging technique that utilizes near-infrared light to excite fluorescent molecules, which experiences less scattering and absorption in tissue and allowing deeper imaging compared to traditional single-photon excitation microscopy. This technique is suitable for imaging EVs in 3D matrices as it can penetrate through thick samples and provides high-resolution imaging in live tissues. Two-photon microscopy enables the visualization of EV dynamics and interactions within 3D environments.

Superresolution microscopy techniques have revolutionized the field of optical imaging by surpassing the diffraction limit of light. They offer enhanced spatial resolution and sensitivity, allowing to analyse the nanoscale features and molecular composition of EVs with unprecedented detail. The most common superresolution techniques include structured illumination microscopy (SIM), stimulated emission depletion microscopy (STED), and direct stochastic optical reconstruction microscopy (dSTORM), that have been utilized for analysis of stem cell-derived EVs (Nizamudeen et al. 2018). However, many of these techniques face certain restrictions when applied to thick samples, they usually require fixation and are time-consuming.

Light-Sheet Fluorescent Microscopy (LSFM) is an imaging technique that uses a thin sheet of light to illuminate the sample. It provides rapid and low phototoxicity imaging of large 3D volumes. Light-sheet microscopy is suitable for imaging EVs in 3D matrices as it can generate high-resolution images with minimal sample bleaching and photodamage. For example, lattice light-sheet fluorescence microscopy with enhanced resolution compared to conventional LSFM has been utilized for imaging intercellular interactions of EVs in a co-culture model of chronic lymphocytic leukaemia and stromal cells (Elgamal et al. 2020).

Time lapse imaging combined to particle tracking techniques can be used to a track kinetics and analyse the trajectories of EVs in 3D cultures or tissues. By tracking the movement of individual EVs over time, parameters such as speed, directionality, and diffusion characteristics can be quantified. Particle tracking algorithms and software can be employed to process the acquired images and extract the trajectories of EVs. For tracking, EVs can be labelled with fluorescent dyes or tagged with fluorescent proteins by expressing them in donor cells. Several studies have utilized fluorescent labelling of EVs as a common approach to track their movement in cells (Liebel et al. 2020) or tissues (Lai et al. 2015).

Intravital imaging techniques allow for the direct visualization and tracking of EV movement within live tissues in animal models. Multiphoton microscopy or spinning disk confocal microscopy can be used to capture high-resolution images of EVs in the intact tissue environment. This approach provides valuable insights into the behaviour of EVs in a physiologically relevant context. For example, the bioluminescent labelling of EVs using different luciferase enzymes tethered to CD63 can be used to monitor distribution of EVs in real time directly after the injection into mice (Gupta et al. 2020).

Also, label-free optical techniques offer a promising approach for EV imaging in vivo. Utilization of third-harmonic generation (THG) have shown promising data to visualize EVs even in the unperturbed human breast tumour microenvironment in ex vivo–resected specimens (Sun et al. 2018), living tumour-bearing animals, and fresh excised human breast cancer tissue (You et al. 2019).

Electron Microscopy (EM) techniques, including transmission electron microscopy (TEM) and scanning electron microscopy (SEM) are powerful imaging techniques providing ultrastructural resolution and label-free visualization of EVs in their native environment. SEM provides information and details on surface morphology of EVs (Fig. 11.1), and TEM offers the highest resolution for the direct visualization of EVs and provides detailed structural information (Fig. 11.2). However, EM techniques are not ideal for high-content imaging, quantification, or analysis of size distribution due to their laborious sample preparation and imaging. Additionally, conventional EM techniques provide only 2D information on the samples.

Electron tomography is a technique that utilizes TEM to reconstruct 3D images of EVs within 3D matrices. It involves acquiring a series of TEM images at different

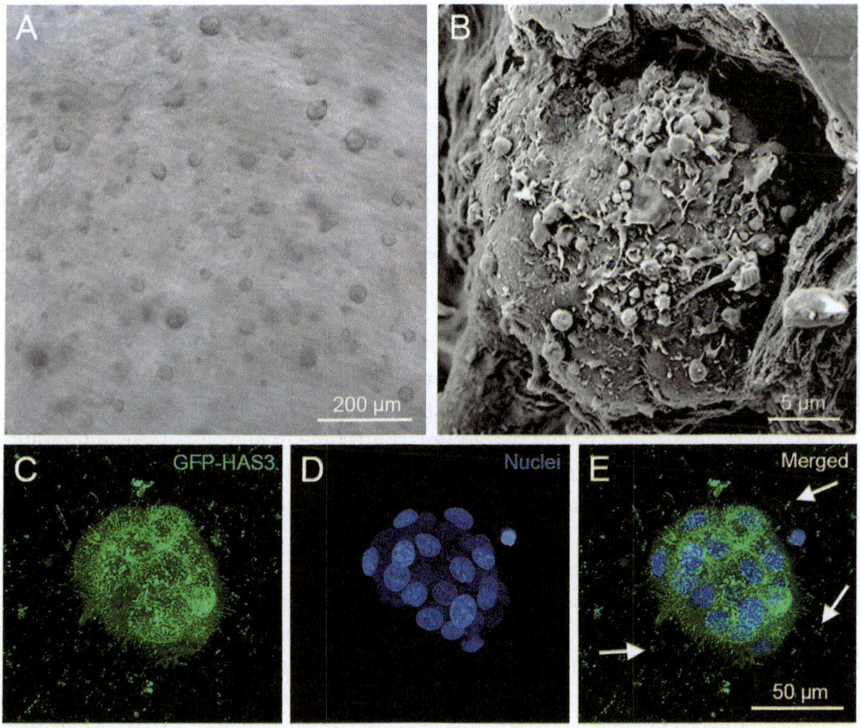

Fig. 11.1 MCF7-GFP-HAS3 spheroids in nanofibrillar cellulose (NFC) cultures. MCF7 breast cancer cells form spheroids during 7 days of culture in 0.5% NFC (**a**). SEM image of a GFP-HAS3-induced spheroid growing in 0.5% anionic NFC (aNFC). (**b**) Shows membrane blebbing and several vesicular structures being attached on the surface of the spheroid. Confocal 3D rendering of a live MCF7-GFP-HAS3 spheroid in 0.33% aNFC stained with NucBlue live nuclear stain (**c**–**e**). The cells form long protrusions and are highly positive for GFP-HAS3 as previously described in literature and secrete EVs into the surrounding NFC matrix (arrows in **e**)

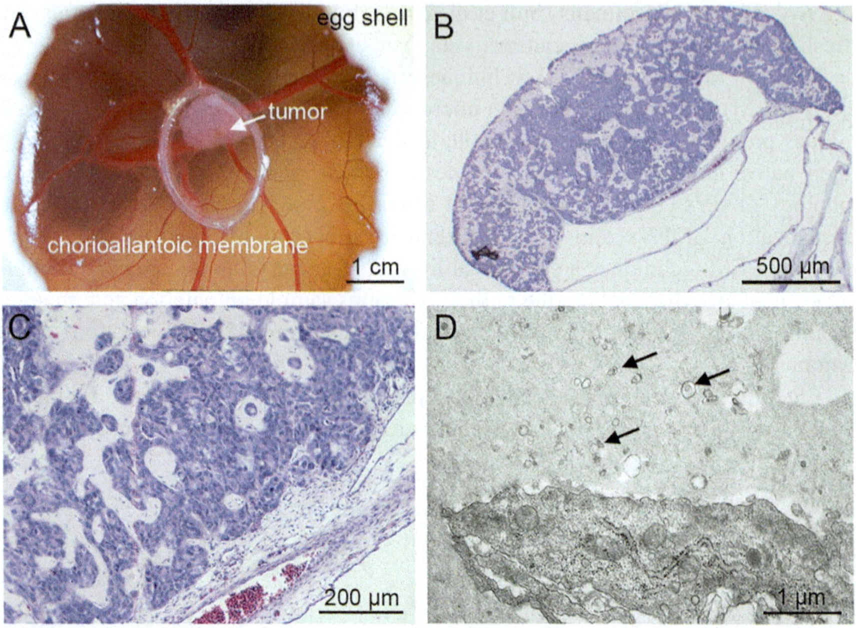

Fig. 11.2 Detection of EV secretion by tumours in chorioallantoic membrane (CAM) cultures. A typical example of MNK74 gastric cancer cell-derived CAM tumour (EDD13) is shown in (**a**). Whole section of HE-stained tumour is shown in (**b**), and higher magnification image from the same tumour in (**c**). TEM reveals EV secretion by tumour cells into the ECM of the tumour (arrows) in (**d**)

tilt angles and computationally reconstructing a 3D volume. Electron tomography provides high-resolution structural information about detailed morphology of EVs, their size distribution, and their interactions within the 3D matrix (Capra et al. 2022). Interestingly, in a recent work Olofsson Bagge et al. used electron tomography for three-dimensional reconstruction of interstitial spherical structures resembling EVs in metastatic and normal human liver tissue samples (Olofsson Bagge et al. 2023).

Correlative light and electron microscopy (CLEM) is a powerful imaging technique that combines the strengths of both light microscopy (LM) and electron microscopy (EM) to correlate functional information obtained from LM with detailed structural information provided by EM. This allows for the precise localization of EVs within cells or tissues and provides information on their subcellular origin and release mechanisms (Hyenne et al. 2019; Verweij et al. 2019). Apart from the growing interest in these techniques in in vitro EV studies (Mittelbrunn et al. 2011; Arasu et al. 2019; Aaltonen et al. 2022), CLEM has been also utilized in in vivo experiments to study distribution and inter-organ communication of tumour-derived EVs (Hyenne et al. 2019) and CD63-pHluorin expressing EVs (Verweij et al. 2019) in zebrafish embryos.

11.6 3D In Vitro and In Vivo Models for In Situ EV Imaging

To understand the biology and function of EVs in health and disease, they need to be studied in their physiological context. Isolation and enrichment of EVs from cell or tissue cultures provides insight on overall EV composition and cargo, but simultaneously loses information about their cellular origin and fate in vivo, including their biodistribution, clearance, and behaviour in circulation or within tissues. Both 3D in vitro and in vivo models can provide information about EV biogenesis and function, while different in vivo model systems are essential in understanding the impact of EVs in development, normal physiology, and pathologies. Multitude of different 3D in vitro and in vivo models for EV imaging have been established (Fig. 11.3), covering a variety of physiological functions and diseases. 3D cell cultures offer a simple and accessible platform to study EV secretion and uptake mechanisms in a more physiological context compared to traditional monolayer cell cultures. Chorioallantoic membrane is a model that combines the simplicity of in vitro and the growth environment of in vivo. *Drosophila* and *Caenorhabditis elegans* have greatly contributed to our understanding of cell biology as genetic model organisms and provided invaluable information about EV signalling and its impact on development and behaviour. Zebrafish is an emerging model for EV imaging and allows real-time tracking of individual EVs from both endogenous and exogenous sources in a vertebrate organ system. Mouse has been used in various applications of in vivo EV imaging and greatly increased our understanding of EV behaviour, such as their biodistribution and organotropism. Finally, human tissue samples are potential models to understand EV function within human tissues. In the following sections these models and their contribution to our understanding of EVs through imaging are discussed further.

11.7 3D Cell Culture

Monolayer cultures have been long utilized in imaging EV biogenesis and their interactions with target cells. The benefit of having a single layer of cells is that it enables microscopic tracking of EV origin, such as shedding or budding from plasma membrane or from tips of cellular protrusions (Rilla et al. 2013), as well as their binding and internalization into receiving cells. With high-resolution imaging it is even possible to track labelled individual EVs inside a single cell (Liebel et al. 2020). It is widely acknowledged, however, that the two-dimensional cellular architecture dramatically affects the morphology, diversity of phenotypes, intercellular interactions, polarization, division, and gene expression of cells, as the natural in vivo organization in tissues is three-dimensional (Kapałczyńska et al. 2018). Therefore, contact with other cells and with the surrounding ECM in tissue has a direct influence on cellular behaviour. EVs are known to reflect the state of the original cell and their biogenesis is affected by culturing cells in 3D (Rocha et al. 2019;

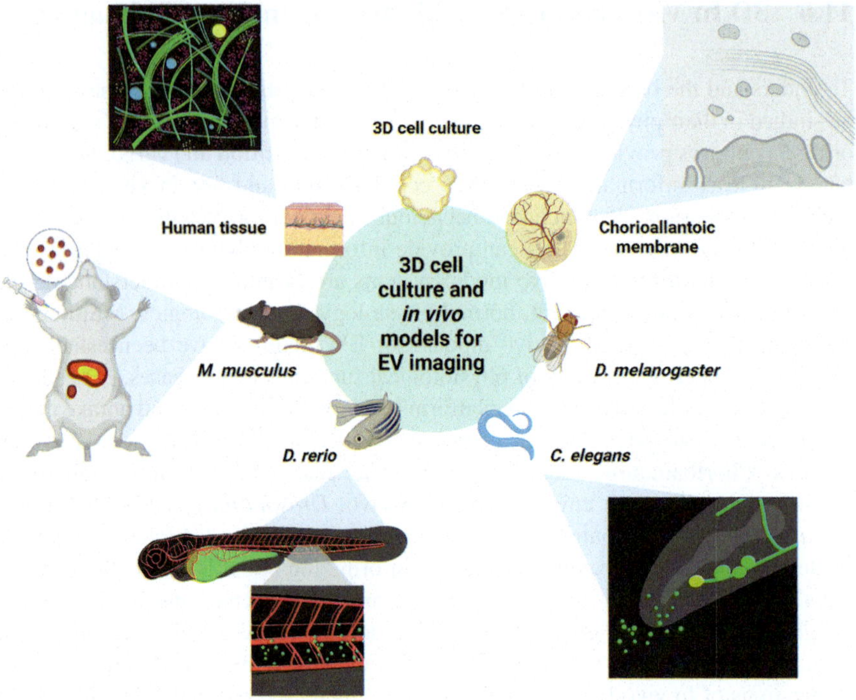

Fig. 11.3 3D cell culture and in vivo models for EV imaging. An illustration of commonly used in vitro and in vivo models for imaging EVs in 3D environment. Examples of EV imaging techniques depict TEM image from EV-secreting cell grown on chorioallantoic membrane, secretion of fluorescently tagged EVs from *C. elegans* neural system by confocal microscopy, biodistribution EVs derived from fluorescent EV-marker expressing tissue in *D. rerio*, biodistribution of exogenous labelled EVs in mouse organs by noninvasive in vivo imaging, and image of human tissue by label-free multiphoton microscopy. Created with BioRender.com

Thippabhotla et al. 2019). Typically, this results in an increase in EV secretion (Rocha et al. 2019; Thippabhotla et al. 2019; Millan et al. 2021), change in EV size, or alterations in EV cargo (Rocha et al. 2019; Thippabhotla et al. 2019; Millan et al. 2021; Villasante et al. 2021). Based on these features, EVs derived from 3D culture have been shown to more closely resemble in vivo EVs (Thippabhotla et al. 2019). Functional differences between 2D and 3D cultured EVs have also been reported (Millan et al. 2021; Kusuma et al. 2022), suggesting that the functionality of EVs may be altered when secreting cells are cultured in 2D.

3D cell cultures can be roughly divided into scaffold-free and scaffold-containing cultures. While both support formation of three-dimensional cell structures, the origin of a single structure differs based on the culture method used. Scaffold-free cultures do not contain any supporting matrix, resulting in spheroids or aggregates that freely float in the culture medium. These kinds of cultures are typically grown by preventing cell adhesion to the culture plate, which can be achieved by using ultra-low attachment plates (Costa Verdera et al. 2017), agarose coating (Millard et al. 2018), nanoculture plates (Taha et al. 2020), or microcarriers (Haraszti et al. 2018). Depending on cell type, spheroidal 3D structures are formed by spontaneous aggregation, or the process can be assisted by using centrifugation or agitation, which typically results in one 3D structure per a single well. While scaffold-free cultures lack the mechanical support of the ECM, 3D cellular architecture was enough to cause changes in gastric cancer cell EV microRNA and protein cargo (Rocha et al. 2019).

Scaffold-free 3D cultures can be utilized to evaluate EV uptake and penetration depth into spheroids (Costa Verdera et al. 2017; Ye et al. 2018; Millard et al. 2018; Taha et al. 2020; Yakavets et al. 2021; Nguyen et al. 2022). In a typical experiment, spheroids are incubated with fluorescently labelled EVs and imaged using confocal microscopy. Internalization of PKH67-labelled A431 human epidermoid carcinoma EV into HeLa cell spheroids was limited to the outer cell layers of the spheroids and was restricted by clathrin-independent endocytosis inhibitor genistein and macropinocytosis inhibitor 5-(*N*-ethyl-*N*-isopropyl)amirolide (Costa Verdera et al. 2017). In another work, natural mouse L929 fibroblast EVs and L929-EVs conjugated with either LDL or Lysine-Leucine-Alanine (KLA)-LDL peptides were stained with PKH26 and incubated with U87 glioma spheroids before confocal imaging. Internalization of all EVs was mainly restricted to the outer cell layers, although the peptide-conjugated EVs seemed to have an increased penetration depth compared to natural EVs (Ye et al. 2018).

Confocal imaging is possible for live or fixed spheroids and enables 3D-visualization of the complete spheroid. On the other hand, frozen-cut spheroid sections can be stained using antibodies to allow more accurate information of the EV fate and allow imaging of larger spheroids where the penetration depth of confocal microscopy is a limiting factor. EV-mediated transfer of MMP3 from rapidly metastatic murine cancer cell line LuM1 to MMP3-knockout LuM1 spheroids was investigated from paraffin-cut thin sections by immunofluorescent stainings (Taha et al. 2020). Human umbilical vascular endothelial (HUVEC) EVs loaded with 3,3′,3″,3‴-(2,3-dihydroporphyrin-5,10,15,20-tetrayl)tetraphenol (mTHPC) demonstrated penetration depth up to 100 μm into HT29 human colorectal adenocarcinoma spheroids, far exceeding that of free mTHPC and mTHPC-loaded liposomes (Millard et al. 2018). Similarly, doxorubicin-loaded EVs from CT26 colorectal cancer cells and 4T1 murine mammary carcinoma cells were able to penetrate deeply into CT26 spheroids (Nguyen et al. 2022). Cell line derived spheroids typically consist of a single type of cells, which poorly represents the cellular heterogeneity of in vivo tissues. Yakavets et al. generated stroma-rich spheroids by co-culturing FaDu squamous carcinoma and MeWo fibroblast cells with liquid overlay technique. HUVEC-EVs loaded with mTHPC were incubated with FaDu and MeWo co-culture spheroids and cryosections from the spheroids were collected and imaged with an upright epifluorescence microscope, and the penetration depth between liposomes and EVs were measured from the fluorescent signal (Yakavets et al. 2021).

In scaffold-containing cultures cells are typically suspended into extracellular-matrix mimicking material to form a culture of multiple spheroids. Common scaffold materials are generally either biologically derived or synthetic hydrogels.

Biologically derived scaffolds consist of molecules commonly found from ECM such as collagen I, hyaluronic acid (Villasante et al. 2021), or fibrin (Bhattacharya et al. 2020). One of the most extensively used biologically derived scaffolds is basement membrane extract isolated from murine Engelbreth-Holm-Swarm sarcoma cells, also commonly referred by one of its commercial names "Matrigel" (Hughes et al. 2010). It is widely used in 3D culture systems due to its high biocompatibility and in vivo basement membrane-like composition. Synthetic hydrogels are composed of cross-linked polymeric material that exhibits tissue-like stiffness, therefore resembling the natural ECM (Caliari and Burdick 2016). An example of a biologically derived but non-mammalian derived hydrogel is nanocellulose (Fig. 11.1) that can be extracted from cellulose-rich plants using mechanical, enzymatical, and chemical methods, or alternatively be produced by bacteria. In aqueous solutions, nano-sized cellulose fibres form a hydrogel with a net-like porous structure (Curvello et al. 2019).

Both biological scaffolds and synthetic hydrogels are typically optically transparent, allowing in situ imaging of cell cultures. EV imaging in matrices requires excellent imaging properties due to the small size of EVs that is often close to the detection limit of light microscopy. Therefore, optimal 3D scaffold allows high-resolution imaging and is free from autofluorescence and any contaminating biological material. Due to the porous structure of many hydrogels, secreted EVs can move through the scaffold, even if the pore size of the material is smaller than that of the diameter of the EV (Lenzini et al. 2020), but especially larger vesicles can get immobilized in the matrix, allowing imaging vesicles around cells (Capra et al. 2022). Scaffolded cultures also make possible to monitor EV biogenesis. An example is EV shedding from the tips of long membrane protrusions (Rilla et al. 2013; Nishimura et al. 2021; Capra et al. 2022). EV uptake can also be studied similarly as in scaffold-free cultures by treating 3D cultures with labelled EVs and imaging the spheroids using confocal microscopy. Interactions of patient adipose tissue-derived mesenchymal cell EVs with human fibroblast-like synoviocytes and articular chondrocytes were demonstrated using fibrin 3D culture on a microfluidic device (Ragni et al. 2020). Additionally, scaffolded cultures make it possible to monitor EV behaviour and functions inside extracellular matrices. It has been demonstrated that cells use EVs as environmental cues for migration (Sung et al. 2020; Aaltonen et al. 2022). Small EV-mediated migration of HT1080 cells was demonstrated using collagen 3D culture (Sung et al. 2020).

11.8 Chorioallantoic Membrane

Chicken chorioallantoic membrane (CAM) is a highly vascularized simple membrane that is formed during chick embryogenesis. It has been utilized as an in vivo model for tumour transplantation for over a century due to its rapid growth, accessibility, cost-effectiveness, and absence of specific immune system on the first 15 days from fertilization. More recently, CAM model has been utilized in the research fields of vascular development and angiogenesis, tumour formation and growth, metastasis, drug delivery and screening, and toxicology. CAM development starts during avian embryogenesis on day 3 from fertilization when allantois starts to extend extra-embryonically from the ventral wall of the embryo's endodermal hindgut. The allantois then partially fuses with chorion between days 5 and 6 from fertilization to form the CAM. The formed CAM rapidly expands to cover the entire inner surface of the shell, while extensive vascular network develops by day 10 and the differentiation into CAM is complete by day 13. CAM is a vastly vascularized multifunctional structure that during embryogenesis enables exchange of respiratory gases, provides calcium supply for osteogenesis via minerals from the egg shell, transports sodium and chloride from the allantoic sac, and stores waste products such as urea (Nowak-Sliwinska et al. 2014; Ribatti 2016).

For CAM assay, fertilized eggs are incubated at 37–38 °C in a humidified incubator under gentle rotation or by occasionally rotating the eggs manually. The embryo can be either kept undisturbed inside the egg for *in ovo* assay or alternatively be transferred to a petri dish at day 3 post-fertilization for further cultivation and an *ex ovo* assay. For *in ovo* assay, the initiation also begins on day 3 post-fertilization, by opening the eggshell with a small hole that can be extended on the day the experiment starts. The benefit of *ex ovo* cultivation is better access to the CAM itself, since performing the assay *in ovo* only allows experimenting and access to the CAM through the small hole in the shell (diameter of around 3 cm). Keeping the embryo *in ovo*, however, drastically improves the survival rate of the embryo and does not alter the physiological environment of the embryo. Starting day for the experiments is dependent on the experimental setup, typically transplantation of cells is done after day 7 when the vasculature of the CAM has formed. However, by day 15 the slowly developing immune system reaches its physiological activity, which may lead to immunological reactions. Depending on the local regulations for use of animals, avian embryos are typically allowed to develop up to day 21 post-fertilization when they reach the hatching state (Nowak-Sliwinska et al. 2014).

For EV research, CAM can offer a simple and accessible model to visualize EVs in vivo. CAM is especially suitable for intravital imaging due to its simple structure, allowing visualization of EVs secreted by labelled cells in live in vivo system. HT1080 human fibrosarcoma cells expressing pHluo_M153R_CD63/mCherry-CaaX were injected into the allantoic vein of *ex ovo* chicks day 10 post-fertilization and the CAM was imaged after 24 h with intravital imaging to visualize secreted labelled small EVs (Sung et al. 2020). EVs can also be imaged from CAM cultivated tumours that have been collected and fixed before preparation into paraffin-cut sections. Hyaluronan-coated EVs were visualized from basement membrane extract-embedded MCF10CA breast cancer tumours, grown implanted onto CAM in *in ovo* assay using immunofluorescent staining of tissue sections and confocal microscopy (Aaltonen et al. 2022). Similarly CD44-positive EVs were detected from fixed and paraffin embedded MKN74 gastric cancer tumours, embedded in matrix and grown on CAM from fixed and histologically stained paraffin tissue sections (Härkönen et al. 2019). Besides light microscopy, CAM tissue biopsies can be imaged using EM, to visualize EVs in higher resolution (Bidarimath et al. 2017). An example of EV detection by EMin CAM tumours is shown in Fig. 11.2.

11.9 *Caenorhabditis elegans*

Caenorhabditis elegans (*C. elegans*) is a nematode with the size of about 1 mm that lives in the soil surface. This simple multicellular eukaryote has been extensively used as a model organism in developmental biology, ageing, and toxicology. Many characteristics make *C. elegans* a powerful model for a variety of research fields. The life cycle of *C. elegans* is short—3 days from egg to adult and 2–3-week lifespan and it rapidly reproduces in great numbers, allowing large-scale production of model organisms (Meneely et al. 2019). The genome of *C. elegans* is well-studied and completely sequenced, with 60–80% of human genes having an orthologue in the genome (Shaye and Greenwald 2011). *C. elegans* is also simple to culture: despite naturally living in soil, it can be grown on small plates with solid medium, or in liquid, containing *Escherichia coli* as a food source (Meneely et al. 2019). Most importantly from EV imaging point of view, *C. elegans* is transparent, which allows monitoring EV biogenesis, secretion, and biodistribution using fluorescent microscopy.

EV research conducted using *C. elegans* has mainly focused on molecular mechanisms of EV biogenesis and release. Transgenic animals expressing fluorescently labelled proteins related to EV biogenesis allow monitoring the importance of known conserved EV biogenesis factors for EV secretion and their effect on animal development and behaviour. In *C. elegans*, EV biogenesis from ciliated sensory neurons is one of the most comprehensively studied processes of EV secretion. Males possess a subset of ciliated sensory neurons where polycystic kidney disease gene product orthologues PDK-2 and LOV-1 are exclusively expressed and localize to the cilia (Barr et al. 2001). Wang et al. investigated the cargo of ciliated sensory neuron EVs via GFP-labelled fusion proteins expressed in the ciliary neurons. Polycystins PKD-2 and LOV-1, as well as cilia-enriched protein CWP-1 were found to be secreted in EVs. Secretion of PKD-2::GFP -positive EVs was independent of endosomal sorting complexes ESCRT-0 and -I, but dependent on intact cilia, demonstrated with mutants of intraflagellar transport polypeptides and kinesin-3 KLP-6 (Wang et al. 2014). More recently, superresolution confocal microscopy revealed that PDK-2 and LOV-1 are co-trafficked and -transported as ciliary EV cargo. The C- and N-terminal fragments of LOV-1 are separated and the C-terminal fragment is co-localized with PDK-2 in the EVs secreted from the ciliary tips (Walsh et al. 2022). Additionally, ciliary protein CIL-7 has been found to regulate secretion of polycystin-containing EVs, shown with GFP-tagged proteins and mutants using light microscopy and TEM (Maguire et al. 2015). Mutations in cephalic male neuron-specific α-tubulin isoform *tba-6* caused abnormal accumulation of PDK-2::GFP and CIL-7::GFP and luminal EVs in the cephalic male neurons, demonstrated by serialTEM. Additionally, *tba-6* mutant animals released less

PDK-2::GFP-containing EVs compared to wild type and the secreted EVs contained KLP-6::GFP which is not EV cargo in the wild types (Silva et al. 2017).

Recently, a comprehensive study of *C. elegans* EV cargo from non-ciliary and ciliary sources identified ciliary-derived EV-enriched proteins: CD9 tetraspanin orthologue TSP-6, double-stranded RNA transporter SID-2, ectonucleotide pyrophosphatase/phosphodiesterase ENPP-1, and minichromosome maintenance protein MCM-3. Their expression, protein localization, and secretion via EVs were investigated using CRISPR/Cas9-inserted mScarlet fluorescent tag, co-expressed with PKD2::GFP in EV-secreting cilial neurons and imaged using superresolution confocal microscopy (Nikonorova et al. 2022). Another study showed that TSP-6 and human CD63 orthologue TSP-7 localize to the cilial region and are a cargo of EVs taken up by the surrounding glial cells by utilizing a similar setting of wrmScarlet tagged fusion proteins and confocal microscopy (Razzauti and Laurent 2021).

While many studies have focused on the biogenesis and cargo of *C. elegans* EVs, transparency of the animal enables studies that follow uptake and biodistribution of exogenous EVs. Uptake and distribution of microalgae derived EVs, called nanoalgosomes, was investigated by administrating Di-8-Anepps-labelled EVs in three different manners: by soaking, *in solido*, and by injection. Whilst fluorescent signal was detected in the intestine of the animals regardless of the administration method, the fluorescence intensity was heavily dependent on the method used (Picciotto et al. 2022).

11.10 *Drosophila*

Fruit fly (*Drosophila melanogaster; Drosophila*) is one of the most utilized in vivo models for genetics and developmental biology. It has also proven to be a valuable model to study the fundamental biology of EVs, including the biogenesis and the role of EVs in development. Despite the external differences between *Drosophila* and vertebrates, many of the organ systems of *Drosophila* are functionally comparable to their vertebrate counterparts. Human and *Drosophila* also share a surprisingly high genetic similarity. From human disease-associated genes, around 75% have a related sequence in *Drosophila* genome (Reiter et al. 2001). Additionally, a comparative analysis of human and *Drosophila* EVs found several features conserved between the species, such as the abundancy of ribosomal RNA and retrotransposons in EV cargo, as well as similarity in size and morphology of EVs (Lefebvre et al. 2016).

In vivo imaging studies of *Drosophila* have revealed the importance of EVs during multiple developmental processes. EVs are required for the long-range signalling of Wnt and Hedgehog morphogens which are conserved proteins in animals and regulate tissue patterning during development §(Matusek et al. 2020). In *Drosophila* larvae during formation of wings, single epithelia layered cells of the imaginal wing disc secrete morphogens that pattern the developing wing tissue in a concentration-dependent manner. Wnt protein Wingless was found to be secreted in

EVs by investigating the *Drosophila* wing formation using confocal microscopy (Gross et al. 2012). Hedgehog-containing EVs have been proposed to be transferred between *Drosophila* wing disc cells via specialized filopodia called cytonemes (Gradilla et al. 2014; González-Méndez et al. 2020). In the neuromuscular junction of developing *Drosophila* larvae, EVs have an important role in synaptic growth by transferring Wingless, which is sorted into EVs by transmembrane protein Evi/Wntless for trans-synaptic transport (Korkut et al. 2009; Koles et al. 2012).

In addition to development, EVs regulate *Drosophila* reproductive behaviour. Confocal imaging of CD63-GFP EVs secreted by epithelial secondary cells of the male reproductive accessory glands demonstrated the fusing of EVs with sperm after mating, as well as their interaction with the epithelial cells of the female reproductive tract. Disrupting the EV release by knockdown of *ALiX* or dominant-negative expression of Rab11 resulted in significantly increased remating behaviour in females (Corrigan et al. 2014). In females, the spermathecal secretary cells that surround the spermatheca capsule where sperm is stored after mating secrete EVs in response to mating. Silencing the expression of *ALiX*, *Hrs*, *Rab11,* and *Rab7* decreased the amount of intracellular CD63-GFP positive spots in the secretary cells in vivo and the number of secreted EVs in an ex vivo system and led to fewer eggs deposited, indicating that mating-induced EV secretion may regulate egg laying (Sanchez-Lopez et al. 2022).

11.11 Zebrafish (*Danio rerio*)

Zebrafish is a versatile organism to study various aspects of physiological mechanisms in health and disease. Zebrafish has a relatively high similarity to human genome, with 71.4% of human genes having at least one orthologue in zebrafish genome and 69% of zebrafish genes having an orthologue in human (Howe et al. 2013). Unlike *C. elegans* and *Drosophila*, zebrafish is a vertebrate, meaning it has a complex organ system that has characteristics of its higher vertebrate counterparts, as well as innate and adaptive immune systems (Goldsmith and Jobin 2012). Zebrafish have several advantages over other common vertebrate models: they are easy to maintain, have external fertilization and breed in high numbers, and can be easily genetically manipulated, making them an ideal model to study development and human pathologies. Zebrafish larvae are naturally transparent and can be maintained transparent up to 14 days post-fertilization. Transparent adult lines have also been generated (White et al. 2008), making zebrafish an excellent in vivo model for imaging experiments. Therefore, in recent years zebrafish has emerged as a novel model to unravel the biology and function of EVs in vivo.

A transparent vertebrate model for EV research is a powerful tool to study the fundamental aspects of in vivo EV biology and the dynamic behaviour of EVs in their physiological context. Genetically encoded fluorophore-tagged EV-enriched proteins, such as tetraspanins can be utilized to label and track endogenous EVs (Verweij et al. 2018, 2019; Jiang et al. 2019) and by mosaic expression, the expression of the fusion protein can be limited to a subset of cells, which provides more detailed information about a subpopulation of EVs (Verweij et al. 2018, 2019). By utilizing fluorophore-tagged EV markers, EVs can be tracked in a live animal in real time and in single EV resolution (Verweij et al. 2019; Jiang et al. 2019; Scott et al. 2021). This can shed light to EV biogenesis and secretion by following the expression patterns of fluorescent EV-related proteins to determine the cell types releasing EVs and even the mechanisms of EV shedding. EVs can be followed in the circulation, which allows tracking of EV life cycle including their half-life, biodistribution, clearance, and organotropism.

Verweij et al. developed a zebrafish model expressing CD63-pHluorin, a pH-sensitive fluorescent reporter which mainly reports the extracellular proportion of CD63 in the physiological pH (Verweij et al. 2018, 2019). A mosaic transient expression of CD63-pHluorin resulted in abundant expression of GFP-tagged CD63 in multiple different cell types of the yolk syncytial layer and allowed tracking of CD63-positive EVs in the bloodstream of live anesthetized animals by spinning disk confocal microscopy. Additionally, GFP-positive EVs were detected from ultrathin cryosections of the zebrafish embryo by immunogold labelling and TEM. The yolk syncytial layer-derived CD63-positive EVs were taken up by endothelial cells of the vein walls and by macrophages. The uptake of pHluorin-CD63 EVs was mediated by scavenger receptors and dynamin-dependent endocytosis, whilst biogenesis of the EVs was dependent on Syntenin, suggesting endosomal origin of the EVs (Verweij et al. 2019). More recently, Scott et al. developed a stable transgenic zebrafish line expressing cell-membrane tethered fluorophores on the inner leaflet of the plasma membrane to label endogenous EVs secreted by endothelial cells and cardiomyocytes in larval and adult zebrafish. High spatiotemporal resolution light-sheet live imaging revealed endothelial cell-derived EVs in the circulation and cardiomyocyte-derived EVs in the pericardial fluid of zebrafish larvae. In adult zebrafish, exchange of EVs from cardiomyocytes to endothelial cells and macrophages and endothelial cells to macrophages was observed from fixed hearts using whole-mount high-resolution imaging (Scott et al. 2021). In addition to observing the behaviour of tissue or cell type specific EVs, protein labelling may allow imaging of specific EV subtypes. Tspan4a-GFP was utilized to investigate migrasome formation during zebrafish gastrulation (Jiang et al. 2019). Another study demonstrated uptake of apoptotic bodies into stem cells in developing zebrafish epidermis (Brock et al. 2019).

While some studies have focused on understanding the biogenesis and fate of endogenous zebrafish EVs, many others have been focusing on exogenous EVs, typically isolated from human cell cultures or secreted by xenotransplanted genetically manipulated cells. Following the behaviour of EVs from external sources can provide information about their organotropism, half-life, and clearance which are especially interesting from the point of view of EV-based therapeutics (Yang et al. 2015; Gong et al. 2023; Heinrich et al. 2023). Some studies have also attempted to understand the behaviour and fate of tumour-derived EVs (Hyenne et al. 2019; Ghoroghi et al. 2021). To track exogenous EVs isolated from cell cultures, either the source cells are required to secrete fluorescently tagged EVs (Hyenne et al. 2019;

Morad et al. 2019) or the EVs must be externally labelled. Different labelling methods used in zebrafish as a model include fluorescent stains Membright (Hyenne et al. 2019; Ghoroghi et al. 2021), DIR (Gong et al. 2023), and Vybrant Dil (Heinrich et al. 2023), but also Rhodamine123-conjucated drugs loaded into EVs have been utilized (Yang et al. 2015). To study the fate of the EVs in circulation, they are usually administered by injection into the cardinal veins, such as the duct of Cuvier (Hyenne et al. 2019; Ghoroghi et al. 2021). Another approach is adding the labelled EVs to the medium of the zebrafish (Gong et al. 2023). However, no study so far has compared the different administration methods and how they might affect the in vivo biodistribution or behaviour of EVs.

Hyenne et al. studied the fate of isolated and Membright-labelled EVs derived from zebrafish melanoma cell line Zmel1 injected into zebrafish embryos and could detect circulating EVs using high speed confocal microscopy. Tumour-EVs were found to have a reduced velocity compared to red blood cells and EVs were found arrested and rolling on the surface of the vessel walls. Majority of the circulating tumour-EVs were taken up by endothelial cells or macrophages (Hyenne et al. 2019), similarly to endogenous EVs (Verweij et al. 2019; Scott et al. 2021). Additionally, using CLEM, Hyenne et al. found that patrolling macrophages capture circulating tumour-EVs with sheet-like protrusions, leading to internalization and targeting of EVs to the endosomal compartments of the cells (Hyenne et al. 2019). In another work, the arrest and internalization of Membright-labelled 4T1 mammary tumour-EVs to endothelial cells was found to be hindered when the EVs derived from RalA or RalB GTPase knockdown cells or when EVs were pretreated with MCAM/CD146 blocking antibody (Ghoroghi et al. 2021). Morad et al. investigated the ability of brain-metastasis forming MDA-MB-231 breast cancer cell EVs to cross the blood-brain barrier in zebrafish brain. Intracardially injected EVs isolated from palmitoylated tdTomato-expressing cells were seen being taken up by cells within brain parenchyma and moving in the endocytic vesicles of endothelial cells before fusing with the membrane, suggesting transcytosis as the process (Morad et al. 2019). Another study suggested drug-loaded EVs being able to deliver anti-cancer drugs into the brain (Yang et al. 2015).

11.12 Mouse

Mouse is the most commonly used in vivo model to study human biology and diseases. Human and mouse genomes are highly conserved and around 80% of the human protein-coding genes have an orthologue in mouse, although there are large number of genes with systemically different expression levels between the species (Breschi et al. 2017). As a mammal, mouse is biologically and anatomically very similar to human and has similar complex organ systems and development. Mouse has also been a common in vivo model to study EVs, especially concerning their biodistribution and organotropism. However, compared to simpler in vivo models introduced previously, mouse models have many limitations considering EV imaging, with the obvious larger size and lack of transparency. Live imaging of EVs in situ with a mouse model is restricted to tissues that are immediately adjacent to the imaging window and the thickness of tissues demands a much higher signal-to-noise ratio to detect the EVs, which most likely favours the detection of large EVs and hinders the ability to detect small EVs. Some endogenous proteins in tissues may also emit autofluorescence, which is a lesser issue in transparent in vivo models. However, amongst common in vivo models, mouse still remains the most representative model for humans, so it has retained its relevancy also in EV imaging studies. Majority of EV studies made using mouse as a model have utilized either in vivo imaging with live animals or ex vivo fluorescent microscopy from fixed tissue samples. Typically used mouse strains in EV imaging studies include BALB/c nude mice (Ye et al. 2018; Ghoroghi et al. 2021; Lázaro-Ibáñez et al. 2021; Nguyen et al. 2022; Ding et al. 2023), an immunodeficient mouse strain and therefore mostly used as a tumour-bearing model, and C57BL/6 mice (Wiklander et al. 2015; Hoshino et al. 2015; Li et al. 2022b), used as a model for EV tracking, biodistribution, and organotropism.

Majority of EV imaging studies made using mice have attempted to understand the biodistribution of exogenous EVs. While the general consensus is that majority of EVs end up in the same organs, mainly liver, spleen, lungs, and kidneys (Kang et al. 2021), the distribution of EVs has been shown to be affected by EV origin, labelling method, dose, targeting, and administration route (Wiklander et al. 2015; Hoshino et al. 2015; Gupta et al. 2020; Lázaro-Ibáñez et al. 2021). Organ distribution of DiR-labelled HEK293T EVs analysed by intravital imaging was dependent on dose of administered EVs and significantly altered between intraperitoneal, subcutaneous, and intravenous administration routes (Wiklander et al. 2015). Bioluminescently labelled EVs from luciferace-CD63-fusion transducted HEK293T cells showed similar pharmacokinetics in plasma when administered intraperitoneally and intravenously but only modest level of luciferase activity was detected from subcutaneously administered NanoLuc-CD63-EVs. On the other hand, NanoLuc-CD63-EVs administrated via intracardiac, intravenous, and intracarotid injections did not exhibit differential biodistribution in organs. However, EVs from luciferase-CD9 transducted cells showed increased accumulation in lung and decreased accumulation on GI tract compared to luciferase-CD63-EVs, indicating subpopulation-dependent biodistribution and organotropism (Gupta et al. 2020). Differential biodistribution of EVs seem to also be dictated by their cellular origin. DiR-labelled EVs from three different murine cell lines showcased different biodistribution in lungs, liver, spleen, and GI tract (Wiklander et al. 2015). Fluorescently labelled metastatic human breast and pancreatic cancer cell line EVs mainly localized to the future metastatic organs of the origin cells (Hoshino et al. 2015).

The biodistribution of EVs may also be influenced by the labelling method. Lazaro-Ibanez et al. investigated the effect of EV labelling to their biodistribution by comparing fluorescent, radioactive, and bioluminescent approaches using in vivo imaging. EVs were labelled either with mCherry-fluorophore fused with CD63, DiR lipophilic fluorescent dye, covalent modification with ^{111}Indium-DTPA radioisotope, or CD63-fused luciferases Fluc and Nluc. From fluorescent labels, DiR dye labelling was better alternative compared to mCherry-fused CD63, as high level of tissue autofluorescence limited the detection sensitivity of mCherry. Radiolabelled EVs imaged using single-photon emission computed tomography coupled with computer tomography showed high specificity and sensitivity in noninvasive and ex vivo imaging. From bioluminescent approaches, Nluc imaging was poorly suitable for noninvasive imaging but ex vivo imaging of organs showed high bioluminescence with increased accumulation in the lungs, in contrast to DiR- or radiolabelled EVs, which were mainly detected in the liver and spleen (Lázaro-Ibáñez et al. 2021).

EV targeting to tumours to act as potential drug delivery vehicles is being widely investigated. Generally, low accumulation of non-targeted EVs into tumours is detected, when compared to major EV uptaking organs (Wiklander et al. 2015; Lázaro-Ibáñez et al. 2021; Nguyen et al. 2022). However, the biodistribution may also be dictated by the EV source. Fluorescent intensity of DiD dye labelled CT26 colorectal cancer cell EVs and 4 T1 murine mammary carcinoma EVs in colorectal tumour-bearing BALB/c mice was higher in spleen, kidney, liver, and lung compared to the tumour, although the CTC26-EVs showed higher accumulation in the tumour region compared to the 4T1-EVs (Nguyen et al. 2022). On the contrary, bone mesenchymal stem cell EVs loaded with photosensitizer chlorin e6 and imaged using multispectral in vivo imaging system showed highest accumulation in mammary tumour, compared to organs, 72 h after injection (Ding et al. 2023). It is most likely that the organotropism of EVs is cell source dependent but can be also affected by external EV labelling.

Surface modifications can be used to increase the targeting of EVs to tumours. Low-density lipoprotein targeted peptide-conjugated EVs were DiR labelled and intravenously injected to glioma-bearing mice and biodistribution of EVs was followed by in vivo imaging. Ex vivo imaging showed no difference between unconjugated and peptide-conjugated EVs in the major organs but an increase of peptide-conjugated EV signal in the brain. Additionally, methotrexate-loaded peptide-conjugated EVs decreased the growth of glioma tumours compared to non-conjugated EVs, free methotrexate and saline (Ye et al. 2018). PEGylated hyaluronic acid surface modified EVs, generated by click chemistry with a covalent Cy5.5 label, were shown to have an increased targeting efficiency to inflamed joints of collagen-induced arthritis nude mouse model and tumours of PC3 tumour-bearing mouse model, compared to bare EVs without hyaluronic acid. Additionally, immunohistochemistry confirmed enrichment of hyaluronic-acid EVs but not bare EVs in the synovium and tumour area (Lim et al. 2021).

Whole-body and ex vivo tissue imaging are powerful methods to study the biodistribution and organotropism of EVs in a mouse model. However, they only grant information about which tissues the EVs are enriched in but not about which cells the EVs are targeted into. Therefore, immunohistochemical sections and stainings are required to determine the cell types that uptake EVs. Despite several publications investigating the biodistribution of exogenous EVs, the EV uptaking cell types within organs are rarely reported. Hoshino et al. showed that EVs derived from tumour cells with specific metastatic sites are directed into those organs to promote the formation of a metastatic niche. Immunofluorescent staining of the different organs revealed distinct cell types that uptake the EVs, associated with the integrin expression patterns of the EVs (Hoshino et al. 2015). Xenotransplantation of fluorescent EV-marker expressing cells can be used to trace in vivo secreted EVs derived from exogenous cells to study EV-mediated crosstalk in cancer or formation of premetastatic niches. Mammary tumour induced by co-culture of GFP-CD63-expressing and RFP-labelled MDA-MB-231 breast cancer cells demonstrated exchange of GFP-CD63 EVs between cancer cells and their microenvironment in the primary tumour and lung metastasis (Suetsugu et al. 2013). Functional RNA, such as Cre mRNA, can be loaded into EVs and transported into the EV uptaking cells in a mouse model. After transplantation of murine glioma cells expressing Cre, subsequent Cre expression within the tumour was mainly found in CD45-positive leukocytes but also in neurons, endothelial cells, and microglia (Ridder et al. 2014). GFP-tagged lung metastatic human melanoma cells and liver metastatic human neuroblastoma cells were orthotopically transplanted to mice to study the EV-mediated premetastatic niche formation. GFP-positive melanoma EVs were detected at day 10 post-implantation in lung, while tumour cells were detected from day 15 onwards. For liver metastatic glioblastoma cells, GFP-EVs were detected in liver on day 14 post-implantation and tumour cells starting from day 28. Resident macrophages were the major EV capturing cell type in metastatic tissues prior tumour cell infiltration, while no GFP signal was detected in the non-metastatic organs, suggesting organ-specific homing of the tumour-derived EVs (Blavier et al. 2023).

Transgenic mouse models can be used to study tissue specific secretion and tropism of endogenously secreted EVs. Several models utilizing various fluorescent EV-related proteins have been established either to study secreted EVs from body liquids (McCann et al. 2020; Li et al. 2022b; Nørgård et al. 2022; Fordjour et al. 2023) or the secretion and uptake of EVs in tissues in situ (Neckles et al. 2019; Men et al. 2019; Li et al. 2022b; Fordjour et al. 2023). An inducible CD9-GFP transgenic mouse model was generated to study the secretion of CD9-positive EVs in the mouse brain during cortical development. Based on immunostained brain sections, astrocytes were found to be the main cell type expressing CD9-GFP and releasing CD9-positive EVs. Additionally, culture astrocyte-derived Dil-labelled EVs injected intraventricularly were found to be taken up by microglia in vivo (Neckles et al. 2019). Another transgenic mouse model with selectively induced CD63-GFPf/f reporter demonstrated secretion of neuron-derived CD63-GFP and their uptake into astrocytes using confocal microscopy and immuno-EM of brain sections (Men et al. 2019). Li et al. generated two transgenic mouse models with tissue-specific expression of CD63-EGFP in liver or intestine and investigated the fate of CD63-EGFP EVs in other tissues. Subsequent CD63-EGFP signal, derived from albumin-CD63-EGFP-mCh expressing liver cells, was mainly detected in kidney and lung. Villin-CD63-EGFP-mCh expressing cells were found mainly from small and large intestine and CD63-EGFP signal was also detected in the kidney. Closer inspection of lung and kidney tissues via immunofluorescently stained tissue sections revealed that liver-derived CD63-EFGP-EVs were mainly taken up by pulmonary surfactant protein C expressing alveolar specific-cells in the lung and nephrin-expressing podocyte-related cells in the kidney. Similarly, majority of intestine-derived EVs were taken up by podocyte-related cells (Li et al. 2022b).

11.13 Human Patient Tissue

While in vivo animal models are required to monitor the biogenesis, movement, targeting, and fate of EVs, imaging human tissue in situ can also provide novel information about the localization and abundance of EVs in tissues, such as in the tumour microenvironment. However, retaining the EVs in the tissue samples can be challenging and may require special fixation methods, as demonstrated with imaging EVs in the vitreous of bovine and human using multiphoton microscopy and TEM (Gupta et al. 2019).

In formalin-fixed paraffin embedded (PPFE) tissue sections, imaged with TEM, EVs of different sizes and morphologies are found in the extracellular space of metastatic melanoma (Jang et al. 2019; Crescitelli et al. 2020) and colon cancer (Crescitelli et al. 2021). 3D reconstruction of immunostained PPFE metastatic colorectal cancer tissue serial sections revealed CD63 and ALIX positive protruding structures in the migrating cancer cells. Further characterization of the structures with superresolution confocal microscopy revealed the shedding of small EV clusters, which were also detected in in vitro cultures using confocal and electron microscopy (Valcz et al. 2019). Recently, transmission electron tomography was utilized to prepare a similar reconstruction of normal and metastatic liver tissue, revealing small spherical structures between cells in both tissues (Olofsson Bagge et al. 2023).

Many tissue imaging methods require extensive sample processing like fixation followed by staining or labelling. However, label-free imaging techniques are a promising approach in visualizing unperturbed tissues and their components and are also applicable to visualize tissue contained EVs. Multicontrast nonlinear imaging of rat mammary tumours revealed an enrichment of various types of EVs in the tumour microenvironment (Tu et al. 2017). Follow-up studies investigated intraoperatively imaged human breast tissue from breast tumours and healthy individuals and observed similar enrichment of EVs in their tumour microenvironment (Sun et al. 2018) as well as greatly increased concentration of NAD(P)H-enriched EVs in the cancer patients compared to healthy individuals, and the amount of NAD(P)H-EVs correlating with the stage of the cancer (You et al. 2019).

Human tissue biopsies can improve our understanding on how EVs are localized and function within tissues, which is essential in understanding the contribution of EVs in human health and disease. While the current knowledge is still limited, developments in imaging techniques are expected to aid understanding the role of EVs in human tissue microenvironment.

11.14 Conclusion and Future Perspectives

Visualization and analysis of EVs by advanced imaging techniques in their natural microenvironment in vitro and in vivo is essential for unravelling their roles in intercellular communication, disease progression, drug delivery, tissue engineering, and diagnostics. However, imaging EVs within 3D matrix networks presents several challenges due to the limitations in methodology, complexity of the environment, and the small size of EVs. Addressing these challenges requires interdisciplinary approaches combining advanced imaging techniques, innovative sample preparation methods, and computational analysis tools. This helps in bridging the gap between in vitro studies and in vivo applications, which is crucial for successful utilization of EVs in various fields including biomedicine and nanotechnology.

References

Aaltonen N, Kyykallio H, Tollis S, Capra J, Hartikainen JM, Matilainen J, Oikari S, Rilla K (2022) MCF10CA breast cancer cells utilize hyaluronan-coated EV-rich trails for coordinated migration. Front Oncol 12:869417. https://doi.org/10.3389/fonc.2022.869417

Altei WF, Pachane BC, dos Santos PK, Ribeiro LNM, Sung BH, Weaver AM, Selistre-de-Araújo HS (2020) Inhibition of αvβ3 integrin impairs adhesion and uptake of tumor-derived small extracellular vesicles. Cell Commun Signal 18:158. https://doi.org/10.1186/s12964-020-00630-w

Arasu UT, Härkönen K, Koistinen A, Rilla K (2019) Correlative light and electron microscopy is a powerful tool to study interactions of extracellular vesicles with recipient cells. Exp Cell Res 376(2):149–158. https://doi.org/10.1016/j.yexcr.2019.02.004

Barr MM, DeModena J, Braun D, Nguyen CQ, Hall DH, Sternberg PW (2001) The Caenorhabditis elegans autosomal dominant polycystic kidney disease gene homologs lov-1 and pkd-2 act in the same pathway. Curr Biol 11(17):1341–1346. https://doi.org/10.1016/s0960-9822(01)00423-7

Bhattacharya S, Calar K, Evans C, Petrasko M, de la Puente P (2020) Bioengineering the oxygen-deprived tumor microenvironment within a three-dimensional platform for studying tumor-immune interactions. Front Bioeng Biotechnol 8. https://doi.org/10.3389/fbioe.2020.01040

Bidarimath M, Khalaj K, Kridli RT, Kan FWK, Koti M, Tayade C (2017) Extracellular vesicle mediated intercellular communication at the porcine maternal-fetal interface: a new paradigm for conceptus-endometrial cross-talk. Sci Rep 7(1):40476. https://doi.org/10.1038/srep40476

Blavier L, Nakata R, Neviani P, Sharma K, Shimada H, Benedicto A, Matei I, Lyden D, DeClerck YA (2023) The capture of extracellular vesicles endogenously released by xenotransplanted tumours induces an inflammatory reaction in the premetastatic niche. J Extracell Vesicles 12(5):12326. https://doi.org/10.1002/jev2.12326

Breschi A, Gingeras TR, Guigó R (2017) Comparative transcriptomics in human and mouse. Nat Rev Genet 18(7):425–440. https://doi.org/10.1038/nrg.2017.19

Brock CK, Wallin ST, Ruiz OE, Samms KM, Mandal A, Sumner EA, Eisenhoffer GT (2019) Stem cell proliferation is induced by apoptotic bodies from dying cells during epithelial tissue maintenance. Nat Commun 10:1044. https://doi.org/10.1038/s41467-019-09010-6

Caliari SR, Burdick JA (2016) A practical guide to hydrogels for cell culture. Nat Methods 13(5):405–414. https://doi.org/10.1038/nmeth.3839

Capra J, Härkönen K, Kyykallio H, Vihinen H, Jokitalo E, Rilla K (2022) Microscopic characterization reveals the diversity of EVs secreted by GFP-HAS3 expressing MCF7 cells. Eur J Cell Biol 101(3):151235. https://doi.org/10.1016/j.ejcb.2022.151235

Corrigan L, Redhai S, Leiblich A, Fan S-J, Perera SMW, Patel R, Gandy C, Wainwright SM, Morris JF, Hamdy F, Goberdhan DCI, Wilson C (2014) BMP-regulated exosomes from drosophila male reproductive glands reprogram female behavior. J Cell Biol 206(5):671–688. https://doi.org/10.1083/jcb.201401072

Costa Verdera H, Gitz-Francois JJ, Schiffelers RM, Vader P (2017) Cellular uptake of extracellular vesicles is mediated by clathrin-independent endocytosis and macropinocytosis. J Control Release 266:100–108. https://doi.org/10.1016/j.jconrel.2017.09.019

Crescitelli R, Lässer C, Jang SC, Cvjetkovic A, Malmhäll C, Karimi N, Höög JL, Johansson I, Fuchs J, Thorsell A, Gho YS, Olofsson Bagge R, Lötvall J (2020) Subpopulations of extracellular vesicles from human metastatic melanoma tissue identified by quantitative proteomics after optimized isolation. J Extracell Vesicles 9(1):1722433. https://doi.org/10.1080/20013078.2020.1722433

Crescitelli R, Lässer C, Lötvall J (2021) Isolation and characterization of extracellular vesicle subpopulations from tissues. Nat Protoc 16(3):1548–1580. https://doi.org/10.1038/s41596-020-00466-1

Curvello R, Raghuwanshi VS, Garnier G (2019) Engineering nanocellulose hydrogels for biomedical applications. Adv Colloid Interf Sci 267:47–61. https://doi.org/10.1016/j.cis.2019.03.002

Deen AJ, Arasu UT, Pasonen-Seppänen S, Hassinen A, Takabe P, Wojciechowski S, Kärnä R, Rilla K, Kellokumpu S, Tammi R, Tammi M, Oikari S (2016) UDP-sugar substrates of HAS3 regulate its O-GlcNAcylation, intracellular traffic, extracellular shedding and correlate with melanoma progression. Cell Mol Life Sci 73(16):3183–3204. https://doi.org/10.1007/s00018-016-2158-5

Ding Y, Ding H, Li H, Yang R, Huang J, Chen H, Wang L, Wang Y, Hu C, An Y, Zhang Z, Yu W, Tang Q, Shao G (2023) Photosensitive small extracellular vesicles regulate the immune microenvironment of triple negative breast cancer. Acta Biomater 167:534–550. https://doi.org/10.1016/j.actbio.2023.06.004

Dong G, Chen P, Xu Y, Liu T, Yin R (2023) Cancer-associated fibroblasts: key criminals of tumor pre-metastatic niche. Cancer Lett 566:216234. https://doi.org/10.1016/j.canlet.2023.216234

Elgamal S, Colombo F, Cottini F, Byrd JC, Cocucci E (2020) Chapter Five - Imaging intercellular interaction and extracellular vesicle exchange in a co-culture model of chronic lymphocytic leukemia and stromal cells by lattice light-sheet fluorescence microscopy. In: Spada S, Galluzzi L (eds) Methods in enzymology. Academic Press, pp 79–107

Fordjour FK, Abuelreich S, Hong X, Chatterjee E, Lallai V, Ng M, Saftics A, Deng F, Carnel-Amar N, Wakimoto H, Shimizu K, Bautista M, Phu TA, Vu NK, Geiger PC, Raffai RL, Fowler CD, Das S, Christenson LK, Jovanovic-Talisman T, Gould SJ (2023) Exomap1 mouse: a transgenic model for *in vivo* studies of exosome biology. Extracell Vesicle 2:100030. https://doi.org/10.1016/j.vesic.2023.100030

Ghoroghi S, Mary B, Larnicol A, Asokan N, Klein A, Osmani N, Busnelli I, Delalande F, Paul N, Halary S, Gros F, Fouillen L, Haeberle A-M, Royer C, Spiegelhalter C, André-Grégoire G, Mittelheisser V, Detappe A, Murphy K, Timpson P, Carapito R, Blot-Chabaud M, Gavard J, Carapito C, Vitale N, Lefebvre O, Goetz JG, Hyenne V (2021) Ral GTPases promote breast cancer metastasis by controlling biogenesis and organ targeting of exosomes. eLife 10:e61539. https://doi.org/10.7554/eLife.61539

Goldsmith JR, Jobin C (2012) Think small: zebrafish as a model system of human pathology. J Biomed Biotechnol 2012:817341. https://doi.org/10.1155/2012/817341

Gong L, Zhou H, Zhang S, Wang C, Fu K, Ma C, Zhang Y, Peng C, Li Y (2023) CD44-targeting drug delivery system of exosomes loading forsythiaside a combats liver fibrosis via regulating NLRP3-mediated pyroptosis. Adv Healthc Mater 12(11):2202228. https://doi.org/10.1002/adhm.202202228

González-Méndez L, Gradilla A, Sánchez-Hernández D, González E, Aguirre-Tamaral A, Jiménez-Jiménez C, Guerra M, Aguilar G, Andrés G, Falcón-Pérez JM, Guerrero I (2020) Polarized sorting of patched enables cytoneme-mediated hedgehog reception in the drosophila wing disc. EMBO J 39(11):e103629. https://doi.org/10.15252/embj.2019103629

Gradilla A-C, González E, Seijo I, Andrés G, Bischoff M, González-Mendez L, Sánchez V, Callejo A, Ibáñez C, Guerra M, Ortigão-Farias JR, Sutherland JD, González M, Barrio R, Falcón-Pérez JM, Guerrero I (2014) Exosomes as hedgehog carriers in cytoneme-mediated transport and secretion. Nat Commun 5:5649. https://doi.org/10.1038/ncomms6649

Gross JC, Chaudhary V, Bartscherer K, Boutros M (2012) Active Wnt proteins are secreted on exosomes. Nat Cell Biol 14(10):1036–1045. https://doi.org/10.1038/ncb2574

Gupta MP, Tandalam S, Ostrager S, Lever AS, Fung AR, Hurley DD, Alegre GB, Espinal JE, Remmel HL, Mukherjee S, Levine BM, Robins RP, Molina H, Dil BD, Kenific CM, Tuschl T, Lyden DC, D'Amico DJ, Pena JTG (2019) Non-reversible tissue fixation retains extracellular vesicles for in situ imaging. Nat Methods 16(12):1269–1273. https://doi.org/10.1038/s41592-019-0623-4

Gupta D, Liang X, Pavlova S, Wiklander OPB, Corso G, Zhao Y, Saher O, Bost J, Zickler AM, Piffko A, Maire CL, Ricklefs FL, Gustafsson O, Llorente VC, Gustafsson MO, Bostancioglu RB, Mamand DR, Hagey DW, Görgens A, Nordin JZ, Andaloussi EL (2020) Quantification of extracellular vesicles in vitro and in vivo using sensitive bioluminescence imaging. J Extracell Vesicles 9(1):1800222. https://doi.org/10.1080/20013078.2020.1800222

Haraszti RA, Miller R, Stoppato M, Sere YY, Coles A, Didiot MC, Wollacott R, Sapp E, Dubuke ML, Li X, Shaffer SA, DiFiglia M, Wang Y, Aronin N, Khvorova A (2018) Exosomes produced from 3D cultures of MSCs by tangential flow filtration show higher yield and improved activity. Mol Ther 26(12):2838–2847. https://doi.org/10.1016/j.ymthe.2018.09.015

Härkönen K, Oikari S, Kyykallio H, Capra J, Hakkola S, Ketola K, Thanigai Arasu U, Daaboul G, Malloy A, Oliveira C, Jokelainen O, Sironen R, Hartikainen JM, Rilla K (2019) CD44s assembles hyaluronan coat on filopodia and extracellular vesicles and induces tumorigenicity of MKN74 gastric carcinoma cells. Cells 8(3):276. https://doi.org/10.3390/cells8030276

Heinrich E, Hartwig O, Walt C, Kardani A, Koch M, Jahromi LP, Hoppstädter J, Kiemer AK, Loretz B, Lehr C-M, Fuhrmann G (2023) Cell-derived vesicles for antibiotic delivery—understanding the challenges of a biogenic carrier system. Small 19(25):2207479. https://doi.org/10.1002/smll.202207479

Hoshino A, Costa-Silva B, Shen T-L, Rodrigues G, Hashimoto A, Mark MT, Molina H, Kohsaka S, Di Giannatale A, Ceder S, Singh S, Williams C, Soplop N, Uryu K, Pharmer L, King T, Bojmar L, Davies AE, Ararso Y, Zhang T, Zhang H, Hernandez J, Weiss JM, Dumont-Cole VD, Kramer K, Wexler LH, Narendran A, Schwartz GK, Healey JH, Sandstrom P, Labori KJ, Kure EH, Grandgenett PM, Hollingsworth MA, de Sousa M, Kaur S, Jain M, Mallya K, Batra SK, Jarnagin WR, Brady MS, Fodstad O, Muller V, Pantel K, Minn AJ, Bissell MJ, Garcia BA, Kang Y, Rajasekhar VK, Ghajar CM, Matei I, Peinado H, Bromberg J, Lyden D (2015) Tumour exosome integrins determine organotropic metastasis. Nature 527(7578):329–335. https://doi.org/10.1038/nature15756

Howe K, Clark MD, Torroja CF, Torrance J, Berthelot C, Muffato M, Collins JE, Humphray S, McLaren K, Matthews L, McLaren S, Sealy I, Caccamo M, Churcher C, Scott C, Barrett JC, Koch R, Rauch G-J, White S, Chow W, Kilian B, Quintais LT, Guerra-Assunção JA, Zhou Y, Gu Y, Yen J, Vogel J-H, Eyre T, Redmond S, Banerjee R, Chi J, Fu B, Langley E, Maguire SF, Laird GK, Lloyd D, Kenyon E, Donaldson S, Sehra H, Almeida-King J, Loveland J, Trevanion S, Jones M, Quail M, Willey D, Hunt A, Burton J, Sims S, McLay K, Plumb B, Davis J, Clee C, Oliver K, Clark R, Riddle C, Eliott D, Threadgold G, Harden G, Ware D, Mortimer B, Kerry G, Heath P, Phillimore B, Tracey A, Corby N, Dunn M, Johnson C, Wood J, Clark S, Pelan S, Griffiths G, Smith M, Glithero R, Howden P, Barker N, Stevens C, Harley J, Holt K, Panagiotidis G, Lovell J, Beasley H, Henderson C, Gordon D, Auger K, Wright D, Collins J, Raisen C, Dyer L, Leung K, Robertson L, Ambridge K, Leongamornlert D, McGuire S, Gilderthorp R, Griffiths C, Manthravadi D, Nichol S, Barker G, Whitehead S, Kay M, Brown J, Murnane C, Gray E, Humphries M, Sycamore N, Barker D, Saunders D, Wallis J, Babbage A, Hammond S, Mashreghi-Mohammadi M, Barr L, Martin S, Wray P, Ellington A, Matthews N, Ellwood M, Woodmansey R, Clark G, Cooper J, Tromans A, Grafham D, Skuce C, Pandian R, Andrews R, Harrison E, Kimberley A, Garnett J, Fosker N, Hall R, Garner P, Kelly D,

Bird C, Palmer S, Gehring I, Berger A, Dooley CM, Ersan-Ürün Z, Eser C, Geiger H, Geisler M, Karotki L, Kirn A, Konantz J, Konantz M, Oberländer M, Rudolph-Geiger S, Teucke M, Osoegawa K, Zhu B, Rapp A, Widaa S, Langford C, Yang F, Carter NP, Harrow J, Ning Z, Herrero J, Searle SMJ, Enright A, Geisler R, Plasterk RHA, Lee C, Westerfield M, de Jong PJ, Zon LI, Postlethwait JH, Nüsslein-Volhard C, Hubbard TJP, Crollius HR, Rogers J, Stemple DL (2013) The zebrafish reference genome sequence and its relationship to the human genome. Nature 496(7446):498–503. https://doi.org/10.1038/nature12111

Hughes CS, Postovit LM, Lajoie GA (2010) Matrigel: a complex protein mixture required for optimal growth of cell culture. Proteomics 10(9):1886–1890. https://doi.org/10.1002/pmic.200900758

Hyenne V, Ghoroghi S, Collot M, Bons J, Follain G, Harlepp S, Mary B, Bauer J, Mercier L, Busnelli I, Lefebvre O, Fekonja N, Garcia-Leon MJ, Machado P, Delalande F, López AA, Silva SG, Verweij FJ, van Niel G, Djouad F, Peinado H, Carapito C, Klymchenko AS, Goetz JG (2019) Studying the fate of tumor extracellular vesicles at high spatiotemporal resolution using the zebrafish embryo. Dev Cell 48(4):554–572.e7. https://doi.org/10.1016/j.devcel.2019.01.014

Jang SC, Crescitelli R, Cvjetkovic A, Belgrano V, Olofsson Bagge R, Sundfeldt K, Ochiya T, Kalluri R, Lötvall J (2019) Mitochondrial protein enriched extracellular vesicles discovered in human melanoma tissues can be detected in patient plasma. J Extracell Vesicles 8(1):1635420. https://doi.org/10.1080/20013078.2019.1635420

Jiang D, Jiang Z, Lu D, Wang X, Liang H, Zhang J, Meng Y, Li Y, Wu D, Huang Y, Chen Y, Deng H, Wu Q, Xiong J, Meng A, Yu L (2019) Migrasomes provide regional cues for organ morphogenesis during zebrafish gastrulation. Nat Cell Biol 21(8):966–977. https://doi.org/10.1038/s41556-019-0358-6

Kang M, Jordan V, Blenkiron C, Chamley LW (2021) Biodistribution of extracellular vesicles following administration into animals: a systematic review. J Extracell Vesicles 10(8):e12085. https://doi.org/10.1002/jev2.12085

Kapałczyńska M, Kolenda T, Przybyła W, Zajączkowska M, Teresiak A, Filas V, Ibbs M, Bliźniak R, Łuczewski Ł, Lamperska K (2018) 2D and 3D cell cultures—a comparison of different types of cancer cell cultures. Arch Med Sci 14(4):910–919. https://doi.org/10.5114/aoms.2016.63743

Koles K, Nunnari J, Korkut C, Barria R, Brewer C, Li Y, Leszyk J, Zhang B, Budnik V (2012) Mechanism of evenness interrupted (Evi)-exosome release at synaptic Boutons. J Biol Chem 287(20):16820–16834. https://doi.org/10.1074/jbc.M112.342667

Korkut C, Ataman B, Ramachandran P, Ashley J, Barria R, Gherbesi N, Budnik V (2009) Trans-synaptic transfer of Wnt signals through release of Evi/Wntless vesicles and trafficking of post-synaptic frizzled-2 receptors. Cell 139(2):393–404. https://doi.org/10.1016/j.cell.2009.07.051

Kusuma GD, Li A, Zhu D, McDonald H, Inocencio IM, Chambers DC, Sinclair K, Fang H, Greening DW, Frith JE, Lim R (2022) Effect of 2D and 3D culture microenvironments on mesenchymal stem cell-derived extracellular vesicles potencies. Front Cell Dev Biol 10. https://doi.org/10.3389/FCELL.2022.819726

Kyykallio H, Faria AVS, Hartman R, Capra J, Rilla K, Siljander PR-M (2022) A quick pipeline for the isolation of 3D cell culture-derived extracellular vesicles. J Extracell Vesicles 11(10):e12273. https://doi.org/10.1002/jev2.12273

Lai CP, Kim EY, Badr CE, Weissleder R, Mempel TR, Tannous BA, Breakefield XO (2015) Visualization and tracking of tumour extracellular vesicle delivery and RNA translation using multiplexed reporters. Nat Commun 6:7029. https://doi.org/10.1038/ncomms8029

Lázaro-Ibáñez E, Faruqu FN, Saleh AF, Silva AM, Tzu-Wen Wang J, Rak J, Al-Jamal KT, Dekker N (2021) Selection of fluorescent, bioluminescent, and radioactive tracers to accurately reflect extracellular vesicle biodistribution in vivo. ACS Nano 15(2):3212–3227. https://doi.org/10.1021/acsnano.0c09873

Lee JH, Won YJ, Kim H, Choi M, Lee E, Ryoou B, Lee S-G, Cho BS (2023) Adipose tissue-derived mesenchymal stem cell-derived exosomes promote wound healing and tissue regeneration. Int J Mol Sci 24(13):10434. https://doi.org/10.3390/ijms241310434

Lefebvre FA, Benoit Bouvrette LP, Perras L, Blanchet-Cohen A, Garnier D, Rak J, Lécuyer É (2016) Comparative transcriptomic analysis of human and drosophila extracellular vesicles. Sci Rep 6(1):27680. https://doi.org/10.1038/srep27680

Lenzini S, Bargi R, Chung G, Shin J-W (2020) Matrix mechanics and water permeation regulate extracellular vesicle transport. Nat Nanotechnol 15(3):217–223. https://doi.org/10.1038/s41565-020-0636-2

Li C, Wei S, Xu Q, Sun Y, Ning X, Wang Z (2022a) Application of ADSCs and their exosomes in scar prevention. Stem Cell Rev Rep 18(3):952–967. https://doi.org/10.1007/s12015-021-10252-5

Li W, Wang J, Yin X, Shi H, Sun B, Ji M, Song H, Liu J, Dou Y, Xu C, Jiang X, Li J, Li L, Zhang C-Y, Zhang Y (2022b) Construction of a mouse model that can be used for tissue-specific EV screening and tracing in vivo. Front Cell Dev Biol 10

Liebel M, Ortega Arroyo J, Beltrán VS, Osmond J, Jo A, Lee H, Quidant R, van Hulst NF (2020) 3D tracking of extracellular vesicles by holographic fluorescence imaging. Sci Adv 6(45):eabc2508. https://doi.org/10.1126/sciadv.abc2508

Lim GT, You DG, Han HS, Lee H, Shin S, Oh BH, Kumar EKP, Um W, Kim CH, Han S, Lee S, Lim S, Yoon HY, Kim K, Kwon IC, Jo D, Cho YW, Park JH (2021) Bioorthogonally surface-edited extracellular vesicles based on metabolic glycoengineering for CD44-mediated targeting of inflammatory diseases. J Extracell Vesicles 10(5). https://doi.org/10.1002/JEV2.12077

Loconte L, Arguedas D, El R, Zhou A, Chipont A, Guyonnet L, Guerin C, Piovesana E, Vázquez-Ibar JL, Joliot A, Théry C, Martín-Jaular L (2023) Detection of the interactions of tumour derived extracellular vesicles with immune cells is dependent on EV-labelling methods. J Extracell Vesicles 12(12):12384. https://doi.org/10.1002/jev2.12384

Maguire JE, Silva M, Nguyen KCQ, Hellen E, Kern AD, Hall DH, Barr MM (2015) Myristoylated CIL-7 regulates ciliary extracellular vesicle biogenesis. Mol Biol Cell 26(15):2823–2832. https://doi.org/10.1091/mbc.E15-01-0009

Matusek T, Marcetteau J, Thérond PP (2020) Functions of Wnt and hedgehog-containing extracellular vesicles in development and disease. J Cell Sci 133(18):jcs209742. https://doi.org/10.1242/jcs.209742

McCann JV, Bischoff SR, Zhang Y, Cowley DO, Sanchez-Gonzalez V, Daaboul GD, Dudley AC (2020) Reporter mice for isolating and auditing cell type-specific extracellular vesicles in vivo. Genesis 58(7):e23369. https://doi.org/10.1002/dvg.23369

Men Y, Yelick J, Jin S, Tian Y, Chiang MSR, Higashimori H, Brown E, Jarvis R, Yang Y (2019) Exosome reporter mice reveal the involvement of exosomes in mediating neuron to astroglia communication in the CNS. Nat Commun 10:4136. https://doi.org/10.1038/s41467-019-11534-w

Meneely PM, Dahlberg CL, Rose JK (2019) Working with worms: Caenorhabditis elegans as a model organism. Curr Protoc Essent Lab Tech 19(1):e35. https://doi.org/10.1002/cpet.35

Millan C, Prause L, Vallmajo-Martin Q, Hensky N, Eberli D (2021) Extracellular vesicles from 3D engineered microtissues harbor disease-related cargo absent in EVs from 2D cultures. Adv Healthc Mater 2021:2002067. https://doi.org/10.1002/ADHM.202002067

Millard M, Yakavets I, Piffoux M, Brun A, Gazeau F, Guigner J-M, Jasniewski J, Lassalle H-P, Wilhelm C, Bezdetnaya L (2018) mTHPC-loaded extracellular vesicles outperform liposomal and free mTHPC formulations by an increased stability, drug delivery efficiency and cytotoxic effect in tridimensional model of tumors. Drug Deliv 25(1):1790–1801. https://doi.org/10.1080/10717544.2018.1513609

Mittelbrunn M, Gutiérrez-Vázquez C, Villarroya-Beltri C, González S, Sánchez-Cabo F, González MÁ, Bernad A, Sánchez-Madrid F (2011) Unidirectional transfer of microRNA-loaded exosomes from T cells to antigen-presenting cells. Nat Commun 2(1):282. https://doi.org/10.1038/ncomms1285

Morad G, Carman CV, Hagedorn EJ, Perlin JR, Zon LI, Mustafaoglu N, Park T-E, Ingber DE, Daisy CC, Moses MA (2019) Tumor-derived extracellular vesicles breach the intact blood-brain barrier via transcytosis. ACS Nano 13(12):13853–13865. https://doi.org/10.1021/acsnano.9b04397

Neckles VN, Morton MC, Holmberg JC, Sokolov AM, Nottoli T, Liu D, Feliciano DM (2019) A transgenic inducible GFP extracellular-vesicle reporter (TIGER) mouse illuminates neonatal cortical astrocytes as a source of immunomodulatory extracellular vesicles. Sci Rep 9(1):3094. https://doi.org/10.1038/s41598-019-39679-0

Nguyen VD, Kim HY, Choi YH, Park J-O, Choi E (2022) Tumor-derived extracellular vesicles for the active targeting and effective treatment of colorectal tumors in vivo. Drug Deliv 29(1):2621–2631. https://doi.org/10.1080/10717544.2022.2105444

Nikonorova IA, Wang J, Cope AL, Tilton PE, Power KM, Walsh JD, Akella JS, Krauchunas AR, Shah P, Barr MM (2022) Isolation, profiling, and tracking of extracellular vesicle cargo in Caenorhabditis elegans. Curr Biol 32(9):1924–1936.e6. https://doi.org/10.1016/j.cub.2022.03.005

Nishimura T, Oyama T, Hu HT, Fujioka T, Hanawa-Suetsugu K, Ikeda K, Yamada S, Kawana H, Saigusa D, Ikeda H, Kurata R, Oono-Yakura K, Kitamata M, Kida K, Hikita T, Mizutani K, Yasuhara K, Mimori-Kiyosue Y, Oneyama C, Kurimoto K, Hosokawa Y, Aoki J, Takai Y, Arita M, Suetsugu S (2021) Filopodium-derived vesicles produced by MIM enhance the migration of recipient cells. Dev Cell 56(6):842–859.e8. https://doi.org/10.1016/j.devcel.2021.02.029

Nizamudeen Z, Markus R, Lodge R, Parmenter C, Platt M, Chakrabarti L, Sottile V (2018) Rapid and accurate analysis of stem cell-derived extracellular vesicles with super resolution microscopy and live imaging. Biochim Biophys Acta Mol Cell Res 1865(12):1891–1900. https://doi.org/10.1016/j.bbamcr.2018.09.008

Nørgård MØ, Steffensen LB, Hansen DR, Füchtbauer E-M, Engelund MB, Dimke H, Andersen DC, Svenningsen P (2022) A new transgene mouse model using an extravesicular EGFP tag enables affinity isolation of cell-specific extracellular vesicles. Sci Rep 12(1):496. https://doi.org/10.1038/s41598-021-04512-0

Nowak-Sliwinska P, Segura T, Iruela-Arispe ML (2014) The chicken chorioallantoic membrane model in biology, medicine and bioengineering. Angiogenesis 17(4):779. https://doi.org/10.1007/s10456-014-9440-7

Olofsson Bagge R, Berndtsson J, Urzì O, Lötvall J, Micaroni M, Crescitelli R (2023) Three-dimensional reconstruction of interstitial extracellular vesicles in human liver as determined by electron tomography. J Extracell Vesicles 12(12):12380. https://doi.org/10.1002/jev2.12380

Perrin L, Bayarmagnai B, Gligorijevic B (2019) Frontiers in intravital multiphoton microscopy of cancer. Cancer Rep (Hoboken) 3(1):e1192. https://doi.org/10.1002/cnr2.1192

Picciotto S, Santonicola P, Paterna A, Rao E, Raccosta S, Romancino DP, Noto R, Touzet N, Manno M, Di Schiavi E, Bongiovanni A, Adamo G (2022) Extracellular vesicles from microalgae: uptake studies in human cells and Caenorhabditis elegans. Front Bioeng Biotechnol 10

Ragni E, Palombella S, Lopa S, Talò G, Perucca Orfei C, De Luca P, Moretti M, de Girolamo L (2020) Innovative visualization and quantification of extracellular vesicles interaction with and incorporation in target cells in 3D microenvironments. Cells 9(5). https://doi.org/10.3390/cells9051180

Razzauti A, Laurent P (2021) Ectocytosis prevents accumulation of ciliary cargo in C. elegans sensory neurons. eLife 10:e67670. https://doi.org/10.7554/eLife.67670

Reiter LT, Potocki L, Chien S, Gribskov M, Bier E (2001) A systematic analysis of human disease-associated gene sequences in Drosophila melanogaster. Genome Res 11(6):1114–1125. https://doi.org/10.1101/gr.169101

Ribatti D (2016) The chick embryo chorioallantoic membrane (CAM). A multifaceted experimental model. Mech Dev 141:70–77. https://doi.org/10.1016/j.mod.2016.05.003

Ridder K, Keller S, Dams M, Rupp A-K, Schlaudraff J, Del Turco D, Starmann J, Macas J, Karpova D, Devraj K, Depboylu C, Landfried B, Arnold B, Plate KH, Höglinger G, Sültmann H, Altevogt P, Momma S (2014) Extracellular vesicle-mediated transfer of genetic information between the hematopoietic system and the brain in response to inflammation. PLoS Biol 12(6):e1001874. https://doi.org/10.1371/journal.pbio.1001874

Rilla K, Pasonen-Seppänen S, Deen AJ, Koistinen VVT, Wojciechowski S, Oikari S, Kärnä R, Bart G, Törrönen K, Tammi RH, Tammi MI (2013) Hyaluronan production enhances shedding of plasma membrane-derived microvesicles. Exp Cell Res 319(13):2006–2018. https://doi.org/10.1016/j.yexcr.2013.05.021

Rilla K, Mustonen A-M, Arasu UT, Härkönen K, Matilainen J, Nieminen P (2019) Extracellular vesicles are integral and functional components of the extracellular matrix. Matrix Biol 75–76:201–219. https://doi.org/10.1016/j.matbio.2017.10.003

Rocha S, Carvalho J, Oliveira P, Voglstaetter M, Schvartz D, Thomsen AR, Walter N, Khanduri R, Sanchez JC, Keller A, Oliveira C, Nazarenko I (2019) 3D cellular architecture affects microRNA and protein cargo of extracellular vesicles. Adv Sci 6(4). https://doi.org/10.1002/advs.201800948

Sanchez-Lopez JA, Twena S, Apel I, Kornhaeuser SC, Chasnitsky M, Miklosi AG, Vega-Dominguez PJ, Shephard A, Hefetz A, Heifetz Y (2022) Male-female communication enhances release of extracellular vesicles leading to high fertility in drosophila. Commun Biol 5:815. https://doi.org/10.1038/s42003-022-03770-6

Scott A, Sueiro Ballesteros L, Bradshaw M, Tsuji C, Power A, Lorriman J, Love J, Paul D, Herman A, Emanueli C, Richardson RJ (2021) In vivo characterization of endogenous cardiovascular extracellular vesicles in larval and adult zebrafish. Arterioscler Thromb Vasc Biol 41(9):2454–2468. https://doi.org/10.1161/ATVBAHA.121.316539

Shaye DD, Greenwald I (2011) OrthoList: a compendium of C. elegans genes with human orthologs. PLoS One 6(5):e20085. https://doi.org/10.1371/journal.pone.0020085

Silva M, Morsci N, Nguyen KCQ, Rizvi A, Rongo C, Hall DH, Barr MM (2017) Cell-specific α-tubulin isotype regulates ciliary microtubule ultrastructure, intraflagellar transport, and extracellular vesicle biology. Curr Biol 27(7):968–980. https://doi.org/10.1016/j.cub.2017.02.039

Suetsugu A, Honma K, Saji S, Moriwaki H, Ochiya T, Hoffman RM (2013) Imaging exosome transfer from breast cancer cells to stroma at metastatic sites in orthotopic nude-mouse models. Adv Drug Deliv Rev 65(3):383–390. https://doi.org/10.1016/j.addr.2012.08.007

Sun Y, You S, Tu H, Spillman DR, Chaney EJ, Marjanovic M, Li J, Barkalifa R, Wang J, Higham AM, Luckey NN, Cradock KA, George Liu Z, Boppart SA (2018) Intraoperative visualization of the tumor microenvironment and quantification of extracellular vesicles by label-free nonlinear imaging. Sci Adv 4(12):eaau5603. https://doi.org/10.1126/sciadv.aau5603

Sun J, Mou J, Zhang Y, Yuan K, Yang Y, Tang Q (2023) Matrix metalloproteinase-2 delivery to extracellular vesicles produced by HPV-positive cervical cancer cells enhances metastasis via the Akt pathway. Cell Biol Int 47(9):1558–1572. https://doi.org/10.1002/cbin.12054

Sung BH, von Lersner A, Guerrero J, Krystofiak ES, Inman D, Pelletier R, Zijlstra A, Ponik SM, Weaver AM (2020) A live cell reporter of exosome secretion and uptake reveals pathfinding behavior of migrating cells. Nat Commun 11:2092. https://doi.org/10.1038/s41467-020-15747-2

Taha EA, Sogawa C, Okusha Y, Kawai H, Oo MW, Elseoudi A, Lu Y, Nagatsuka H, Kubota S, Satoh A, Okamoto K, Eguchi T (2020) Knockout of MMP3 weakens solid tumor organoids and cancer extracellular vesicles. Cancers (Basel) 12(5):1260. https://doi.org/10.3390/cancers12051260

Thippabhotla S, Zhong C, He M (2019) 3D cell culture stimulates the secretion of in vivo like extracellular vesicles. Sci Rep 9(1):1–14. https://doi.org/10.1038/s41598-019-49671-3

Tian T, Yang Z, Li X (2021) Tissue clearing technique: recent progress and biomedical applications. J Anat 238(2):489–507. https://doi.org/10.1111/joa.13309

Tu H, Liu Y, Marjanovic M, Chaney EJ, You S, Zhao Y, Boppart SA (2017) Concurrence of extracellular vesicle enrichment and metabolic switch visualized label-free in the tumor microenvironment. Sci Adv 3(1):e1600675. https://doi.org/10.1126/sciadv.1600675

Valcz G, Buzás EI, Kittel Á, Krenács T, Visnovitz T, Spisák S, Török G, Homolya L, Zsigrai S, Kiszler G, Antalffy G, Pálóczi K, Szállási Z, Szabó V, Sebestyén A, Solymosi N, Kalmár A, Dede K, Lőrincz P, Tulassay Z, Igaz P, Molnár B (2019) En bloc release of MVB-like small extracellular vesicle clusters by colorectal carcinoma cells. J Extracell Vesicles 8(1):1596668. https://doi.org/10.1080/20013078.2019.1596668

Verweij FJ, Bebelman MP, Jimenez CR, Garcia-Vallejo JJ, Janssen H, Neefjes J, Knol JC, de Goeij-de HR, Piersma SR, Baglio SR, Verhage M, Middeldorp JM, Zomer A, van Rheenen J, Coppolino MG, Hurbain I, Raposo G, Smit MJ, Toonen RFG, van Niel G, Pegtel DM (2018) Quantifying exosome secretion from single cells reveals a modulatory role for GPCR signaling. J Cell Biol 217(3):1129–1142. https://doi.org/10.1083/jcb.201703206

Verweij FJ, Revenu C, Arras G, Dingli F, Loew D, Pegtel DM, Follain G, Allio G, Goetz JG, Zimmermann P, Herbomel P, Del Bene F, Raposo G, van Niel G (2019) Live tracking of inter-organ communication by endogenous Exosomes in vivo. Dev Cell 48(4):573–589.e4. https://doi.org/10.1016/j.devcel.2019.01.004

Verweij FJ, Balaj L, Boulanger CM, Carter DRF, Compeer EB, D'Angelo G, El Andaloussi S, Goetz JG, Gross JC, Hyenne V, Krämer-Albers E-M, Lai CP, Loyer X, Marki A, Momma S, Nolte-'t Hoen ENM, Pegtel DM, Peinado H, Raposo G, Rilla K, Tahara H, Théry C, van Royen ME, Vandenbroucke RE, Wehman AM, Witwer K, Wu Z, Wubbolts R, van Niel G (2021) The power of imaging to understand extracellular vesicle biology in vivo. Nat Methods 18(9):1013–1026. https://doi.org/10.1038/s41592-021-01206-3

Villasante A, Godier-Furnemont A, Hernandez-Barranco A, Coq JL, Boskovic J, Peinado H, Mora J, Samitier J, Vunjak-Novakovic G (2021) Horizontal transfer of the stemness-related markers EZH2 and GLI1 by neuroblastoma-derived extracellular vesicles in stromal cells. Transl Res 237:82–97. https://doi.org/10.1016/j.trsl.2021.06.006

Walsh JD, Wang J, DeHart M, Nikonorova IA, Srinivasan J, Barr MM (2022) Tracking N- and C-termini of C. elegans polycystin-1 reveals their distinct targeting requirements and functions in cilia and extracellular vesicles. PLoS Genet 18(12):e1010560. https://doi.org/10.1371/journal.pgen.1010560

Wang J, Silva M, Haas L, Morsci N, Nguyen KCQ, Hall DH, Barr MM (2014) C. elegans ciliated sensory neurons release extracellular vesicles that function in animal communication. Curr Biol 24(5):519–525. https://doi.org/10.1016/j.cub.2014.01.002

White RM, Sessa A, Burke C, Bowman T, LeBlanc J, Ceol C, Bourque C, Dovey M, Goessling W, Burns CE, Zon LI (2008) Transparent adult zebrafish as a tool for in vivo transplantation analysis. Cell Stem Cell 2(2):183–189. https://doi.org/10.1016/j.stem.2007.11.002

Wiklander OPB, Nordin JZ, O'Loughlin A, Gustafsson Y, Corso G, Mäger I, Vader P, Lee Y, Sork H, Seow Y, Heldring N, Alvarez-Erviti L, Smith CE, Le Blanc K, Macchiarini P, Jungebluth P, Wood MJA, Andaloussi SE (2015) Extracellular vesicle in vivo biodistribution is determined by cell source, route of administration and targeting. J Extracell Vesicles 4. https://doi.org/10.3402/jev.v4.26316

Wu B, Liu D-A, Guan L, Myint PK, Chin L, Dang H, Xu Y, Ren J, Li T, Yu Z, Jabban S, Mills GB, Nukpezah J, Chen YH, Furth EE, Gimotty PA, Wells RG, Weaver VM, Radhakrishnan R, Wang XW, Guo W (2023) Stiff matrix induces exosome secretion to promote tumour growth. Nat Cell Biol 25(3):415–424. https://doi.org/10.1038/s41556-023-01092-1

Yakavets I, Francois A, Lamy L, Piffoux M, Gazeau F, Wilhelm C, Zorin V, Silva AKA, Bezdetnaya L (2021) Effect of stroma on the behavior of temoporfin-loaded lipid nanovesicles inside the stroma-rich head and neck carcinoma spheroids. J Nanobiotechnol 19(1):3. https://doi.org/10.1186/s12951-020-00743-x

Yang T, Martin P, Fogarty B, Brown A, Schurman K, Phipps R, Yin VP, Lockman P, Bai S (2015) Exosome delivered anticancer drugs across the blood-brain barrier for brain cancer therapy in Danio Rerio. Pharm Res 32(6):2003–2014. https://doi.org/10.1007/s11095-014-1593-y

Ye Z, Zhang T, He W, Jin H, Liu C, Yang Z, Ren J (2018) Methotrexate-loaded extracellular vesicles functionalized with therapeutic and targeted peptides for the treatment of glioblastoma multiforme. ACS Appl Mater Interfaces 10(15):12341–12350. https://doi.org/10.1021/acsami.7b18135

You S, Barkalifa R, Chaney EJ, Tu H, Park J, Sorrells JE, Sun Y, Liu Y-Z, Yang L, Chen DZ, Marjanovic M, Sinha S, Boppart SA (2019) Label-free visualization and characterization of extracellular vesicles in breast cancer. Proc Natl Acad Sci USA 116(48):24012–24018. https://doi.org/10.1073/pnas.1909243116

The manufacturer's authorised representative in the EU is Springer Nature Customer Service Centre GmbH, Europaplatz 3, 69115 Heidelberg, Germany. If you have any concerns regarding our products, please contact ProductSafety@springernature.com

Printed and bound by CPI Group (UK) Ltd, Croydon, CR0 4YY

30/09/2026

02225593-0007